THE ECOLOGY OF SANDY SHORES
ANTON MCLACHLAN AND ALEC BROWN

Cover photographs
Front: An intermediate beach, Ras Madrakah, Oman (photo by A. McLachlan)
Back: A reflective beach, Seychelles (top, photo by T. Donn), and a dissipative beach, South Australia (bottom, photo by A. McLachlan)

The Ecology of Sandy Shores

A. McLachlan
College of Agricultural and Marine Sciences,
Sultan Qaboos University,
Oman

A. C. Brown
Zoology Department,
University of Cape Town,
South Africa

GOVERNORS STATE UNIVERSITY
UNIVERSITY PARK
IL 60466

Second English Edition of *The Ecology of Sandy Shores* by A. C. Brown and
A. McLachlan

ELSEVIER

AMSTERDAM • BOSTON • HEIDELBERG • LONDON
NEW YORK • OXFORD • PARIS • SAN DIEGO
SAN FRANCISCO • SINGAPORE • SYDNEY • TOKYO
Academic Press is an imprint of Elsevier

Academic Press is an imprint of Elsevier
30 Corporate Drive, Suite 400, Burlington, MA 01803, USA
525 B Street, Suite 1900, San Diego, California 92101-4495, USA
84 Theobald's Road, London WC1X 8RR, UK

This book is printed on acid-free paper. ∞

Library of Congress Cataloging-in-Publication Data
Application Submitted

British Library Cataloguing-in-Publication Data
A catalogue record for this book is available from the British Library.

ISBN 13: 978-0-12-372569-1
ISBN 10: 0-12-372569-0

For information on all Academic Press publications
visit our Web site at www.books.elsevier.com

Printed in the United States of America
06 07 08 09 10 9 8 7 6 5 4 3 2 1

Working together to grow
libraries in developing countries
www.elsevier.com | www.bookaid.org | www.sabre.org

ELSEVIER   BOOK AID International   Sabre Foundation

Acknowledgements

The first edition of *Sandy Shores* grew largely from work undertaken at the Universities of Port Elizabeth and Cape Town in South Africa in the 1970s and 1980s. Since then I moved to Sultan Qaboos University in Oman and Alec Brown retired as Emeritus Professor in Zoology in Cape Town. While the research support we received from these two South African institutions was fundamental in providing the base for our sandy-beach studies and we have leaned heavily on the work of colleagues and students there, this second edition draws much more widely. The sandy-beach research community has broadened and strengthened, especially following international symposia on sandy-beach ecology in Chile in 1994 and Italy in 2001, as well as a meeting in Poland in 2004 focusing on related topics. We are indebted to many colleagues around the globe for their stimulating research and diverse inputs to the content of this book.

In particular, for this edition we would like to recognize Nancy Maragioglio from Academic Press for supporting our proposal for a second English edition and for her encouragement. We would also like to thank Andy Richford, Cindy Minor, and Carl M. Soares, for their support and professional assistance. Support from Sultan Qaboos University is gratefully acknowledged. Beth Umali, as well as Elise Eisma and Fe Alcachupas, contributed many hours in careful typesetting and organizing figures and tables, and Atsu Dorvlo plotted several figures. Reg Victor and Bill Huguelet read the entire draft and provided valuable comments on the language, layout, and general content. The following colleagues made useful and insightful specialist contributions in reviewing sections of the book: David Schoeman, Omar Defeo, Jenny Dugan, Felicita Scapini, Andy Short, Derek du Preez, Tris Wooldridge, Karl Nordstrom, Tom Gheskiere, Nadine Strydom, Patrick Hesp, Yosuke Suda, and Anvar Kacimov. David Hubbard willingly provided sketches of beach birds and clams. Prof. Yasuhiro Hayakawa kindly translated the second appendix from the Japanese edition into English. The advice and suggestions of all of these colleagues have added greatly to the final product. Any remaining errors and omissions are entirely my responsibility.

Anton McLachlan   *February 2006*

Sadly, Alec Brown died in March of 2005,
before this second edition was completed.
In our last discussions on the nature
and content of the book, we both knew
that we would like to dedicate
this final joint endeavor
of our sandy-beach research
to our wives,

Hester and Rosalind

Contents

Introduction

Ocean sandy beaches are dynamic environments that make up two-thirds of the world's ice-free coastlines. Here sea meets land, and waves, tides, and wind engage in a battleground where they dissipate their energy in driving sand transport. The alternating turbulence and peace of the beach environment enhance its scenic and aesthetic appeal, while its relative simplicity provides an ideal template for research. This should attract the student of coastal ecology. However, the biological study of sandy beaches has traditionally lagged behind that of rocky shores and other coastal ecosystems (Fairweather 1990). A few farsighted workers produced seminal papers — for example, Bruce (1928) in England, Stephen (1931) in Scotland, Remane (1933) in the North Sea, and Pearse *et al.* (1942) in the USA — but for the most part scientific investigation (as opposed to casual observation and intermittent beachcombing) began only some 50 years after the first intensive studies of rocky shores. The reasons for this early neglect of sandy-beach ecosystems are not far to seek.

Unlike sandy beaches, rocky shores teem with obvious life. Many of the plants and animals are large and highly colored, and when the tide is out the life in rock pools may be observed without undue effort. Sessile (slow-moving) animals are not only easy to collect but the biologist may enclose or exclude species from an area to study biological interactions and recolonization. None of these things is true of any but the most sheltered sandy beaches. To the casual observer, exposed intertidal sands may seem almost devoid of life. There are no attached plants intertidally. The majority of the animals are too small to be seen conveniently with the naked eye and most macrofaunal invertebrates are cryptic, hiding within the sand and emerging only when necessary to feed or to perform other vital functions — often when covered by the tide. On a sheltered beach, the openings of burrows may give visible evidence of the life within the sand, but on beaches exposed to heavy wave action the sand is far too unstable to support burrows intertidally. As the sand surface is in constant movement while covered by the tide, so must the animals themselves be highly mobile in order to maintain their positions on the beach or to regain them if swept out to sea. This mobility, coupled with a semicryptic mode of life, renders the observation of sandy-beach animals *in situ* far more difficult than for other shore types. There simply appears to be little there. Indeed, sandy beaches have been likened to marine deserts.

Yet the ocean beach is teeming with life, microscopic and macroscopic. The spectrum of life in the sand includes clams, whelks, worms, sand hoppers, crabs, sea lice, sand dollars, and a host of smaller animals — as well as protozoans, microscopic plants, and bacteria. In addition to these residents of the intertidal beach, a variety of species move up over the beach from the surf zone on the rising tide, and others descend onto the beach from the dunes on the falling tide. All of these components interact in a trophic network to create the open ecosystem of the sandy beach, which exchanges materials with sea and land. Increasingly, we have begun to realize that sandy beaches are not marine deserts but are interesting and often productive ecosystems. And so, fortunately, the earlier neglect of sandy beaches by researchers has been quite strongly addressed in the past few decades, providing the material for this book.

Whereas most early work on sandy-beach ecology was descriptive, this changed in the 1970s and 1980s. The complexity of the interactions among the surf-zone fauna and flora, the animals of the intertidal slope, and the backshore biota were brought home to workers in this field at the first international symposium on sandy beaches in 1983 (McLachlan and Erasmus 1983), where sandy-beach ecology emerged for the first time as a distinct field of coastal science. In the decade following that symposium, systems energetics was a major theme among sandy-beach researchers. Many ideas emerging from that phase of sandy-beach research were covered in the first edition of this book (Brown and McLachlan 1990). In the 16 years since the first edition, emphasis has shifted to macrobenthic population and community ecology.

The aim of this second edition of *Sandy Shores* is again to present an integrated account of sandy-shore ecology, including the surf zone, the intertidal slope, the back beach, and the dunes. Since the first edition, our understanding of the ecology of intertidal macrofauna has advanced considerably and this second edition includes major additions to this aspect of beach ecology. As before, we consider beaches as ecosystems, and human impacts and conservation and management of these systems are stressed. Accordingly, this edition retains much of the first edition. However, new sections have been added, based on literature and reviews that have been published since 1990. For a full listing of the sandy-beach ecological literature prior to 1990 the reader is referred to McLachlan and Erasmus (1983) and Brown and McLachlan (1990).

This book is unashamedly biased in favor of exposed beaches of pure sand. We have omitted consideration of estuarine sand flats, and such sheltered environments are mentioned only in passing, attention being concentrated on the world's open oceanic beaches. Because these are dynamic physically controlled systems, our account begins with two chapters appraising the physical environment of the sandy beach. It is essential for any beach ecologist to have a sound understanding of the main features and processes of the physical environment of the sandy beach. Following this are chapters on the main components of flora and fauna. Thereafter, beach and dune systems are considered as a whole. Because sandy coastlines in general (and especially their associated dune systems) are relatively fragile environments facing many threats — and because most are eroding — they require conservation and special management techniques if they are to continue to function ecologically and provide for quality recreation. Appropriate management and successful conservation can only be achieved if the complex ecology of these areas is understood. Furthering this understanding is also the task of this book.

Ocean sandy beaches are wonderful venues for recreation for everyone. They are also fascinating and important ecosystems for the student of coastal ecology. They are magnets that draw our attention by their dynamic beauty and their contrasting restlessness and tranquillity. And they still hold many secrets awaiting discovery.

Willard Bascom's (1964) epilogue is as true today as it was nearly half a century ago: "Fortunately the beaches of the world are cleaned every night by the tide. A fresh look always awaits the student, and every wave is a masterpiece of originality. It will ever be so. Go and see." We hope the chapters that follow will encourage the reader to do just that.

The Physical Environment

2

2.1   Introduction

Sandy coastlines are dynamic environments where the physical structure of the marine habitat is determined by the interaction among sand, waves, and tides. Sandy beaches constitute one of the most resilient types of dynamic coastline because of their ability to absorb wave energy. This wave energy is expended in driving surf-zone water movement, which carries sand offshore during storms and moves it back onshore during calms. The beach is characterized by wave-driven sand transport and by aeolian (wind) transport in the backshore and dunes. Most beaches are backed by dunes and interact with them in terms of sediment budgets by either supplying or receiving sand. This sediment transport, in the surf zone by wave action and in the dunes by wind action, has both a shore normal and a longshore component. On many coasts, longshore transport accounts for vast volumes of sand. The sandy beach is thus an extremely dynamic environment where sand and water are always in motion. Before considering the overall interactions of these parameters, it is appropriate to examine the characteristics of the defining elements — sand, waves, and tides. Thereafter, we explore the types of beaches that result from these interactions and some of their key processes. Two useful general references covering physical processes and features of beaches are Komar (1998) and Short (1999).

2.2   Sand

Particle Size

Sand originates mainly from erosion of the land and is transported to the sea by rivers. Beaches may, however, also receive sand from biogenic sources in the sea such as animal skeletons, and from sea cliff erosion. The two main types of beach material are quartz (or silica) sands of terrestrial origin and carbonate sands of marine origin. Quartz sands have a slightly lower density ($2.66\,\mathrm{g\cdot cm^{-3}}$) than carbonate sands (2.7 to $2.95\,\mathrm{g\cdot cm^{-3}}$ for calcite and aragonite), and quartz particles tend to be more rounded. Despite their higher density, calcium carbonate particles sink more slowly in water due to their more irregular shapes. Other materials that may contribute to beach sands include heavy minerals, basalt (volcanic rock), and feldspar. The most important feature of sand particles is their size. Particle size is generally classified according to the Wentworth scale, in phi units, where $\varphi = -\log_2$ diameter (mm). This classification is summarized in Table 2.1.

Analysis of sand particle size can be accomplished using either a settling tube or a nest of sieves. Use of a settling tube is based on Stokes' law, which defines the rate of sinking of particles in water (Figure 2.1). Sieving involves passing a sample of wet or

Table 2.1.   Wentworth size scale for sediments.

	Generic name	Wentworth scale size range (ϕ)	Particle diameter (mm)
Gravel	Boulder	<−8	>256
	Cobble	−6 to −8	64 to 256
	Pebble	−2 to −6	4 to 64
	Granule	−1 to −2	2 to 4
Sand	Very coarse	0 to −1	1.0 to 2.0
	Coarse	1 to 0	0.50 to 1.0
	Medium	2 to 1	0.25 to 0.50
	Fine	3 to 2	0.125 to 0.25
	Very fine	4 to 3	0.0625 to 0.125
Mud	Silt	8 to 4	0.0039 to 0.0625
	Clay	>8	<0.0039

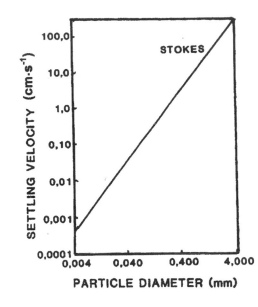

Figure 2.1. Graphical illustration of Stokes' law of falling quartz spheres in water.

dry sand through a series of sieves whose mesh corresponds to 1φ or 0.5φ intervals. For beach sands, 50 g of sand and 15-cm-diameter sieves of the following mesh sizes are usually used: 2 mm, 1.41 mm, 1 mm, 710 μmm, 500 μmm, 250 μmm, 177 μmm, 125 μmm, 88 μmm, and 63 μmm. In some sheltered beaches there may be a significant silt/clay component that warrants the use of finer sieves.

Following laboratory analysis of the weights of sand falling within each size fraction, further graphical analysis is required. This is done by plotting cumulative curves on probability paper (Figure 2.2) and calculating the following parameters (Folk 1974).

- Measures of average size:
  - Median particle diameter (Mdφ) is the diameter corresponding to the 50% mark on the cumulative curve ($\varphi50$).
  - Graphic mean particle diameter (M_z), where

$$M_z = (\varphi16 + \varphi50 + \varphi84)/3$$

- Measures of uniformity of sorting:
  - Phi quartile deviation (QDφ), where

$$QD\varphi = (\varphi75 - \varphi25)/2$$

  - Inclusive graphic standard deviation ($\sigma1$), where

$$\sigma1 = \frac{(\phi84 - \phi16)}{4} + \frac{(\phi95 - \phi5)}{6.6}$$

- Measures of skewness:
  - Phi quartile skewness (Skqφ), where

$$Skq\varphi = (\varphi25 + \varphi75 - \varphi50)/2$$

  - Inclusive graphic skewness (Sk_1), where

$$Sk_1 = \frac{(\phi16 + \phi84 - 2\phi50)}{(2(\phi84 - \phi16))} + \frac{(\phi5 + \phi95 - 2\phi5)}{(2(\phi95 - \phi5))}$$

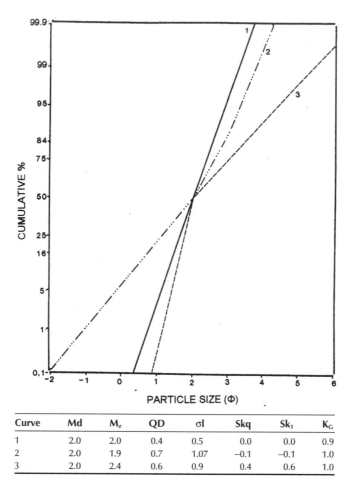

Figure 2.2. Cumulative curves of three beach sands. All three have medians of 2ϕ, or 250 μm. (1) Well sorted, symmetrical, and slightly peaked. (2) Moderately to poorly sorted, slightly negatively skewed, and with normal peakedness. (3) Moderately sorted, positively skewed, and showing normal peakedness. Values for the various characteristics displayed by these curves are given in the following table in ϕ units.

Curve	Md	M_z	QD	σI	Skq	Sk_1	K_G
1	2.0	2.0	0.4	0.5	0.0	0.0	0.9
2	2.0	1.9	0.7	1.07	−0.1	−0.1	1.0
3	2.0	2.4	0.6	0.9	0.4	0.6	1.0

- Measure of kurtosis or peakedness:

$$K_G = \frac{(\phi 95 - \phi 05)}{2.44(\phi 75 - \phi 25)}$$

Unless sands are badly skewed, median and mean particle diameters are very similar and for most ocean beaches are in the range of fine to coarse sand (0 to 2.5φ or 180 to 1,000 μm). The inclusive graphic standard deviation is the best measure of sorting. Values below 0.5 indicate good sorting, values between 0.5 and 1.0 moderate sorting, and values above 1.0 poor sorting, where a wide range of particle sizes are evident. The best sorting attained by natural sands is about 0.2.

Skewness measures the asymmetry of the cumulative curve, and plus or minus values indicate excess amounts of (or tails on the sides of) fine and coarse material, respectively. The inclusive graphic skewness is the best measure of this. Values between +0.1 and −0.1 indicate near symmetry, values above +0.1 indicate fine skewed sand, and values below −0.1 indicate coarse skewed sands. Values seldom exceed +0.8 or −0.8.

Kurtosis is not often considered by ecologists. For normal curves, K_G is 1.0, whereas curves with a wide spread have values over 1.0 and curves with little spread have values below 1.0. Some representative curves are shown in Figure 2.2. In addition to analyses of particle size, the degree of particle roundness may be estimated by comparison with a pictorial scale, and the calcium carbonate content by acid digestion and gravimetry or titration.

Porosity and Permeability

Porosity is the volume of void space in the sand, usually expressed as a percentage of total sand volume. Thus, the porosity of a sediment is the volume of water needed to saturate a given weight of dry sand. It may, however, be expressed on either a volume or mass basis. For most sands, porosity is about 30 to 40% of the total volume, or 20 to 25% of the total mass of wet sand. Generally, the finer a sand the greater its porosity, despite the decreasing size of individual pore spaces. Crisp and Williams (1971) studied pore spaces using thin sections and showed that mean pore diameters were 30 to 40% of particle diameters in almost all uniform sands and 15 to 20% of particle diameters in poorly sorted shell gravels. Whereas pore sizes may be estimated by direct measurement of thin sections of resin casts, porosity is usually measured gravimetrically by determining water loss. A speedy moisture tester is a rapid way of measuring the moisture content of sands based on acetylene gas pressure generated by moisture in the sand interacting with calcium carbide. Porosity is important in determining the moisture-holding capacity of sand.

Whereas porosity is the total pore space or volume in a sand, permeability refers to the rate of flow or drainage of water through the sand. Fine sands, although holding more water than coarse sands, have lower permeabilities due to their smaller pore sizes. Permeability is important in determining the amount of flushing through and drainage of a sand. It can be measured by passing a known volume of water through a known depth and area of sand under a set head and noting the time taken (see Section 3.2).

Penetrability

Penetrability of sand is related to particle size and porosity, but is also dependent on other factors. Penetrability can be important to the macrofauna of sandy beaches, as all species must be able to burrow into the substratum. The proportion of clay and silt, as well as the water content of the sand, plays vital roles in determining its penetrability, as well as its resistance to erosion. In water-saturated sand, ease of penetration also depends markedly on the manner in which the penetration force is applied — a sudden pressure causing dilatancy and increased resistance, gentle probing encouraging thixotropy and decreased resistance.

Penetrability is best measured by employing a spring-loaded piston of a known cross-sectional area. The piston is forced into the sand to a standard depth, the pressure exerted compressing the spring and causing a pointer to move along a scale on the side of the instrument. Calibration is thus simple, and the effective range of the instrument can be extended by exchanging the piston for others presenting a greater or lesser contact area with the sand. The readings obtained are normally expressed in $kg \cdot cm^{-2}$. Multiplication by a factor of 10 expresses the force more correctly in Newtons. It should be noted that the results gained with a spring-loaded piston sunk to a standard depth cannot be correlated with results from other types of penetrometer in which a spike or rod is driven into the substratum to a variable depth using a set force.

There can be considerable longshore variation in penetrability. In addition, the crests of the ripples display greater resistance to penetration than the troughs between them. Resistance to penetration can increase with depth below the surface of the sand and decrease with increasing angle from the vertical.

2.3   Waves

In the context of sandy beaches, we are mainly concerned with surface gravity waves, although internal and tidal waves may also be important (see Section 2.4). Surface gravity waves and the secondary currents they induce constitute the driving forces behind most processes occurring on open sandy beaches. Waves are generated by wind stress on the water surface, with friction between air and water causing a viscous drag, which stretches the surface like an elastic membrane. This distortion by wind and restoration by surface tension causes undulations (or waves). If the wind is strong and the waves grow, gravity replaces surface tension as the restorative force and the waves move off before the wind. Waves thus transfer energy from winds at sea to the coastal zone.

Basic wave features are illustrated in Figure 2.3. Wavelength (L) is the horizontal distance between successive crests, and wave height (H) is the vertical height of the wave from trough to crest. The time required for successive crests to pass a fixed point is the period (T). The wave steepness is H/L, and wave speed or celerity-C = L/T. The height and period of a wave are related to the strength, time, and fetch of the wind that generated it. The stronger the wind, the longer it blows, and the greater the fetch (or distance over which the wind blows) the larger are L and T.

For waves with short fetches, wave height increases directly as a function of wind velocity, but for waves with long fetches wave height is lower. Steep waves and confused seas occur where strong winds blow. However, the more regular swell that moves off across the ocean to break on distant beaches is not steep, typical steepness values being 0.002 to 0.025.

Water particles in a wave oscillate in a circular path, returning to their original positions after one complete cycle (or wavelength) has passed. The velocity and radius of the circle decline with depth (Figure 2.4) until the particle no longer describes a circle but moves back and forth horizontally. At a depth of one-half the wave length, orbital motion becomes negligible. Thus, whenever water depth is less than L/2 the wave feels bottom and begins to change.

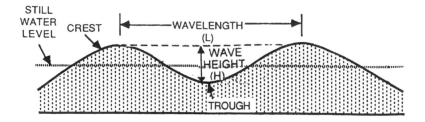

Figure 2.3.   Features of a progressive wave.

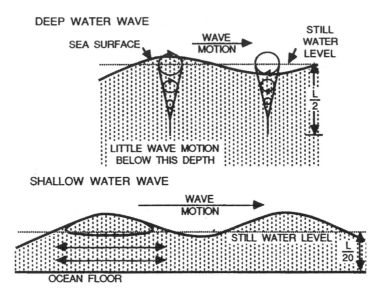

Figure 2.4. Particle motion in deep- and shallow-water waves.

Types of Waves

Waves that do not feel bottom are called deep-water waves and their speed is given by $C = 1.56\,T\,m\cdot s^{-1}$. Thus, their speed is governed by their period, with long-period waves traveling the fastest. For such waves, the group speed is half the wave speed, because waves in front of the train continually decay and new ones are generated behind.

Where the water depth is between 1/2L and 1/20L, waves are transitional and bottom effects become significant. Here, C is determined partly by T and partly by water depth. For most wind waves, this occurs at periods of 10 to 12 s and depths less than 100 m.

Where water depth is less than 1/20L, wave speed is controlled by depth and the waves are called shallow-water waves. Here waves shorten, steepen, and eventually break. In these cases, $C = \sqrt{g \cdot d}$, where g = gravity = $9.1\,m\cdot s^{-1}\cdot s^{-1}$ and d = depth in m. Particle motion here takes the form of a very shallow ellipse approaching horizontal oscillation. For these waves, group speed = C (or wave speed).

Wave Energy

Waves contain two types of energy: kinetic (the energy of particle motion) and potential (the displacement of the sea surface related to wave height). As wave height determines both the orbital diameter (or kinetic energy) and the amplitude (or potential energy), wave energy is proportional to the square of wave height. High-energy coasts dissipate considerable amounts of wave energy.

Refraction

Waves feeling the bottom decelerate. Such a change in speed of one section of a wave causes it to change direction. This refraction, or bending, of waves as they approach the shore tends to align them with the contours of the coastline. It also tends to focus wave energy on headlands and dissipates it in bays (Figure 2.5). Convergence of wave energy

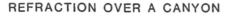

REFRACTION OVER A CANYON

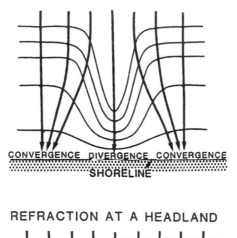

Figure 2.5. Refraction of waves approaching a shoreline caused by deeper water over a canyon and shallower water off a headland.

also occurs over raised areas of the bottom, such as reefs or bars. This convergence results in most damage on headlands during storms.

Shoaling and Breaking

When depth decreases, speed slows. T is conserved, causing L to decrease and wavelength to shorten. As the wave translates into shallow water, crests become more pronounced. The ratio H/L thus increases until the wave breaks, where H/L = 1/7 and the water depth = 1.3 H. H here is the breaking height, which is generally greater than the deep-water height. Breaking occurs in two main ways (Figure 2.6).

- *Plunging.* Wave speed decreases as shallower water is entered, while the orbital velocity of the particles increases as the wave steepens, until a point is reached where the maximum orbital velocity exceeds wave speed. The water particles under the wave crest are traveling faster than the wave crest itself, and faster still than the trough (which is in shallower water). The crest then plunges into the wave trough ahead as a water jet. This is a plunging breaker.
- *Spilling.* The maximum vertical acceleration in the wave motion increases until it exceeds the downward gravitational acceleration. Water particles then start popping out of the wave surface, forming a spilling breaker.

The type of breaker is determined by two factors: the deep-water wave steepness (H/L) and the beach slope. Spilling breakers occur when steep waves reach a gently sloping beach, whereas plunging breakers occur on all slopes but with a lower wave steepness.

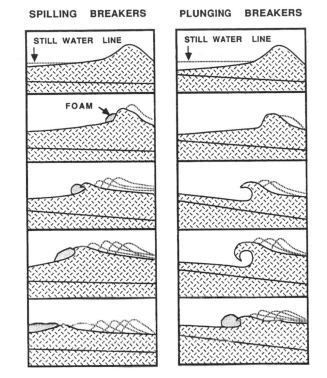

SPILLING BREAKERS   PLUNGING BREAKERS

Figure 2.6. The two main types of breaking waves.

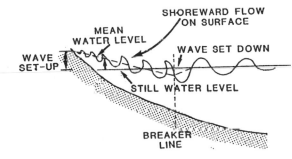

Figure 2.7. Wave setup and set-down for spilling breakers. The elevation of the water line at the shore and lowered water level just outside the breakers due to shoreward water transport by the breakers (after Swart 1983).

A third type of breaker also occurs; namely, a surging breaker (with very low wave steepness and steep beach slope). Here, the wave does not break but surges up the beach face and is partly reflected back to sea. In actual fact there is not a sharp transition between these breaker types, as they tend to grade into each other. Waves may break where the water depth is between $2\,H$ (spilling waves) and $0.8\,H$ (plunging breakers).

Wave energy is dissipated in the breaker zone. Spilling breakers dissipate their energy gradually, whereas plunging breakers dissipate it rapidly. The dissipated energy sustains a setup (or rise) of the mean water level inside the breaker zone (Figure 2.7). At the surface (within the breaker zone) there is a mass flow of water shoreward, causing a drop in water level at and just outside the breakers, called set-down. Water accumulated against the beach by waves is discharged out of the surf zone by rip currents or bed return flow (see material following).

In the case of perpendicular wave attack, the dissipated energy goes mainly toward sustaining a higher mean water level inside the breakers than outside the breakers. For plunging breakers, wave setup may increase abruptly. What usually happens, however, is that waves break initially as plunging breakers and then reform and shoal landward as spilling breakers. Maximum wave setup at the mean water line can be 20 to 50% of the outer breaker height. Thus, 2-m breakers can result in a maximum elevation of the water level inside the surf zone of more than 0.4 m above sea level. These water level variations are the driving force for secondary surf-zone circulations (see Section 2.9).

When a wave breaks, a cavity (the primary vortex) is formed and the plunging jet of the wave may form more than one such vortex. Collapse of these vortices may lead to spouts of water erupting from just behind the breaker front. After breaking, the wave changes to a bore (a type of spilling breaker), which advances toward the beach. The speed of movement of such a bore is constrained by the water depth. Once the sand is reached, the bore collapses to form a thin layer of swash that runs rapidly up the beach face. Because much wave energy may be consumed in the surf zone, not every incident wave necessarily results in a swash reaching the beach face.

Bound and Infragravity Waves: Surf Beat

In nature, many gravity wave trains with different properties impinge on a coast at any one time. These wave trains interact or interfere with one another to produce bound waves called infragravity waves. For example, if two wave trains having periods of 7 s and 10 s interfere, they will result in a bound wave of 70-s period (that is, with crests 70 s apart). The presence of a bound wave component in shallow-water waves means that the water level will fluctuate with a period of usually between 1 and 10 minutes. These bound waves manifest themselves in variations in water level with periods longer than those of the observed breaking gravity waves. Excess water brought into the surf by this process will be returned seaward by pulsating rip currents or bed return flow (see Section 2.9). This variation in breaker height with periods of a few minutes is known as surf beat.

If wave approach is oblique, the bound waves will propagate alongshore, resulting in a system of rip currents moving along the shore. This is similar to the effects of edge waves (see material following), which always propagate alongshore (even when wave attack is perpendicular).

Edge Waves

Edge waves are waves running along the shore and contained within the surf zone. They are formed by reflection of incident and infragravity waves off the beach and their refraction and entrapment within the surf zone. They are still little understood but are thought to be responsible for most longshore rhythmic topography (see Section 2.8). Edge waves have a longshore periodicity and amplitude decaying exponentially offshore, their energy being trapped against the shore by refraction. They absorb energy from the incoming surface waves. The compound water-level fluctuations at the water line are the result of gravity waves, bound waves, edge waves, and tides. Edge waves can cause rhythmic variations along the shoreline, such as beach cusps.

2.4   Other Drivers of Water Movement

Tides

Tidal currents are usually much less important than wave-induced currents in surf zones. However, in macrotidal areas the reverse may be true. Furthermore, tides limit wave height by affecting nearshore water depth and are important in determining the volume of water within the surf zone. The largest waves usually occur at high tide.

Tides are normally observed against coastlines as a periodic rise and fall of the sea surface. The maximum elevation of the tide is known as high tide, and the minimum elevation as low tide. On most coastlines, two high tides and two low tides occur each day — the vertical difference between them being the tidal range. This varies on open coasts from a few centimeters in the Mediterranean Sea to nearly 10 m. Tides are generated by the gravitational attraction of the moon and the sun on the oceans. According to Newton's law of universal gravitation, *the gravitational attraction between two bodies is directly proportional to their masses and inversely proportional to the square of the distance between the bodies*. The moon, therefore, exerts twice as much tide-generating force as does the much larger sun because the latter is much more distant.

The moon orbits the earth each lunar month (27.5 days). To maintain this orbit, the gravitational attraction between the earth and moon exactly balances the centrifugal force holding the bodies apart. Together, these two opposing forces create two tide-producing forces at the earth's surface. If the earth were completely covered with water, two bulges of water (or lunar tides) would pile up — one on the side of the earth facing the moon and the other on the opposite side (Figure 2.8).

Because the earth makes a complete rotation every 24 hours, a point on the earth's surface will experience two high tides and two low tides each day. However, during that rotation the moon advances in its own orbit and thus an additional 50 minutes of the earth's rotation is required to bring a point on its surface directly in line with the moon again. Therefore, a reference point on the earth's surface experiences only two equal-high and two equal-low tides every 24 hours and 50 minutes (a lunar day).

The sun earth system generates similar tide-producing forces that yield a solar tide about one-half as large as the lunar tide. The solar tide is experienced as a variation on

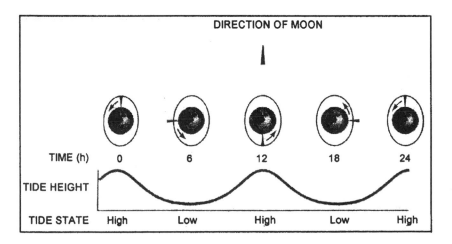

Figure 2.8. Each day as the earth rotates, a point on its surface (indicated by the marker) experiences high tides when under tidal bulges and low tides when at right angles to the tidal bulges (after Sumich 1999).

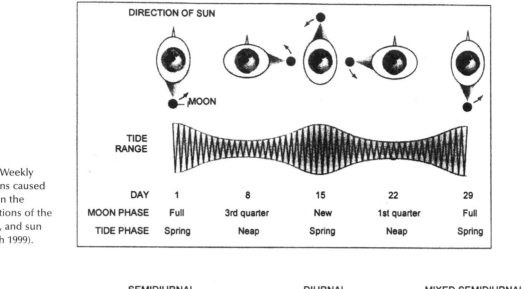

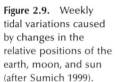

Figure 2.9. Weekly tidal variations caused by changes in the relative positions of the earth, moon, and sun (after Sumich 1999).

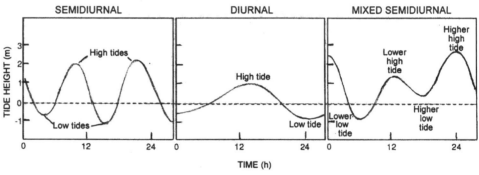

Figure 2.10. Three common types of tides (after Sumich 1999).

the basic lunar tidal pattern, not as a separate set of tides. When the sun, moon, and earth are in alignment (at the time of the new and full moon), the solar tide complements the lunar tide, creating extra high tides and very low tides, collectively called spring tides. A week later, when the sun and moon are at right angles to each other, the solar tide partially cancels the lunar tide to produce smaller tides known as neap tides. During each lunar month, two sets of spring tides and two sets of neap tides occur. That is, a period of 14 days (Figure 2.9).

In the world's oceans, the continents act to block the westward passage of the tidal bulges as the earth rotates under them. When they cannot move freely around the globe, these tidal impulses generate complex patterns within each ocean basin that may differ greatly from the tidal patterns of adjacent ocean basins or other regions of the same ocean basin.

Figure 2.10 shows types of tides experienced on ocean beaches. In semidiurnal tides, the two high tides are quite similar to each other, as are the two low tides. Where there is a single tidal cycle each day, this is a diurnal (or daily) tide. A third pattern consists of two high tides and two low tides each day, but successive high tides are quite different from each other. This type of tidal pattern is a mixed semidiurnal tide. Figure 2.11 shows the geographical distribution of diurnal, semidiurnal, and mixed semidiurnal tides.

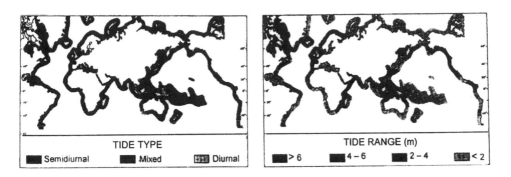

Figure 2.11. The geographical occurrence of the three types of tides and tidal ranges (after Davies 1977).

Tide range is considered microtidal if less than 2 m, mesotidal if 2 to 4 m, and macrotidal if 4 to 8 m. In the deep ocean, tide range tends to be smaller toward the equator and larger toward the poles. In general, it is tide range rather than tide type that is most important in sandy-beach ecology.

Internal Waves

Gravity waves occur at the interface between water and air, whereas internal waves occur at the boundary between water layers of different densities. Because the difference in density between two such layers is much less than between water and air, greater wave heights (up to 30 m or more) can be supported. Internal waves may be caused by tides, underwater avalanche, or wind. They move slowly, with periods of 5 to 8 minutes and wavelengths of 0.5 to 1 km. Internal waves may cause fluctuations in water level in the surf zone.

Wind

The shear stress on the sea surface caused by wind blowing over it induces a water current in the same direction as the wind. This flow decreases rapidly below the surface and is deflected due to Coriolis forces, to the left in the Southern Hemisphere and to the right in the Northern Hemisphere. Generally, surface water movement is at about 2% of wind speed. Winds are important in shaping wave characteristics. Strong onshore winds increase wave height and the tendency for spilling breakers, thereby increasing the size of the surf zone. Offshore winds flatten the surf and increase the tendency for plunging breakers. Winds are also responsible for sediment transport between the beach and dunes (see Chapter 13).

2.5 Sand Transport

Water movement results in shear stress on the sea bed. This may move sand off the bed into the water, whereupon it can be transported. The coarsest sands occur around the break point, and sands generally become finer offshore and onshore corresponding to the distribution of current velocities.

As shear stress on the bed increases with a shoaling wave, a point is reached where the drag on sand particles becomes sufficient to rock them to and fro. Closer inshore this movement is accentuated. Cyclic water movement leads to the formation of ripples in the sand. The size of the ripples increases with increasing particle size and wave height or current speed, up to a maximum current speed of $1 \, \text{m·s}^{-1}$, above which the ripples become washed out until the bed becomes smooth.

Sand can be transported in two modes: as bed load and as suspended load. Suspended load is that part transported in the water column above the bed. Oscillating flow over ripples sets up eddies in the lee of the ripples, which explode when the flow is reversed (and material is ejected beyond the crest of the ripples). Gravity pulls these particles downward, whereas turbulence carries them upward. A balance is reached, with an equilibrium profile of material suspended at various levels in the fluid. Sediment may also be suspended by plunging breakers. Bed load is defined as that part of the total volume of material moving close to the bed and not much above ripple height. Coarser material is mainly carried as bed load.

This transport may be in a longshore, as well as in an on-offshore, direction. During oblique storm wave attack, large amounts of sediment may move alongshore — mostly in the outer surf zone, where turbulence is greatest. Longshore transport is one of the most important processes occurring on exposed beaches and is the focus of much attention by coastal zone managers (Chapters 14 and 15). Longshore sand transport along exposed beaches can exceed $100,000 \, \text{m}^3$ per year, and blocking this with coastal engineering structures can cause accretion and erosion problems of great magnitude.

Return flow of water set up in the surf zone by breakers may be in the form of either rip currents or bed return flow. Relative to these conditions, sand movement may be onshore or offshore. The steeper the waves and the beach the greater will be the tendency for accretion. Beaches go through cycles of erosion and accretion coupled to changes in wave energy. During storms beaches are eroded and flattened, whereas during calm periods sand moves slowly shoreward, causing accretion. This results in the range of beach states defined in the next section.

2.6 Interaction Among Beach Slope, Waves, Tides, and Sand

The slope of a beach face depends on the interaction of the swash and backwash processes planing it. Swash running up the beach carries sand with it and therefore tends to cause accretion and a steep beach face. Backwash has the opposite effect. If a beach consists of very coarse material, such as pebbles, uprunning swashes tend to drain into the beach face, thereby eliminating backwash. Sand or pebbles are thus carried up the beach but not back again, resulting in a steep beach face. Fine-sand beaches, on the other hand, stay waterlogged because of their low permeability — so that each swash is followed by a full backwash, which flattens the beach by removing sand suspended by the swash. Thus, the coarser the sand the steeper the beach face for a given regime of wave action.

If sand particle size is kept constant and wave height increased, the beach will flatten. This is because bigger waves result in larger swashes, which cause a greater amount of waterlogging of the sand and greater erosion in the strong backwash. Beach slope is therefore not merely a function of particle size. This important relationship between beach slope, sand particle size, and wave action is illustrated in Figure 2.12.

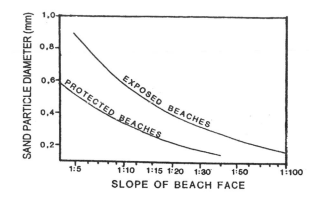

Figure 2.12. The general relationships among sand particle size, exposure to wave action, and beach face slope.

Tides also influence slope, in a fashion similar to that of waves (i.e., beaches become wider and flatter as tide range increases). Storms generally move sand off the beach and expand the surf zone, whereas calm conditions have the opposite effect. The range of beach morphodynamic states resulting from the relationship among sand, waves, and tides is covered in Section 2.8.

2.7   Beach Indices

Various indices have been used to characterize beach type. The most useful are as follows.

- DFV $(\Omega) = H_b \cdot 100 / W \cdot T$
- RTR $= \text{tide} / H_b$
- BI $= \log_{10} (\text{sand} \cdot \text{tide} / \text{slope})$
- Slope* $= 1 / \text{beach face slope}$

Here, DFV is dimensionless fall velocity, H_b is significant breaker height (m), W is sand fall velocity (cm·s^{-1}), T is wave period (s), RTR is relative tide range, *tide* is maximum spring tide range (m), BI is beach index, slope is beach face slope, and *sand* is mean sand particle size in phi units + 1 (McLachlan and Dorvlo 2005; see Appendix A for a list of other indices). Tables from Gibbs et al. (1971) are used to calculate settling velocities based on particle size. The indices are all dimensionless except for BI (log phi·m).

The DFV (Ω, also referred to as Dean's parameter) is based on a measure of sand transport potential and wave energy and is essentially an index of the ability of waves to move sand. High values (> 5) indicate much erosion of the beach by waves and hence a flat (or dissipative) beach. Conversely, low values (< 2) indicate limited ability of waves to erode and hence the beach is more accretional (or steeper).

RTR is a measure of the relative importance of waves and tides in influencing the beach morphology. Low values (< 3) indicate wave-dominated beaches, values in the range 3 to 12 indicate tide-modified beaches, and values > 12 indicate tide-dominated beaches fronted by sand flats. This index is also dimensionless.

BI combines slope, sand, and tidal values into a single measure that can characterize a beach and enable ecologists to compare beaches of differing tide range. It ranges 0 to 4, from beaches with coarse sand, small waves, and small tides to beaches of fine sand, big waves, and large tides.

Slope* is a simple measure of the integrated effects of sand, tides, and waves and is especially useful when comparing beaches subject to similar tide range. The most convenient way of expressing slope is to take the reciprocal of beach face slope (i.e., 1/slope). This generally ranges between 5 and 100 for ocean beaches.

2.8   Beach Types

Microtidal Beaches

The six major microtidal (or wave-dominated) beach types are illustrated in Figure 2.13. The two extremes in this system are the dissipative-beach/surf-zone and the reflective-beach/surf-zone, with a series of intermediate states. The reflective end of the scale occurs when conditions are calm and/or the sediment is coarse. Here, all of the sediment is stored on the intertidal beach and backshore. There is no surf zone and waves surge directly up the beach face. Cusps, or short longshore undulations caused by edge waves, are a typical feature of such beaches. Usually the tidal range is also small. The beach face is characterized by a step on the lower shore (where incoming waves and backwash collide and deposit sediment) and by a berm (or platform) above the intertidal slope. Wave energy is reflected from such a beach face.

Figure 2.13. Microtidal beach types, ranging from reflective (lower right) to dissipative (top left), with four intermediate states characterized by bars moving further offshore toward the dissipative state (after Short and Wright 1983).

As bigger waves cut back a beach and spread out its sediments to form a surf zone, such a reflective beach gives way to a series of intermediate forms. If wave action is strong enough and/or sediment particle size fine enough, the fully dissipative state may be reached. Here, the beach is flat and maximally eroded, and the sediment is stored in a broad surf zone that may have multiple bars (sandbanks) parallel to the beach. Waves

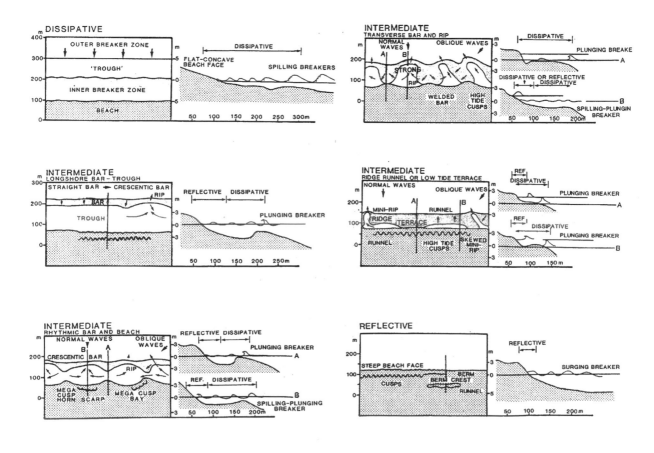

tend to be spilling and break a long way from the beach, often reforming and breaking again. In this way, most wave energy is consumed in the surf zone before reaching the beach. Swash along the shoreline may be gentle, although there are usually pronounced infragravity waves. Landward water flow occurs as surface bores, while return flow is mainly in the form of bed return flow (although some widely spaced rip currents may occur). Wave energy is thus dissipated in the surf zone rather than reflected from the beach face.

Between these two extremes, four intermediate states may be recognized. Intermediate beaches are characterized by high temporal variability, sand storage both on the beach and in the surf zone, and bars and troughs in a surf zone (usually supporting well-developed rip currents). Consider a sandy beach after a violent storm, when it has been planed down to a dissipative state. As conditions become calm, sediment starts to move shoreward. The next recognizable state is referred to as a longshore bar-trough system. Here, the outer surf-zone bars weld to form a single bar onto which the waves first break. Waves then reform over a deeper trough and break again on the shore, where the inner bars may have welded with the beach. Small rip currents may lead from the shore to the trough, while larger rips break through the outer bar. The beach face itself may be reflective. If still calmer conditions occur, the bar moves further shoreward and starts to undulate as a result of edge waves, forming a rhythmic bar and beach system. Here, the beach face also develops a series of rhythmic megacusps. Rip currents are well developed, broaching the bar between its horns.

As conditions become still calmer, the horns of the bar migrate to weld with the beach in places, forming a crescentic (or rhythmic-shaped) bar. Under still calmer conditions the rhythmic bar may disappear between these horns. The result is a series of shore perpendicular transverse bars with well-developed rips in between. Progressing to even calmer conditions, all of the bars move in to weld with the beach and form a low-tide terrace in which ridges and runnels may occur as remnants of the bars and troughs. If the sediment from this terrace moves onto the beach, the system has returned to a reflective state. Complete transition from fully dissipative to fully reflective conditions, where sand storage shifts from surf zone to beach, is unlikely to occur on any one beach. Transition upward toward the dissipative state during storms may occur rapidly, whereas transition back toward the reflective state is much slower during calms and may take weeks.

Dissipative beaches usually occur where waves exceed 2 m and sands are finer than 200 μm, whereas reflective beaches are usually found where waves are less than 0.5 m and sands are coarser than 400 μm. The morphodynamic state of a microtidal beach can be described by DFV (see Section 2.7; DFV = Hb/W·T, where Hb is the breaker height, T the wave period, and W the fall velocity of the sand for Stokes' law, Figure 2.1). This generally resolves into a scale of 1 to 6, which agrees closely with the six states from reflective to dissipative. It is thus possible, simply by using this parameter, to estimate the state of a wave-dominated beach for any given wave height.

Most of the previously cited observations can only be relied upon where the tidal range does not exceed 2 m. Features of such microtidal beaches are summarized in Table 2.2. Where the tide range is significantly greater than 2 m the picture becomes more complex. Generally, the greater the tidal range the flatter and more concave a beach, as increased tidal range allows greater input of water into the sand, making the lower shore more waterlogged and susceptible to erosion by backwash. On fine-grained macrotidal beaches the lower tidal zones are often flat and extremely dissipative, whereas the high

Table 2.2.   General features of different beach types.

Modal beach type	Dissipative	Intermediate	Reflective
Energy source	Infragravity standing waves and bores	Gravity and infragravity waves, rips	Gravity and edge waves
Morphology	Flat, with multiple bars	Variable bars	Deep water inshore
Sand storage	Stored in surf zone	Shifts between surf zone and beach	Stored on beach
Dunes'	Usually large	Intermediate	Usually small
Filtered volume*	Small	Intermediate	Large
Residence time*	About 24 h	6 to 24 h	About 6 h
Surf circulation	Vertical, bores on surface, undertow below	Horizontal cells	No surf zone, swash circulation within cusps

' See Chapter 13.
* Filtered volume is the volume of seawater flushed daily through the intertidal sand; residence time is the time it takes to percolate (see Chapter 3).

tidal zones are reflective. This is the condition of many Northern European and British beaches. Tidal effects can be taken into account by considering both DFV (Ω) and RTR (Short 1996).

Tidal Effects

The beach states described previously apply to microtidal conditions. Increasing tide range makes the picture more complex (Figure 2.14). When RTR < 3, the three microtidal beach types (reflective, intermediate, and dissipative) apply. Landward of the breaker zone these beaches are dominated by surf- and swash zone processes. As tide range increases, the impact of both the swash and surf-zone processes decreases and shoaling waves become more important. The swash zone is limited to the high-tide beach, whereas the surf zone is initially detached from the swash zone by an intertidal zone and is located at the low-tide level. With increasing tide dominance, shoaling waves are more important than broken waves and wave shoaling controls the intertidal and subtidal morphology. This smooths out the intertidal beach profile.

When RTR is between 3 and 12, three beach types can be identified: namely, the reflective type with a low-tide terrace (RLT), the low-tide bar and rip type (LBR), and the ultradissipative (UD) type. When DFV < 2 and low waves prevail, the beach face is steep and reflective, but rather than a step the lower beach face has a low-tide terrace that may be continuous or cut by rips. This beach is called the reflective plus low-tide terrace beach type. Rips may occur when the RTR < 7 and waves exceed 1 m. If RTR increases, the width of the low-tide terrace also increases. If wave height increases, raising DFV to between 2 and 5, the low-tide bar and rip type dominates. Here, a steeper high-tide reflective beach face is fronted by a wide low gradient intertidal zone, which may contain a low swash bar (ridge and runnel) at low tide. When RTR > 7, higher waves (> 2 m) and fine sand are needed to maintain the bar/rip type. Otherwise, it becomes ultradissipative.

If higher waves and/or finer sands raise DFV above 5, the entire intertidal and subtidal profiles becomes flat and relatively featureless, called ultradissipative. Cusps may be present in the spring high-tide zone, and weak swash bars may form in the intertidal

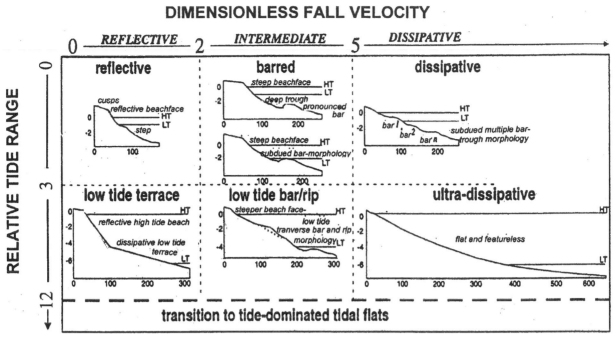

DIMENSIONLESS FALL VELOCITY

zone, but the overall profile is low gradient and concave without clear bars or rips at low tide. The intertidal width of these beaches increases with tide range and may reach several hundred meters.

A characteristic of beaches in areas of high-tide range is a distinct break in slope between the dry high-tide beach and the saturated lower-tide beach. It usually occurs on the reflective/low-tide terrace beach but becomes less prominent on more dissipative beaches. Where the high-tide beach is steep, the break in slope may be sharp and marked by an abrupt change to finer sediment where a low-gradient low-tide dissipative beach is formed. The elevation of the break in slope increases with finer sand, decreases with coarser sand, and usually coincides with the point of low-tide water discharge from the beach (the effluent line).

When RTR exceeds 12, beaches are fully tide dominated and tend toward low-energy high-tide beaches fronted by tidal flats. Overall, the impact of increasing tide range is to raise the dominance of wave shoaling, particularly at the expense of the surf zone, and thereby restrict swash dominance to the narrow high-tide zone. Thus, increasing tide range produces beaches with a steeper swash-dominated high-tide swash zone, usually composed of coarser sediment. The surf zone shifts with the tide, and only controls the morphology at low tide, during the turn of the tide, or not at all. The intertidal profile is low in gradient and concave in profile. The exposed surface is flat at low tide but covered by shoaling-wave ripples at high tide and groundwater discharge at low tide.

Figure 2.15 illustrates the environmental range of macro- to microtidal beaches, showing the convergence of the boundaries in areas of both low waves and low tide. Thus, tide-dominated beach types can exist in areas of low tide range if waves are low enough. Small changes in wave height and tide range may have a pronounced impact on beach morphology in these low wave-tide environments, such as in estuaries. Longshore changes in the level of wave exposure, however, can lead to rapid spatial changes in beach

Figure 2.14.
Conceptual model covering beaches of all tide ranges, based on the dimensionless fall velocity (Ω) and the relative tide range (RTR) (after Short 1996 and personal communication). When RTR < 3 and Ω < 2, the microtidal beach types dominate. When RTR is between 3 and 12, tide range increasingly modifies the wide intertidal beach, with cusps restricted to high tide (and bars and rips, when present, to low tide). When RTR > 12, the transition to tide-dominated beaches is entered.

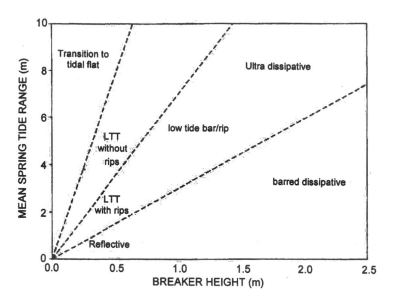

Figure 2.15. The location of the beach types presented in Figure 2.14 based on tide range and breaker height, in the case of a wave period (T) of 8 seconds and grain fall velocity (Ws) of 0.04 m/sec (after Short 1996). The boundaries are indicated by dotted lines, as their position will shift with changes in T and Ws. In areas of low waves or low tides, small changes in tide range and/or wave height will produce large shifts in beach type.

and/or tidal flat morphology. Relative tide range is important in determining beach types. When RTR is low, the surf zone (dominated by breaking waves) largely controls beach morphology. However, as RTR increases the zone dominated by shoaling waves determines beach and intertidal morphology.

2.9   Circulation Cells and Mixing

The interaction of surface-gravity waves moving toward the beach and edge waves moving alongshore produces alternating zones of high and low waves that determine the position of rip currents. The classical pattern that results from this is the horizontal eddy known as the nearshore circulation cell (Figure 2.16). This nearshore circulation system produces a continuous interchange between surf zone and offshore waters and acts as a dispersing mechanism. The outer limits of the cells are usually about double the width of the surf zone. These cells will be symmetrical if wave approach is shore normal and asymmetrical if wave approach is oblique. Much of the water carried offshore by rip currents recirculates with the breakers, so that mixing between adjacent cells may be greater than mixing or exchange between surf-zone and offshore water at the outer limit of rip heads.

During oblique wave attack, a longshore current is generated that may be roughly twice the width of the breaker zone offshore. For spilling breakers, maximum longshore current velocity occurs inside the breakers, whereas for plunging breakers it is at the breaker line. Where there are troughs in the bed of the surf zone, there may be high longshore current velocities. Generally, bed resistance (bottom friction) determines current velocity, whereas lateral mixing determines the shape of the current profile. Maximum longshore velocities can reach 1 to 2 m·s.$^{-1}$, and mean values across the surf zone are generally about a third of this.

Where there is a longshore variation in wave height, water inside the breakers tends to flow toward areas of lower water level. A rip current will then develop there, as a low wave height results in a lower setup and water from the higher setup flows in. If the wave

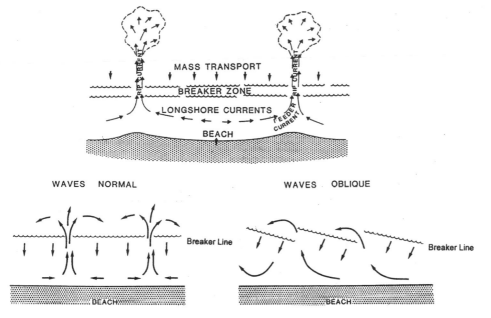

Figure 2.16. Nearshore circulation with mass transport by breakers leading to longshore currents feeding into rip currents. Discharge of rips outside the breakers results in surf circulation cells that differ under normal and oblique wave attack.

approach is oblique, the rip current will be superimposed on a longshore current system. If the incident wave pattern changes with time, the entire meandering current pattern migrates alongshore.

2.10   Embayments and Headlands

The foregoing applies to long beaches without boundary effects. However, headlands, rocks, and other structures impact the beach and surf zone through their influence on wave refraction and attenuation, and by limiting the development of longshore currents, rips, and rip feeder currents. As wave height increases and shoreline length decreases, a threshold is reached where the beach model is increasingly modified. On beaches without headlands, normal surf zone circulation prevails. When headlands are present but widely spaced and/or the beach receives low waves, the beach becomes transitional, as the headlands impact surf zone circulation only locally. As wave height increases and/or the headlands are closer together, the entire beach circulation may become impacted. At this stage, topographically controlled cellular circulation develops, including megarips under high wave conditions (Short 1996).

Normal circulation occurs when beaches are long enough not to be affected by headlands. Transitional circulation occurs when the embayment size and shape begin to increasingly influence the surf-zone circulation, initially by causing longshore currents to turn and flow seaward against each headland, while still maintaining some normal beach circulation away from the headlands.

Cellular circulation occurs when the headlands control the circulation within the entire embayment. Longshore flow dominates within the embayment, with strong seaward-flowing topographic rips occurring at one or both ends of the embayment. In longer embayments, megarips may also occur away from the headlands.

Figure 2.17. Breaker wave height plotted against shoreline length to indicate the range of the normal, transitional, and embayment beach models (broken lines) and the predicted megarip shoreline spacing (solid lines with distances in m) (after Short 1996). The largest megarips are produced in the longer embayments by the higher waves, whereas small embayments will shift to embayment circulation under relatively small waves.

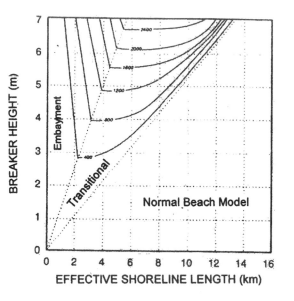

Figure 2.17 illustrates the combination of embayment dimensions and breaker wave heights that produce transitional and embayment circulation, together with the approximate spacing for the megarips. Small embayments (< 2 km) may shift to embayment circulation when waves exceed 3 m, but long embayments (> 5 km) require waves to exceed 6 m to reach embayment circulation with rips spaced 2 km apart.

Embayments and the megarips they produce also influence beach erosion and the seaward extent of surf-zone circulation. Whereas normal beach rips usually begin to dissipate seaward of the breaker zone, megarips may flow at high velocity (~2 to 3 m/s) up to 1 km seaward of the breakers. This has important implications for beach erosion and for seaward transport of sediments, nutrients, and organisms. The net result of embayment circulation is more rapid and severe beach erosion on such beaches, with the eroded sediment taking longer to return to shore (commonly 2 to 5 years following severe erosion).

2.11 Swash Climate

We have seen that waves transform into breakers, then into bores, and finally into swashes as they approach the shore. Although all these stages influence overall beach morphodynamics, the macrofauna organisms living in the intertidal zone mostly experience wave-driven water movement as swash. Swash climate on the beach face therefore deserves some attention.

Swash climate is closely coupled to beach type and can be quite well predicted by beach face slope (McArdle and McLachlan 1991, 1992). Traces of swash movement over 15-minute periods on reflective, intermediate, and dissipative beaches are shown in Figure 2.18. The following are some features ecologists can measure as the swashes run over the beach face.

- Swash length: the distance from the point of bore collapse to the upper limit of the swash on the beach face.

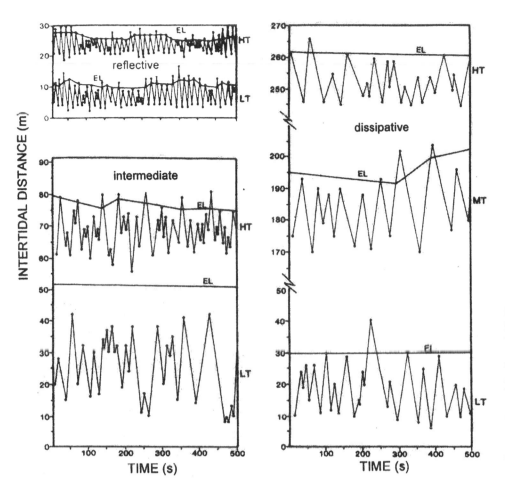

Figure 2.18. Swash profiles over 8 minutes for a typical reflective, intermediate, and dissipative beach (after McArdle and McLachlan 1991). Distance is relative to low-water mark. HT, MT, and LT refer to periods of high, mid, and low tide, respectively. EL = effluent line.

- Swash period: the average time between swashes, which can be divided into upwash time and backwash time.
- Swash speed: the swash length divided by upwash time.
- Effluent line crossing: refers to swashes running above the water table outcrop (i.e., reaching unsaturated sand).

Steep, narrow reflective beaches have swashes that are brief (i.e., short periods and short lengths), whereas dissipative beaches show the opposite trend (i.e., long swashes of long period). In the reflective situation, every incoming wave surges up the beach face to generate a swash, whereas in the dissipative situation much wave energy is consumed in the surf zone so that only longer-period infragravity swashes or bores reach the beach face. Further, because of the low position of the effluent line (water table outcrop) on reflective beaches most swashes reach above this, moving over unsaturated sand and flushing water into the sand. On dissipative beaches, by contrast, the saturated nature of the sand and high-water table (effluent line) results in most swash activity moving over saturated sand and hence little water input. Swash activity above the effluent line, prevalent on reflective beaches, may cause difficulties for macrofauna (which may not be able to burrow into unsaturated sand).

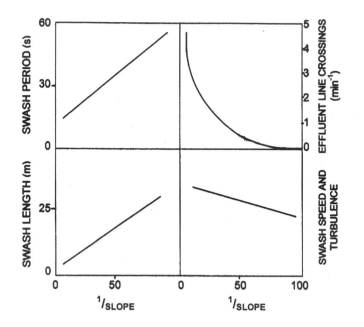

Figure 2.19. Generalized relationships between some swash features and the beach type as indicated by the reciprocal of beach face slope (after McArdle and McLachlan 1991, 1992).

Table 2.3.   Swash characteristics of different beach types.

Beach type →	Reflective	Intermediate	Dissipative
Beach slope	Steep	Moderate	Flat
Swash period	Short	Intermediate	Long
Swash length	Short	Medium	Long
Swash speed	Fast	Fast	Variable
Effluent line	Low	Intermediate	High
Turbulence	High	Intermediate	Low

Changes in swash climate toward more dissipative beaches include longer periods, fewer effluent line crossings, less turbulence, and more variable swash speeds (Figure 2.19). This is summarized in Table 2.3. Swash climate correlates closely with beach face slope. Overall, Figure 2.19 and Table 2.3 indicate a more benign swash climate toward dissipative conditions (i.e., flatter slopes). Denny (1988) has shown for rocky shores that swash run-up speed increases as the slope steepens for any height on the shore, so that organisms living on a gently sloping beach are subject to less energetic water flows than those on steeper beaches.

2.12   Slope

We have seen that beach face slope (for example, the average gradient between the high-water drift line and the low-tide swash zone during spring tides), can be related to most aspects of sandy-beach morphodynamics. Not only does slope predictably flatten from reflective to dissipative beaches but slope flattens in response to finer sand, larger waves, and larger tides (Figure 2.20) and is indicative of swash climate. Beach face slope generally ranges from 1/10 (5.7 degrees) for steep reflective systems to 1/100 (0.6 degrees) for very dissipative beaches. Values steeper than 1/10 generally indicate very coarse sand or gravel, and values flatter than 1/100 indicate the transition to tidal flats. Thus, beach

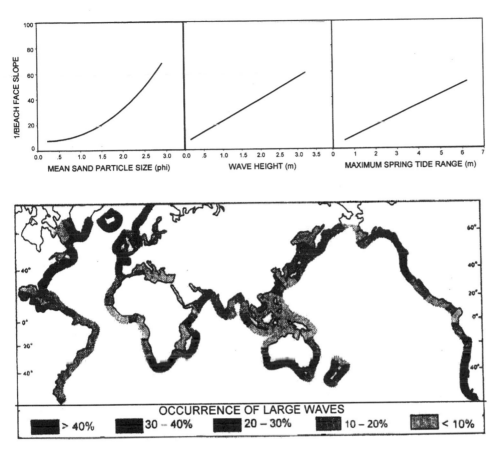

Figure 2.20. The response of beach face slope to changes in sand particle size, breaker height, and tide range. Slope is plotted as the reciprocal of beach face slope. Data obtained from McLachlan and Dorvlo (2005).

Figure 2.21. Global distribution of wave energy around the coastlines of the world (after Davies 1977). Percentage occurrence in at least two quarters of the year of waves 2.5 m and higher.

face slope (easily measured by ecologists) is a useful index of beach type. It is often convenient to use the reciprocal of slope, which is usually in the range 10 to 100. Beach face slope is determined by surveying the beach profile, usually from above the drift line down to the extreme low-water swash zone during spring tides. Surveying can be done simply using a ranging pole, tape measure, and the horizon (or more accurately with a level).

2.13 Latitudinal Effects

Latitudinal changes in climate cause changes in physical factors that impinge on sandy beaches. The most significant change is an increase in wave energy from the tropics toward cold temperate areas (Figure 2.21), which causes a shift in beach types. Tropical areas tend to have a greater proportion of marine biogenic calcareous sands, which are often coarser than the fine sands of terrestrial origin on temperate beaches. As a consequence of the lower waves and coarser sand, tropical areas tend to have a preponderance of reflective beach types, whereas temperate areas with higher waves and finer sand have a greater proportion of intermediate to dissipative beaches (Soares 2003, McLachlan and Dorvlo 2005).

2.14 Conclusions

The general features of sandy beaches are summarized in Figure 2.22. An ecologist studying a sandy shore should consider a number of physical parameters when assessing and

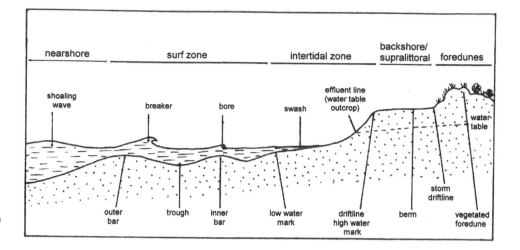

Figure 2.22. Features of a typical sandy beach at mid tide.

describing the environment. It is suggested that any description of a beach/surf-zone study area should include at least the following.

- Sediment parameters from samples taken at the drift line, midshore and lower shore, and from the surf zone if this is possible.
- A description of wave climate, prevailing winds, and tidal regime.
- A description of modal (and range of) morphodynamic states, including DFV (Ω), RTR, slope, and BI. This requires knowledge of sand grain size, tidal regime, breaker height (wave climate), and beach face slope.
- General description of geomorphology of the foredunes, berms, cusps, drift line, beach length, nearby rocks, and other relevant features.
- Where feasible, swash climate parameters can be measured for 15-minute periods during high, mid, and low tide.
- Depth of reduced layers, if present, in the sand (see Chapter 3).

The sandy beach and its surf zone constitute a physically controlled environment where wave energy is the driving force for most physical, chemical, and biological processes. It is therefore essential that the ecologist have a good understanding of the physical structure and dynamics of these systems. This means familiarity with the concepts outlined in this chapter and the ability to recognize these features and processes in the field.

The Interstitial
Environment

3

Chapter Outline

3.1   Introduction

The sand body of a beach consists of sediment grains and the pore spaces between them, the latter constituting the interstitial system. This lacunar system is important as a habitat for organisms and for the filtration of seawater. Sediment particles that define the interstitial system are usually mixtures from several sources, often having been reworked and previously incorporated through geological time in a number of coastal environments. Various factors control the dimensions of the interstitial spaces. Properties of the sediment and resulting sand bodies important for defining the interstitial environment include grain size, sorting, shape, packing, porosity, pore size, permeability, and possibly thixotropy/dilatancy. These are discussed in turn in the sections that follow. This chapter is based on the review of McLachlan and Turner (1994).

3.2   Characteristics of the System

Grain Size

Sand grain size is most commonly described by the Wentworth scale (defined in Section 2.2). This logarithmic phi scale is preferred by sedimentologists and geologists because of the convenience with which most common sediment sizes less than 1 mm can be referred to. The linear millimeter scale may be more useful for engineers and other fields, where quantitative physical processes are the focus. Biologists have tended to use both scales interchangeably. On most ocean beaches, the majority of particles falls in the range 0.1 to 1.0 mm (i.e., fine to coarse sand).

Mineralogy

Most sand grains on ocean beaches fall into two mineral categories: quartz fragments (along with other particles originating from the weathering of rocks) and calcium carbonate fragments of biogenic origin (the proportion of the latter generally increasing toward the tropics and in arid temperate latitudes). Quartz occurs in granites and other igneous rocks, and upon weathering forms grains generally < 1 mm in diameter. Calcium carbonate grains are usually larger and more angular than quartz particles. Like quartz, most other sedimentary particles are derived from the silica tetrahedron with various additions and modifications (e.g., feldspars and clays). Beach sands are usually mixtures of grains of more than one mineral type.

Sorting

Sandy beaches are not composed of uniform sediment of a single size. Thus, a measure of sorting provides insight into the distribution of grain sizes present within a sample. Sediment sorting is defined as the standard deviation of grain sizes about the mean. Hence, a small value for sorting indicates relative uniformity of sediment, whereas a large value for sorting indicates poor sorting (i.e., that a wide range of grain sizes is present). Other statistical measures of sand samples include skewness and kurtosis, which (respectively) are measures of the asymmetry and abundance of extreme particle sizes within a sediment distribution. Beach sands, because of the strong sorting action of waves, tend to be well sorted with limited skewness or kurtosis. These factors were covered in Section 2.2.

Grain Shape

Due to their origin as fragments of crystals, which in turn are chips of amorphous material of geological or biogenic origin, sand grains are rarely spherical. There is little agreement as to an appropriate nomenclature to describe the diverse range of grain shapes found on sandy beaches, both sphericity and roundness being useful measures. Sphericity is a measure of how closely the volume of a particle compares to the volume of a circumscribing sphere. Roundness refers to the outline of the grain and compares the average radius of corners to the radius of the maximum inscribed circle. These two measures are often described jointly — in that sphericity is unlikely to change greatly by abrasion during transport and is therefore a fairly fundamental property of the grain, whereas roundness can be altered significantly during transport. The quantitative analysis of grain shape involves a tedious measurement of individual grain dimensions. As a result, visual comparison to standard charts is often more practical. In general, quartz sands are more spherical and rounded than carbonate sands, and roundness tends to increase landward across the beach and toward finer grades.

Porosity

An accumulation of sand comprises discontinuous particles (which may touch at points) and continuous channels and lacunae (the interstitial system). The ratio of total void volume to the total sediment volume (both grains and voids) defines sediment porosity. The porosity of a sediment depends on the arrangement of individual grains (sediment packing). The packing of a sediment is in turn related to both the sorting and shape of sand grains, and to the nature of sediment deposition. For the simple case of uniform spheres, there are four ways of packing grains, the densest having a porosity of 26% and the most loosely packed a porosity of 48%. Close packing can take the form of four or six spheres enclosing voids that will be tetrahedrons or cubes with sizes 0.225 or 0.414 of the sizes of the spheres, respectively. For an infinite lattice of spheres, there will be twice as many voids as spheres (and half will be tetrahedrons and half will be cubes). If smaller spheres are added, they can fill the voids and reduce them or push the spheres further apart, depending on their size and packing. This applies to the ideal situation of perfect spheres. Natural sands are not uniform, nor are they arranged in this manner.

For natural sands — depending on size, sorting, and packing — porosity may range 20 to 50% by volume, with a mean around 37% for well-sorted beach sands. This porosity can be measured in thin sections or as the amount of water needed to saturate the sand (on a mass or volume basis). When measured on a mass basis, porosity generally falls in the range of 15 to 25% and must be converted to volume using the densities of water and sand (e.g., quartz = 2.65) at the appropriate temperature. For graded sands, porosity is typically 16 to 25% by mass and 30 to 40% by volume.

Maximum porosity of natural sands is obtained with well-rounded grains and very well-sorted (nearly uniform) sand. Porosity decreases as grains become more angular and varied in size, and lowest porosity occurs where very coarse and fine sand mix — resulting in the most efficient packing because fine particles fill the voids between the large

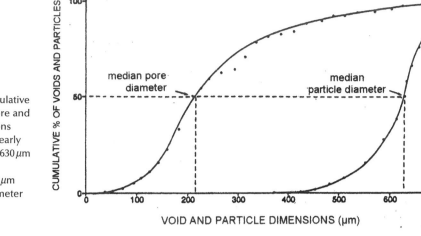

Figure 3.1. Cumulative curves of void/pore and particle dimensions for an artificial, nearly uniform, sand of 630 μm median particle diameter and 220 μm median pore diameter (after Crisp and Williams 1971).

ones. Porosity is thus determined by the size and shape of the dominant grain size fraction and the sizes and proportions of other fractions mixed with it. There also tends to be an increase in porosity with a decrease in the mean grain size of the dominant fraction, in that the finer the sand the less the chance of still finer particles filling the voids.

Pore Size

The minimum number of pores leading to a dilation or cavity is four for tetrahedral packing, but much higher numbers can be obtained (and six may be a good average). Measuring this in thin sections gives values of 0.15 to 0.20 for poorly sorted natural shell gravels and 0.3 to 0.4 for nearly monometric uniform spheres (Crisp and Williams 1971, Figure 3.1). For natural beach sands, mean pore size may be 0.2 to 0.4 mean particle size (so that for fine to medium beach sands most pores will be in the range 10 to 200 μm).

Permeability

The porosity or total void volume and the rate at which water can flow through a sandy beach are important characteristics of the interstitial environment. The discharge rate of a liquid through a body of sand depends on grain and liquid properties as well as on the pressure head during the flow. The expression for the discharge of water (Q) moving a distance (L) through a cross-sectional area (A) in time t due to a hydraulic head (h) is known as Darcy's law and is given by either of the following.

$$Q = KAht/L \, (m^3 \cdot s^{-1})$$

$$K = QL/Aht \, (m \cdot s^{-1})$$

Here, K is the hydraulic conductivity (or permeability) of the sediment with units of meters per second. There is often confusion between hydraulic conductivity and specific permeability (k), whose dimensions are length squared. Specific permeability is a property of porous media only, and is related to hydraulic conductivity by

$$K = kgP/v,$$

where v and P are the kinematic viscosity and density of the fluid and g is acceleration due to gravity.

The hydraulic conductivity of beach sands typically ranges from 10^{-2} to $10^{-6}\,m\cdot s^{-1}$. Well-sorted coarse sands have the highest permeabilities, and poorly sorted fine sands have the lowest. Permeability decreases dramatically with the addition of very fine sand, especially silt, because of its effect in filling and blocking the voids. Permeability also depends on fluid viscosity, which is influenced by temperature. Water, for example, is twice as viscous at 5° C as at 30° C, causing seasonal changes in the permeability of fine sands in temperate climates.

The porous system of the sand includes both capillary space and cavity space, the ratio between the two defining the specific permeability defined previously. Water flows through pores and is stored in cavities. Thus, this ratio gives an idea of movement relative to stagnant water in the interstitial system and may be used for the study of interstitial fauna. This index is specific for a particular lattice, irrespective of the degree of compaction, and is therefore fairly constant for any sand.

Moisture Content

Pore spaces within a sandy beach may be entirely filled with water, may contain a combination of air and water, or may be devoid of moisture. Capillary forces (the mutual attraction of water molecules and the attraction of water molecules to sand grains) can draw water 4 to 50 cm up a column of sand, depending on particle size and sorting (typical values for beach sands being 20 to 30 cm). Capillarity increases with decreasing grain size (Figure 3.2). Capillary forces can supersaturate sand, pushing the grains apart and increasing porosity.

When water leaves the sand it is replaced by air, which can remain behind after the sand is inundated again — thereby reducing the permeability, although not changing capillary lift. About 85% of the air in the intertidal sand body of a beach may be displaced from the sand when the tide comes in.

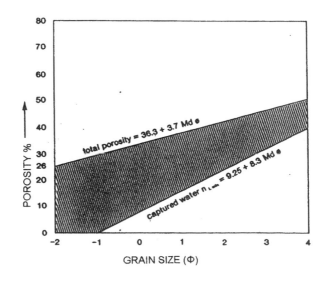

Figure 3.2. Porosity and capillarity (captured water) as a function of sand grain size (after Sakamoto 1990). Shift porosity (shaded) is the proportion of total porosity that can gain and lose water.

In moist sand, surface tension holds grains together with rings of moisture at their points of contact, forming a firm lattice. This pendular water, occupying about 20% of void space, gives moist sand the cohesion needed for making sand castles!

Capillary rise fills about 95% of void space just above a water table, but only 25% of void space at its upper limit where there are more air bubbles in the sand. A cross section from the sand surface to the water table shows the following in terms of moisture: dry sand, moist sand retaining pendular moisture, sand wetted by capillary rise, and finally saturated sand at the water table (Webb 1991).

A volume of beach sand may thus contain sand particles (about 60 to 65% of volume), captured water (e.g., 10 to 20%), and air bubbles (e.g., 15 to 20%) alternately filled with water and air (Sakamoto 1990). A relation between porosity and mean particle size showing both captured (capillary and/or pendular) water and "shift" porosity (or space available for new water) is depicted in Figure 3.2.

Thixotropy and Dilatancy

Thixotropy is the term given to the reduction in resistance of sand with increased rate of shear, as opposed to dilatancy (where increasing shear force causes increased resistance). This is of particular significance for animals that burrow in sand, in that dilatancy makes burrowing impossible. Thixotropy is mainly dependent on the water content of sand, although the fluidity of the sand is also a function of the viscosity and density of the liquid filling the interstices. Saturated fine sand exhibits maximum thixotropy. In extreme cases, this takes the form of a tendency to liquefy like quicksand — where the sand is supersaturated due to expansion of the lattice by capillarity, thereby reducing the contact between grains and increasing the number of floating grains. This is more important for macrofauna, which burrow in the sand, than for interstitial fauna.

3.3   Processes of Water Input

We have seen that the interstices of sandy beaches vary in size and shape due to grain characteristics and the arrangement of the sand matrix, and that the moisture content within this environment will range from dry to saturated conditions. The input of water to the system is via a combination of terrestrial, marine, atmospheric, and biological processes. These processes may include precipitation, groundwater discharge, tides, wave run-up, subtidal wave pumping, and bioturbation.

The activities of animals, such as burrowing, can irrigate the sediment. This can be important in deep fine sediments where other input mechanisms are absent or in dense beds of thalassinid prawns on sand flats, but it is relatively insignificant on ocean beaches. Rain also represents a small and sporadic input of freshwater, but this is very limited in comparison to the other inputs. We shall therefore not discuss precipitation or bioturbation further.

Groundwater Discharge

Sandy coasts generally act as unconfined aquifers connected hydraulically to the sea through permeable beach sediments. Their hydraulic heads are above the sea and they

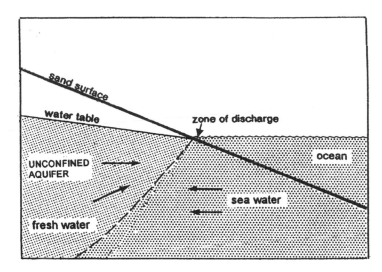

Figure 3.3.
Diagrammatic representation of an unconfined aquifer discharging through a beach system.

are therefore characterized by groundwater discharge to varying extents. The groundwater discharge rate depends on the hydraulic head (height of the water table above sea level) and the permeability (hydraulic conductivity) of the sand.

This water may discharge on the beach or in the subtidal, in that it typically flows out between the fresh/seawater interface and the water table (effluent line) outcrop on the beach. The discharge to the sea will be brackish because of mixing with saltwater in the zone of diffusion. The thickness of the discharge zone is proportional to the volume of freshwater flow. A wedge of intruding saltwater typically underlies the aquifer along the coast, impeding the downward mixing of less dense freshwater and forcing the aquifer to discharge close to shore (Figure 3.3).

The discharge zone may be in the intertidal during low tide and inundated during high tide. In exceptional cases, a high-water table can result in the entire beach being the zone of discharge. Water in the main aquifers typically moves slowly seaward at 0.0001 to $0.01\,\mathrm{m \cdot h^{-1}}$. However, discharge at the beach is faster (typically $0.1\,\mathrm{m \cdot h^{-1}}$) because it is focused through a narrow zone. On the global scale, groundwater discharge to the sea is much less important than river discharge, but it is significant as a supplier of nutrients to coastal waters and for its role in water filtration, its influence on interstitial salinities, and its enhancement of beach face erosion.

Tides

The rise and fall of the tides across the intertidal region of a sandy beach produces an alternately landward-directed then seaward-directed hydraulic gradient at the frequency of the local tides. This necessitates the flow of water into and out of the beach. Due to the ability of water on the rising tide to infiltrate almost vertically into a beach much more rapidly than it can drain out nearly horizontally on the falling tide, there is the tendency for the super-elevation of the beach water table above the mean sea level. Water input therefore only occurs when the elevation of the tide exceeds the elevation of the beach water table. Thus, input occurs on the rising tide and discharge mainly on the outgoing tide. This can be discerned as a rising and falling water table, including a tidal wave that decays back into the beach sand body. Although output may occur all of the time, input only occurs when the tide rises above the water table outcrop.

Simple patterns would result from tidal inputs alone: the sand would fill with water up to the high-tide mark on the rising tide, resulting in a water table sloping off landward. This would be followed by drainage on the falling tide, initially both landward and seaward and then only seaward. The rates of these processes would depend on the beach face slope, permeability, and tide range.

Beach Face Wave Run-up

In addition to tides, ocean beaches are subject to wave effects. Water inputs to the intertidal occur when swashes run up the beach face, cross the water table outcrop, and infiltrate water into unsaturated sand. Wave run-up on the beach face raises the local water level, which in itself drives flow into the beach and sometimes an offshore bottom current (bottom return flow) in the surf zone. Because the frequency of waves and swashes on ocean beaches is nearly 10^3 times that of the tides, they represent a considerable amount of hydrodynamic energy and are usually more important than the tides as an input mechanism, especially on coarse sand beaches that drain more rapidly.

Riedl (1971) described the mechanism of seawater input to the beach face by swash action in terms of filling wedges (Figure 3.4). The beach surface below the drift line shows two boundaries: (1) the water table outcrop (or effluent line), below which the surface sand is saturated and glassy in appearance, and (2) the water's edge (or swash line), which pulses up and down the beach as swash and backwash. Above the water table outcrop the water table slopes landward into the sand, drawing air into the capillary network. Thus, a section cut through the beach perpendicular to the shoreline will show a wedge-shaped body of unsaturated sand between the water table and the surface above

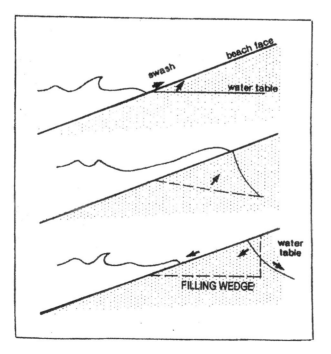

Figure 3.4.
Representation of a filling wedge in the beach face (after Riedl 1971). Arrows indicate direction of water table movement.

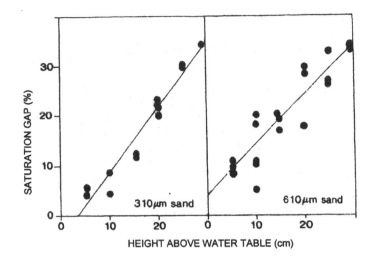

Figure 3.5. Saturation gaps or percentage air space for two different sands (after McLachlan *et al*. 1985).

the outcrop. Close to the water table (in the area of capillary rise), the air space in the sand will be small, but above this there will be increasing air space. The sharp seaward edge of this wedge will migrate seaward as water seeps out of the beach, surface drainage reducing pore pressure and causing a loss of the glassy appearance even though sand just above the water table may remain saturated. Whenever a swash overruns the water table outcrop and the top of the saturated capillary fringe (generally 5 to 20 cm above the outcrop, depending on particle size), it will fill this wedge as far up as it reaches. This input water will displace a corresponding volume of air. Once the backwash retreats, the outcrop will again migrate seaward as water drains out of the sand. This cycle of alternate landward (after a swash) and seaward (after a backwash) migration of the water table outcrop will be repeated for every wedge-filling swash.

The dimensions of a filling wedge will depend on the slope of the beach and the depth of the water table, the length of the swash above the outcrop, and the amount of air space in the sand (the saturation gap). Because of capillary forces, saturation gaps will generally be zero just above the water table, but may increase to as much as 30% 20 to 30 cm above the water table (Figure 3.5). However, using a neutron probe technique Turner (1993) has indicated that saturation gaps may be lower than estimated by other techniques and that the saturated zone may extend tens of centimeters above the water table. The wedge is filled with water not only from the surface, but also by capillary rise and lateral seepage. This especially occurs when the beach is flat and the water table responds and rises ahead of the advancing swash. Conversely, on steep coarser-grained beaches the water table may be left behind and air may be trapped between water infiltration from above and the water table rising from below. The wedge in fact is not flat. Its bottom may be curvilinear and not sharply defined (Riedl 1971). The slope of the filling wedge can range from 1/10 to 1/80, being steepest toward high tide.

Subtidal Wave Pumping

Water input through the bed in the subtidal zone occurs as a result of wave pumping (i.e., it is driven by the pressure differences between wave crests and troughs passing overhead). This inflow and outflow can operate over much of the beach during high tide and through all states of the tide in the subtidal. In the intertidal, wave pumping will often be working against the water that is draining out of the sand due to the water table head. However, pumping currents may have twice the amplitude of gravity drainage currents (Riedl et al. 1972). The energy consumed in the process of water percolating into and out of the bed results in wave energy attenuating as waves propagate in toward the shore. Unlike the intertidal swash filtration process, this subtidal pumping is oscillatory flow, with input and output approximately balanced.

3.4   Water Filtration

As outlined in Section 2.3, the driving forces for water filtration through beach sands are hydraulic gradients resulting from tides, waves, and aquifer recharge. The moisture content of beaches varies both spatially within a beach and temporally over the tidal cycle. Having covered the mechanisms responsible for water input to the interstitial system of sandy beaches and the distribution of this water within the beach sand matrix, we now examine the filtration process and quantities of water filtered.

Volumes and Residence Times of Tide and Wave-driven Inputs

The inputs of seawater through intertidal sandy beaches have been estimated by several authors since Riedl (1971) and summarized by McLachlan and Turner (1994). On an intermediate beach in North Carolina (mean wave height of 1 m and tide range of just over 1 m, slopes 1/15 to 1/25, and mean grain size 250 μm), an average filtered volume was estimated at 6 $m^3 \cdot m^{-1} \cdot d^{-1}$ (Riedl and Machan 1972). Input ranged from 0.09 to 0.7 $m^3 \cdot m^{-1} \cdot d^{-1}$ over the tidal cycle, lowest values being just after low tide. It was estimated that tides alone could account for about 25% of the filtered volumes, and mean percolation path length (the filtration distance through the sand from input to discharge) was 24 m (or 35% of the intertidal distance) and mean percolation time was 22 h.

Over a range of exposed intermediate to dissipative South African beaches, filtered volumes ranged from 1 to 12 $m^3 \cdot m^{-1} \cdot d^{-1}$, being highest where beaches had steepest slopes and small to moderate tides. There was a clear pattern, with greatest inputs on the late incoming tide (Figure 3.6) for beaches of 200- to 300-μm sand and a maximum tide range of 2 m.

Microtidal reflective beaches near Perth, Australia, with medium to coarse sand, modal wave heights of 0.4 m, and a maximum diurnal tide range of 0.9 m had filtered volumes of 19 to 92 $m^3 \cdot m^{-1} \cdot d^{-1}$ on two wave-exposed beaches, indicating high input volumes on beaches with steep slopes, coarse sand, and small tides. A third beach, covered in algal wrack that filtered out swash effects, received < 1 $m^3 \cdot m^{-1} \cdot d^{-1}$. Residence times for this filtered water in the beach sediments were estimated at 1 to 7 h and > 90 h on the wrack-covered beach. These are matched by mean percolation paths of 2 to 5 m. These beaches clearly had water input driven mainly by waves (> 90%), with tides playing a small role (McLachlan et al. 1985).

Two dissipative meso/macrotidal beaches in Oregon with fine to medium sands, flat slopes, mixed tides up to 3.6 m, and subject to large waves (the modal breaker height being 1 to 2 m in summer and 3 to 4 m in winter) had low filtered volumes (0.1 to 7 $m^3 \cdot m^{-1} \cdot d^{-1}$) and long resident times (15 to 400 days). During a major storm with 7-m waves, one beach filtered 31 $m^3 \cdot m^{-1} \cdot d^{-1}$. These inputs were almost entirely accounted

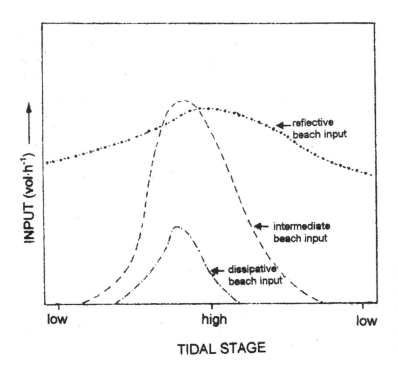

Figure 3.6. Changes in water input to reflective, intermediate, and dissipative beaches by swash and tide action over a tidal cycle (after McLachlan 1982).

for by tidal effects, most wave energy being filtered out by the dissipative surf zones except during storms (McLachlan 1989). However, because of the wide saturated sand surface typically associated with dissipative beaches they would be expected to filter significant volumes by wave pumping during the high tide. Sakamoto (1990) estimated water input to a Japanese beach by tides alone at $6\,m^3{\cdot}m^{-1}{\cdot}d^{-1}$ during a 1.95-m spring tide and $0.7\,m^3{\cdot}m^{-1}{\cdot}d^{-1}$ during a neap tide with a range of 0.31 m.

A plot of input or filtered volume, residence time, and relative importance of waves versus tides is shown in Figure 3.7 for different beach types. In reflective beaches, large volumes of water are filtered rapidly and input is driven primarily by waves. On dissipative beaches, smaller volumes are filtered slowly and input is driven mainly by tides. Intermediate beaches are intermediate in all of these parameters, and both waves and tides play a role. Where waves have direct access to the beach face and are not too greatly dissipated in wide surf zones, they greatly increase input volume by swashes above that accounted for by tides alone. Turner (1993), however, suggested that saturation gaps might be overestimated (i.e., by the presence of the saturated capillary fringe) by the methods employed and consequently that filtered volumes may be more conservative than indicated in Figure 3.7.

Flow Patterns and Interstitial Climate

Seawater infiltration into and through sandy beaches is greatest toward the late incoming tide (Figure 3.6), when the swash zone is over the unsaturated upper shore and the water table is rising rapidly. It is a complex process driven by tidal and swash effects. If wave pumping below the swash zone is ignored, this input by filling wedges consists of two types of flow: (1) slow smooth gravity flows that may drain landward at and above the input zone on the incoming tide and seaward below this and throughout the beach on

INPUT MECHANISM

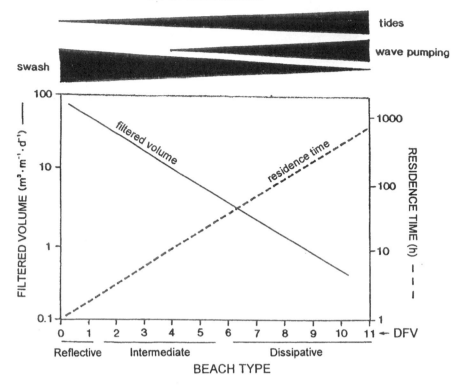

Figure 3.7. Graphic model of the volume of water filtered and its mean residence time as a function of beach type for microtidal beaches (after McLachlan and Turner 1994). The relative importance of tides, wave swash, and wave pumping over saturated sand in driving input is indicated.

the outgoing tide, and (2) more complex pulsing flows resulting directly from the swash inputs (Riedl and Machan 1972). The frequency of the latter is 2 to 4 orders of magnitude higher that that of the gravity flows, which are primarily tidal. Tidally driven gravity flows occur throughout the sand body, whereas swash-driven pulsing flows only occur near the surf and swash (i.e., the input zone).

In the absence of waves, only gravity flows will occur. Seaward gravity flows dominate at all times except on the mid to late incoming tides, when input is high and landward flows are important (especially at and above the input zone). A point on the midshore will therefore experience a sequence of gravity currents over a tidal cycle as illustrated in Figure 3.8.

Swash pulsing is driven by swashes filling wedges above the water table outcrop, this input extending downward and sideways from the wedge and spreading predominantly in the cross-shore direction. Swash pulsing is superimposed on the more constant gravity flows. Therefore, the pulsing flows will be retarded on one side of the swash zone (e.g., landward if water is flowing seaward) and boosted on the other (e.g., seaward), where they run against and with the gravity currents, respectively. Current speeds in the interstices can be up to $370 \, \mu\text{m·s}^{-1}$ and can even reach $2 \, \text{mm·s}^{-1}$ in the filling wedge. These are high speeds for interstitial animals.

Rhythmic discharge of the head of water in the wedges influences flow far down into the interstices. Riedl and Machan (1972) called this region of pulsing flow the filling bag and defined it as the area where pulsing currents exceed $10 \, \mu\text{m·s}^{-1}$ (Figure 3.9). Greatest velocities and velocity changes are encountered at the wedge/bag boundary, and these

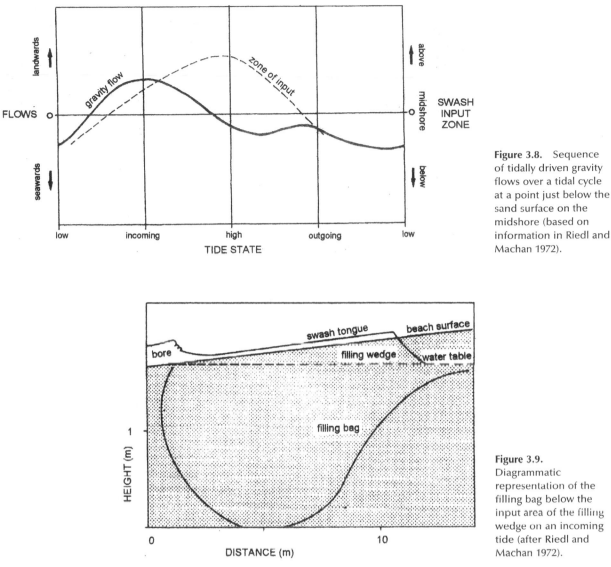

Figure 3.8. Sequence of tidally driven gravity flows over a tidal cycle at a point just below the sand surface on the midshore (based on information in Riedl and Machan 1972).

Figure 3.9. Diagrammatic representation of the filling bag below the input area of the filling wedge on an incoming tide (after Riedl and Machan 1972).

drop off deeper into the bag, which can be as much as 1 m into the sediment. Input thus proceeds from the filling wedge to the bag, and then drains into the rest of the porous system. Whereas the input zone migrates over the shore with the tides, the output area (extending seaward from the bag) expands when the tide rises and shrinks as the tide falls. The mean length of the percolation pathway is about 35% of the intertidal width. Paths would therefore be shorter in narrow reflective beaches and longer in wide dissipative beaches.

Surf pulsing occurs in the sand below waves and is caused by pressure and motion fields, which are generated by waves, penetrating into the sediment. It is thus strong in the submerged part of the intertidal (Riedl and Machan, 1972). On the beach this is superimposed on the seaward gravity current, which has a speed just more than half the amplitude of the surf pulsing current. Thus, the gravity current is continuous but varies with the amplitude of the surf pulsing current.

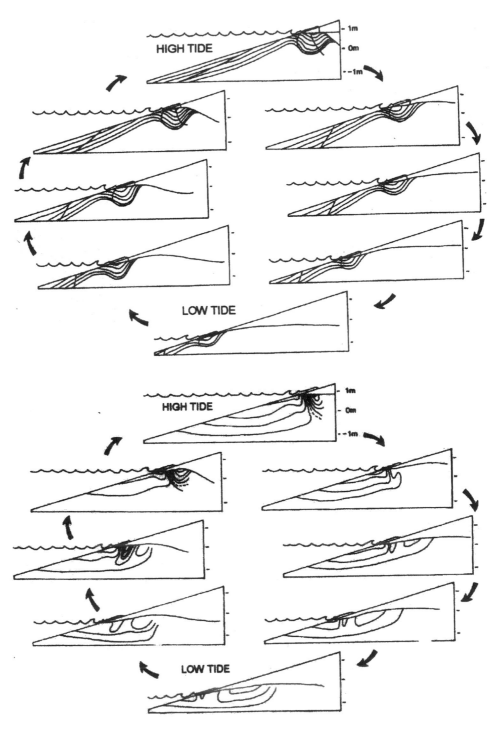

Figure 3.10. Velocity field of pulsing (swash) flows over a tidal cycle (after Riedl and Machan 1972). Contours (isotachs) represent flows of 10 to 1,000 $\mu m \cdot s^{-1}$ and show most vigorous interstitial water movement at and below the input zone during the high and incoming tides.

Figure 3.11. Velocity field of tidally driven gravity flows over a tidal cycle (after Riedl and Machan 1972). Contours represent flows of 10 to 500 $\mu m \cdot s^{-1}$ and show strongest flows during the late incoming and high-tide times.

The previously cited flows occur in various combinations and will vary in proportions and timing between beach types. Movement of the velocity field for interstitial currents driven by swash and surf pulsing over a tidal cycle is shown in Figure 3.10, and that for gravity flows is shown in Figure 3.11.

Interstitial climatic parameters derived from a synthesis of the previously cited processes and currents are shown in Figure 3.12. For a typical beach, vigorous input

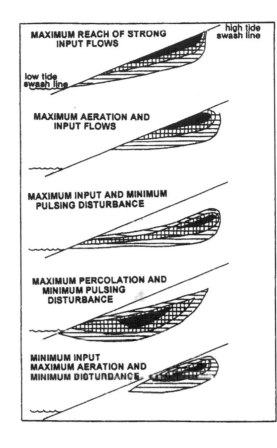

Figure 3.12. Diagrammatic representation of interstitial climatic parameters on an intermediate sandy beach (after Riedl and Machan 1972, McLachlan and Turner 1994).

occurs on the upper shore (where aeration and infiltration are strong to some depth in the sand). Just below this and extending downshore is a stratum characterized by considerable water input and circulation but low pulsing disturbance. The area of maximum percolation but minimum disturbance occurs below the surface on the lower shore, whereas maximum input, aeration, and disturbance of the sand occur near the surface at mid to upper tide levels. Thus, for interstitial fauna requiring inputs of water, oxygen, and food materials optimum conditions may occur in a layer extending from just below the surface near mean tide level landward to well below the surface toward the high-tide level. Above this, disturbance and aeration may become extreme, and below this conditions will tend toward stagnation because of restricted flow and net output.

Subtidal Wave Pumping: Input Volumes and Flow Patterns

Although wave pumping is predominantly a feature of the subtidal, it will also occur in the intertidal during much of the high tide, especially on dissipative beaches. Currents in the interstitial system under this situation have two components: gravity currents draining water seaward from the intertidal, and superimposed on these currents regular variations at wave frequency and amplitude, dependent on wave energy. Flow speeds through the interstices may be in the range 0 to $200\,\mu m \cdot s^{-1}$. The more efficient this pump the greater will be the irrigation and oxygenation of submerged sands and the deeper the reduced layers will be driven. In the shallow subtidal of exposed beaches this process can account for filtered volumes of 0.01 to $1.0\,m^3 \cdot m^{-2} \cdot d^{-1}$ (McLachlan and Turner 1994).

3.5 Water Table Fluctuations

The dynamics of water tables within sandy beaches are critical in distinguishing zones of contrasting interstitial moisture content, potential regions of water inflow and outflow to the interstitial environment, and hence water flow patterns through the beach. Beach water tables are dynamic and respond to, and modify, wave and tide action.

The water table defines the upper boundary of the zone of groundwater, where pore water pressure equals atmospheric pressure, whereas below the water table pore pressure is greater than atmospheric. This definition allows a clear distinction of the water table from the capillary fringe — that region above the water table that is also saturated due to capillary rise — in that the only distinction between these two zones is pore pressure.

Tidal Effects

Water tables in marine sandy beaches fluctuate in response to the tides. The water table in an intermediate beach responds to the tide with a landward-moving pulse that decays back into the beach with a lag of 1 to 3 h over a distance of 10 to 15 m landward of the water line (Figure 3.13). The beach acts as a tidal filter, with water table fluctuations both diminishing and changing phase in the landward direction. The water table slopes landward during the rising tide and seaward on the falling tide. Lag times increase toward the backshore and are greatest on the falling tide because drainage rates are significantly slower than rates of rising-tide infiltration. Because of this asymmetry, the water table surface within a beach averaged over an entire tidal cycle will stand above the mean ocean level.

Groundwater and Swash Effects

Groundwater discharge through sandy beaches can significantly affect water tables. Tidally induced fluctuations in the beach water table are superimposed on the effects of both beach morphology and groundwater recharge from the backshore. The water table

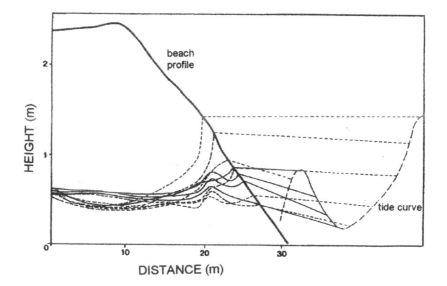

Figure 3.13.
Fluctuations of the water table in a sandy beach over a tidal cycle (after Emery and Foster 1948).

near the swash zone responds to individual waves, but the beach face and sediment matrix effectively act as a filter to these higher-frequency fluctuations, which are minimal above the swash zone.

Influence on Beach Face Erosion/Accretion

The elevation of the beach water table influences the onshore/offshore movement of sand on the beach face. A dry beach facilitates accretion by swash infiltration and hence a reduction in backwash, whereas on a saturated beach outflow supplements backwash, dilates the sand, and encourages erosion. On the rising tide, run-up advances above the water table and onshore transport is promoted due to reduced backwash. In contrast, during the falling tide the water table lags behind the tide and hence backwash is enhanced by water table outflow, resulting in erosion (Figure 3.14).

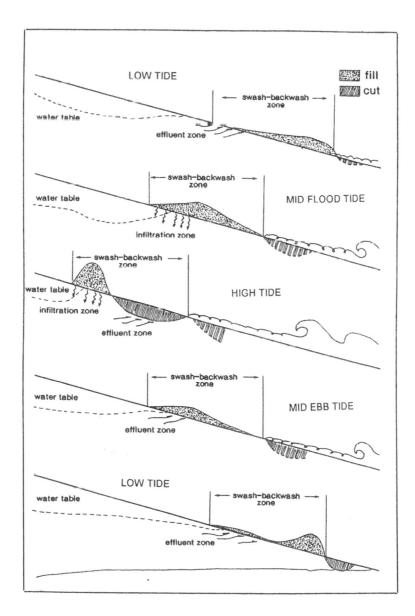

Figure 3.14. Sediment distribution and swash-backwash accretion and erosion on the beach face over a tidal cycle (after Duncan 1964).

On beaches exposed to large tide ranges, the intertidal outcrop of the water table around low tide distinguishes an upper region of the beach that is unsaturated during the rising tide (promoting accretion and beach face steepening) from a lower region that is permanently in a saturated state (enhancing erosion and beach face lowering). The distinct intertidal break in slope often observed on macrotidal beaches usually marks the time-averaged position of this point of divergent sediment transport, often with coarser sediment on the upper beach and finer sediment on the saturated lower beach.

Zones of Interstitial Moisture

Webb (1991) described the moisture zones in beach sand from the surface downward as dry sand, moist sand retaining pendular moisture, sand near the water table wetted by capillary rise, and saturated sand at and below the water table. Much earlier, however, Salvat (1964) had recognized similar zones, which he used to describe animal zonation. His scheme, initially defined for the surface of sandy beaches, has been expanded to describe strata encompassing the sand body. Combining the schemes of Salvat (1964), Pollock and Hummon (1971), McLachlan (1980a), and the information in the foregoing sections, the following stratification of water content can be described from the top of the shore downward (Figure 3.15).

- *A stratum of dry sand* above neap high tides loses capillary and pendular moisture during the low tide and at neap tides. It undergoes strong thermal fluctuations and is immersed irregularly.
- *A stratum of retention* underlies the former zone. Here, the sand remains moist during low tide. This stratum is wetted by all tides, but loses capillary water and retains only pendular moisture during low tide. Loose packing and high porosity and permeability allow extensive percolation of interstitial water. On the surface, most water input occurs here during the mid and later part of the incoming tide (Figure 3.6). Because it does not retain capillary moisture, its lower limit is defined by the upper limit of capillary rise above the water table during low tide, typically

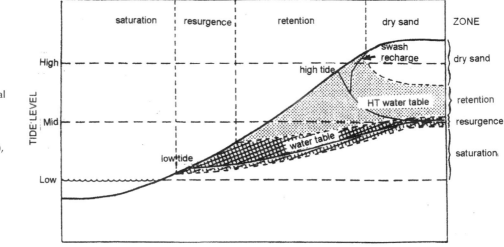

Figure 3.15. Interstitial moisture strata on an intermediate sandy beach. Derived from models in Salvat (1964), Pollock and Hummon (1971), and McLachlan (1980a). (Note: this is expanded to include reflective and dissipative types in Figure 3.17.)

20 to 30 cm above the low-tide water table. It is often the most extensive stratum, especially on well-drained beaches. Temperature variations may be moderate in the upper layers.

- *A stratum of resurgence* consists of wet sand holding capillary moisture and is marked by intense circulation of interstitial water. Much gravitational water drains out through this zone on the ebb tide, this resurgence continuing through the low tide. This discharge consists of water that was input to the retention zone and has circulated through it. Oxygen tensions here are thus slightly lower than in the zone above. The sand is more compact in this stratum, which is underlain by the water table. In steep reflective beaches it will be narrow, but it will widen considerably in flat dissipative beaches. On steep coarse-grained reflective beaches it may extend seaward over the next stratum as a thin veneer reaching to the low-water mark. Temperature variations are small.

- *A stratum of saturation* at and below the permanent water table, where percolation is slow and groundwater discharge occurs. The sand is generally compact and circulation is sluggish. On fine-sand beaches it will be a surface zone, but on very coarse-grained beaches it may not surface at all in the intertidal zone. Where this stratum extends out seaward beyond the resurgence stratum, its surface is subject to the wave pumping described previously and may thus be well oxygenated. Other than this, however, there may be a tendency for stagnation with low interstitial oxygen tensions because water circulating through this stratum generally comes via the retention and resurgence strata, where interstitial fauna utilize much of the oxygen and organic materials. Temperature variations are minimal.

These strata appear as surface zones during low tide, and all except the dry sand stratum become saturated during high tide. In terms of water filtration, the retention stratum is the main receiver of wave- and tide-driven inputs, and the resurgence and saturation strata are the main discharge sites. The dry sand stratum only receives water inputs during spring high tides. Coupled to changes in water flow through these strata are variations in oxygen tensions, interstitial chemistry, and interstitial fauna (see Chapter 9). If these strata are defined by oxygen and water circulation alone, the resurgence stratum would always overlie the saturation stratum. However, if they are defined by moisture (as previously) the saturation stratum will emerge on the lower shore in most cases. The subsurface parts of the retention stratum generally correspond to the area with optimal conditions for interstitial life (see Figure 3.12).

3.6   Interstitial Chemistry

Temperature

Maximum temperature variations in intertidal sandy beaches occur on the surface at upper tide levels, and temperatures become more stable toward the sea and down into the sediment. Extreme temperatures are generally only encountered in the upper few centimeters at mid- to high-tide levels during the low-tide period. Otherwise, most of the sand body takes on temperatures close to those of the adjacent sea. Temperature changes will affect the viscosity of pore water and thereby influence flow rates and capillary forces. Temperature also influences the rates of chemical processes.

Groundwater Inputs

Fresh groundwater discharging from land carries elevated levels of inorganic nutrients, resulting both in inputs to the adjacent sea and elevated nutrient levels in the interstitial system. Groundwater discharge can also influence oxygen levels. Interstitial nutrient levels are usually well above those of the sea, depending on organic load and water circulation time in the interstitial system. Highest nutrient levels occur with high organic loads and long residence times.

Salinity

The salinity of interstitial water is determined both by the salinity of seawater and the extent and distribution of fresh groundwater seepage. Because most beaches experience some fresh groundwater seepage, and this groundwater usually underlies the intertidal sand wedge, salinities will decrease from seawater values in the input zone landward and possibly deeper into the sediment.

Organic Inputs

Organic inputs to beaches occur as dissolved organic matter (DOM) or particulate organic matter (POM). DOM inputs depend on primary production levels in the adjacent seawater and usually exceed POM inputs. This DOM is carried into the interstitial system by water filtration and is capable of supporting diverse interstitial fauna. POM inputs include larger debris cast ashore as well as fine particulate matter that may be carried directly into the interstices. The POM in larger debris will enter the interstitial system after breakdown and consumption by the macrofauna. On beaches adjacent to kelp beds, seagrass meadows, or other areas of macrophytes, this input can be substantial. Benthic microflora constitutes a resident source of organic input from primary production in the interstices. This is generally insignificant in the intertidal zone of exposed beaches, but becomes increasingly important with shelter and distance seaward. Some details of chemical reactions based on organic materials in sediments are outlined in Appendix B.

Oxygen Concentrations

Exposed sandy beaches have been termed *high-energy windows* because, in contrast to most other sediments (which are anoxic below the surface), strong hydrodynamic forces keep them oxygenated (Fenchel and Riedl 1970). Oxygen availability generally decreases into the sand. However, oxygen can be absent a few centimeters below the surface in low-energy fine-sand beaches but near saturation more than 1 m below the surface in well-drained coarse-grained beaches. The availability of oxygen in the interstitial system is crucial in determining the redox status of nutrients, the vertical distribution of redox conditions, and levels of microbiological activity.

Fine-sand beaches tend to be dissipative, with low permeability and filtered volumes. These features and the high biological oxygen demand in fine sands (with greater surface area for bacterial attachment) contribute to low oxygen availability. Every cubic meter of sand of $250\,\mu m$ and 38% porosity provides 1.5 ha of sand grain surface for bacterial colonization. The finer the sand the greater this surface area and the greater the micro-

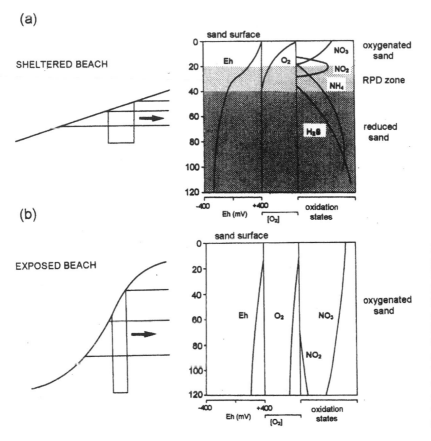

Figure 3.16. Chemical gradients in contrasting beach types: (a) reduced layers in a fine sand beach of low wave energy (the low-energy low-oxygen dissipative type) and (b) the fully oxygenated case in a well drained beach of the high-energy intermediate or reflective type (after McLachlan and Turner 1994).

bial populations. Thus, fine sands tend to develop higher microbial oxygen demands than coarse sands. When such beaches receive high organic loads, reducing conditions will occur. The three redox zones defined by Fenchel and Riedl (1970) will then become evident (Figure 3.16): (1) an upper oxygenated layer with "yellow" sand; (2) a gray transition zone or redox potential discontinuity (RPD), where oxygen becomes limiting and a switch occurs from oxidizing to reducing conditions; and (3) a reduced or black layer discolored by iron sulfides and distinguished by toxic reduced compounds such as H_2S and NH_4. The reduced layer is not black in carbonate sands because of the absence of iron sulfides. Some details of the chemistry associated with these layers and the chemical processes involved in the conversion of organic matter to inorganic compounds are provided in Appendix B.

The amount of oxygenation of the interstitial water, and consequently the depth of reduced layers, depends primarily on the balance between organic input and oxygen input. The depth of the black layers shows an inverse correlation with sand permeability. Low permeability decreases oxygen inputs by diffusion or flushing and brings reduced layers toward the surface. These reduced layers migrate and may move down when storms stir the sediment and introduce extra oxygen; or they may move up when higher summer temperatures raise benthic metabolism and cause more rapid oxygen depletion.

In contrast, high-energy coarse-sand beaches (especially in reflective situation) tend to be well flushed with seawater, so that oxygen may be supplied in excess of the beaches' digestive requirements. Here, aerobic conditions will occur to great depth in the sand. The contrasting vertical gradients in oxygen and nutrient associated with these different beach types can be referred to as the low oxygen dissipative and well-flushed reflective extremes (Figure 3.16).

Beaches have been considered to be in equilibrium with organic inputs from the ocean. Swash and tide effects move water and oxygen through the sand, causing vertical movements of redox boundaries (as do groundwater discharge and gravity currents). Swash input of oxygenated water drives the reduced layers deeper, whereas gravity drainage of filtered water with lower oxygen tensions raises them. Vertical migration of reduced layers in the sand can thus be caused by (1) increasing or reducing wave energy, (2) daily or seasonal temperature changes modifying respiration rates, (3) varying levels of organic input, (4) sunlight causing photosynthesis at the surface, and (5) changing rates of groundwater discharge.

Nutrients

The interstitial systems of sandy beaches have been described as "great digestive and incubating systems" (Pearse *et al.* 1942) because of their role in mineralizing organic materials and recycling nutrients. This refers especially to nitrogen (as nitrate or ammonia) and phosphorus (as phosphate). Much release of nutrients may be regular and governed by water output and diffusion, but storms reworking the sediment are also important forces for episodic release of stored nutrients. Most mineralization of organic material occurs in the first part of the filtration process, and the more refractory components are broken down more slowly. Thus, microbial activity is concentrated near the input zone where interstitial percolation paths start (i.e., the retention stratum in most cases).

Prograding beaches could act as nutrient sinks by storing nutrients derived from mineralization in microbial biomass or adsorbed to the sand. In the long term, however, beaches must be in equilibrium and return to the sea all nutrients they receive. Changing organic inputs can change the equilibrium, but self-regulation is the rule unless assimilation capacities are exceeded by very high loads. Exposed sandy beaches will have highest capacities to assimilate organic matter, and because interstitial biological oxygen demand (BOD) is usually proportional to flow reflective beaches with high flow rates will generally be able to process DOM and fine POM fastest. Phosphate may, however, not be recycled but rather adsorbed on carbonate sands, depending on redox state.

Nutrient concentrations in interstitial waters are generally several times higher than in overlying waters and can be exceptionally high in some cases: NO_3-N levels can exceed $5\,mg \cdot l^{-1}$ in areas of groundwater discharge, and H_2S levels up to $700\,mg \cdot l^{-1}$ can occur in reduced zones. The more vigorous the interstitial water circulation and the more rapid the flushing rate the lower nutrient concentrations will be. Sheltered situations will exhibit the highest concentrations. In low-energy beaches, nutrient concentrations and distribution in the interstitial system may be controlled by wave action in the top layers, by waves and diffusion at intermediate depths, and by diffusion alone in deeper layers. In situations of high wave energy, water filtration paths will be more important and distribution patterns more complex.

3.7   The Interstitial Environment

The interstitial environment of sandy beaches may be conceptualized as spanning a continuum (Figure 3.17). At one extreme is the coarse-grained reflective beach that filters large water volumes. Here, water circulation is rapid and pulsing currents dominate. Filtered water has a low residence time in the interstices, which are well flushed (by wave action), well drained during low tide, and highly oxygenated. This is a physically controlled system dominated by vigorous interstitial circulation and with limited chemical gradients. It supports a deep-dwelling interstitial fauna adapted to powerful hydrodynamic forces and large pore sizes. This fauna has a deep vertical distribution, and recognizable strata can penetrate several meters into the sand.

At the other extreme, the fine-grained dissipative beach filters small volumes of seawater. Because wave effects are largely filtered out in the surf zone, this is driven mainly by tidal action, so that slow gravity currents dominate. Long residence times of interstitial water and stagnant interstitial conditions result in low oxygen tensions and steep vertical chemical gradients. This is principally a chemically controlled system with most interstitial life confined to the oxygenated upper few centimeters. Black reduced layers may occur deeper in the sand. The distribution patterns of interstitial fauna are thus mainly horizontal across the beach in zones that cover only the upper 5 to 20 cm of sand.

Intermediate situations displaying some elements of both of these extremes are most common on open coasts. In general, upper shores will tend to have more physical control driven by wave flushing because most beaches tend to be more reflective toward high tide and this is the area where inputs occur. Lower shores will tend to be more chemically controlled. Truly reducing conditions will only occur in the most sheltered and organically loaded sandy beaches. Nevertheless, low oxygen tensions will occur in the deeper sediment in most cases.

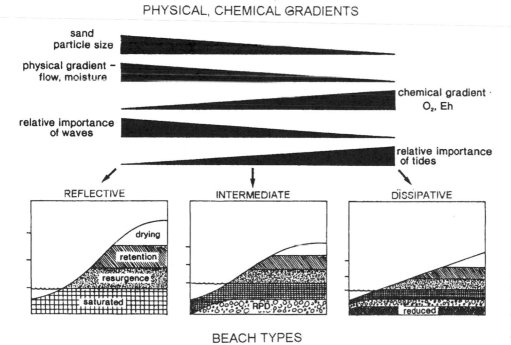

Figure 3.17.
Conceptual model of main gradients defining the interstitial system in different beach types (after McLachlan and Turner 1994).

Across the entire continuum, groundwater seepage also has an influence as it raises water tables and thus causes beaches to become flatter, filter less water, and tend toward stagnation. Groundwater will also lower salinities and elutriate the fauna. Thus, although discharge of groundwater high in nutrients may contribute to elevated surf-zone productivity its direct effects in the intertidal zone are less likely to be beneficial to the interstitial fauna.

Between the physical and chemical extremes on this continuum lies the ideal situation for interstitial life in terms of taxonomic diversity and abundance (see Chapter 9). The optimum may be a low intermediate beach of medium sand (300 to 400 μm), where the role of tides in water input is less than that of waves and the oxygen and organic inputs are sufficiently in balance to just prevent the development of reducing conditions in the deeper sediment. Here, gradients will be physical in the upper layers but tending toward chemical in the deeper sediment. In warm climates where high temperatures elevate metabolic rates, such beaches should be capable of processing considerable quantities of organic materials without becoming anaerobic.

3.8 Conclusions

The interstitial system has its physical structure and dimensions defined by the granulometry of the sand but its dynamics controlled by the process of water filtration through the beach face. Water filtration is driven by waves and tides. Reflective beaches typically consist of coarser sand with high permeability. They filter large volumes of water at fast rates and are consequently well flushed and oxygenated. Dissipative beaches, by contrast, consist of finer sands with lower permeabilities and filter smaller volumes of water at low rates. Waves are thus more important in reflective systems and tides in dissipative cases. Interstitial chemistry is primarily determined by a response to the balance between inputs of oxygen and organic materials, with the former more vigorous in reflective situations. Surplus organic input will lead to development of anaerobic conditions. This is more likely to develop in dissipative situations, whereas reflective beaches are more likely to be physically dynamic with strong interstitial flows and desiccation during low tide. It may be concluded that optimum conditions for the development of an abundant interstitial fauna, rich in species of diverse taxa, are likely to occur in intermediate beaches.

Beach and Surf-zone Flora

4

Chapter Outline

4.1   Introduction

Sandy beaches are normally devoid of living aquatic macrophytes, although seagrasses may occur in sheltered situations. Sandy-beach flora typically consists of benthic microalgae and surf-zone phytoplankton, both components often dominated by diatoms. On sheltered beaches the benthic microflora is usually relatively abundant, whereas on exposed dissipative beaches surf-zone diatoms may be far more important. In intermediate conditions, both may be present in roughly equal proportions or neither group may be well represented. Wave action, and the vertical mixing it causes, limits the growth of diatoms attached to sand grains. Consequently, they tend to be absent from wave-battered beaches but may then be more abundant offshore, in quieter water, where they concentrate toward the sand surface. Seagrasses grow in the subtidal zones of sheltered shores and may in some cases reach the lower intertidal of sandy beaches.

4.2   Benthic Microflora

The benthic microflora of marine sands includes bacteria, blue-green bacteria (cyanobacteria), autotrophic flagellates, and diatoms. Those attached to sand grains are generally known as epipsammon (Figure 4.1). Where there is vigorous wave action, living diatoms may be mixed to considerable depth in the sediment, whereas under more sheltered conditions or in the sublittoral zone they tend to concentrate toward the surface. The photic zone within the sand increases in depth with increasing particle size but does not normally exceed 5 mm (Figure 4.2). Long wavelengths penetrate the deepest because the refractive index of quartz increases with decreasing wavelength. The total surface area of the sand grains increases with decreasing particle size (Figure 4.3), providing increased space for attachment by the microflora but decreased pore space. In sediments that are not well oxygenated, different components (e.g., algae, photosynthetic bacteria,

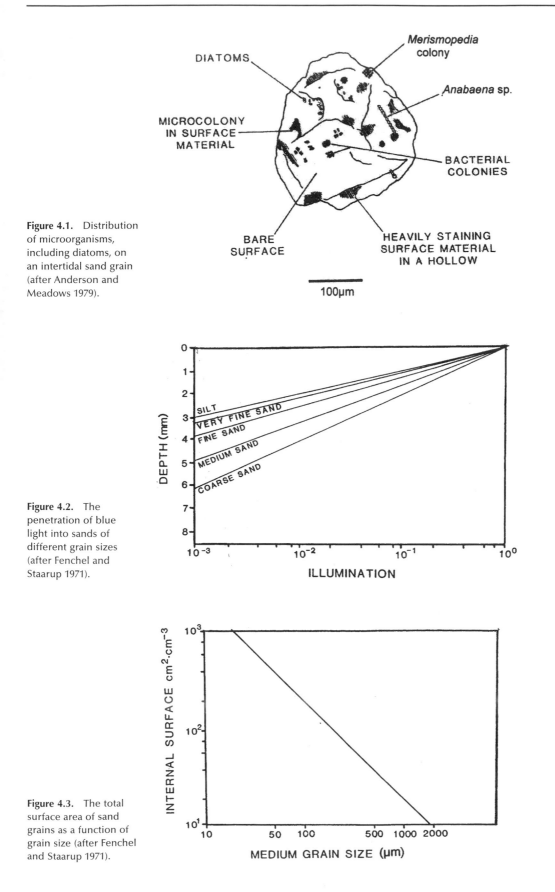

Figure 4.1. Distribution of microorganisms, including diatoms, on an intertidal sand grain (after Anderson and Meadows 1979).

Figure 4.2. The penetration of blue light into sands of different grain sizes (after Fenchel and Staarup 1971).

Figure 4.3. The total surface area of sand grains as a function of grain size (after Fenchel and Staarup 1971).

flagellates) may form layers at different depths within the sand. This is less marked, however, in exposed situations.

Some microalgal assemblages move actively through the sand, whereas those attached to sand grains are of necessity relatively immobile. Pennate diatoms (elongate rod-like) move very slowly from grain to grain, whereas centric (disc-shaped) forms are usually permanently attached. These epipsammic species generally adhere to the grains by means of mucus, which may be secreted by a groove or raphe on the valves or through pores on the frustule or skeleton. Only the forms near the surface of the substratum move fast enough to display diurnal or tidal migrations.

Populations of the pennate diatom *Hantzschia* undergo rhythmic vertical migrations associated with tidal and light cycles. On sheltered sands in Massachusetts, during daytime low tides, the cells move to the surface of the sand, returning to the subsurface interstitial spaces before the incoming tide reaches them. Similar vertical migrations have been reported for several other benthic diatoms and are a characteristic feature of surf-zone species, from which the intertidal benthic forms may have evolved. Benthic diatoms are, however, mostly pennate, whereas surf-zone species may be pennate or centric.

There have been few estimates of benthic microfloral densities from marine sands, but values are known to reach 10^3 cells cm^{-3} under optimal conditions. Many forms may extend into the subtidal, and where abundant can be a significant source of primary production.

4.3   Surf-zone Phytoplankton

Rich accumulations of diatoms are a typical feature of the surf zones of many exposed beaches. These accumulations are composed in most cases of a single species of *Aulacodiscus*, *Attheya*, *Asterionellopsis*, or *Anaulus* (Figure 4.4). They were first thought to be blooms and have an appearance like oil slicks in the surf. They have been shown not

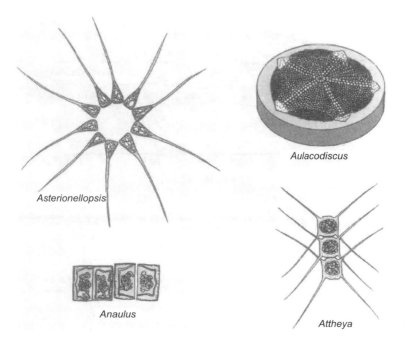

Asterionellopsis

Aulacodiscus

Anaulus

Attheya

Figure 4.4.   Common surf-zone diatom genera.

Table 4.1.   The earliest record for each locality at which surf water discoloration due to diatom accumulations has been reported. The first record for each species is also given. Based on Campbell (1996) and du Preez (pers. comm) which must be consulted for original references.

Locality	First report	Species
Zaire	Van Heurck 1896	*Aulacodiscus africanus*
USA: Washington & Oregon	Thayer 1935	*Aulacodiscus kittonii*
		Asterionellopsis socialis
Nicaragua	Thayer 1935	Unknown
New Zealand: Ninety Mile Beach	Rapson 1954	*Attheya armata*
		Asterionellopsis glacialis
New Zealand: North Island west coast	Cassie & Cassie 1960	*A. armata*
Argentina: North east coast	Kuhnemann 1966	*A. glacialis*
USA: Washington & Oregon	Lewin & Norris 1970	*A. armata*
Marquesas Archipelago	Sournia & Plessis 1974	*A. africanus*
South Africa: Cape Town, Port Elizabeth	McLachlan & Lewin 1981	*Anaulus australis*
Brazil: Rio Grande	Gianuca 1983	*A. glacialis*
New Zealand: North Island east coast	Kindley 1983	*A. kittonii*
		A. armata
Tasmania: Strahan Beach	McLachlan & Hesp 1984	*A. armata*
		A. australis
Brazil: Southeast coast	Lewin & Schaefer 1983	*A. glacialis*
Costa Rica & Panama	Lewin & Schaefer 1983	*A. africanus*
Australia: Warren Beach	McLachlan & Hesp 1984	*A. australis*
Australia: Goolwa Beach, Coorong	McLachlan & Hesp 1984	*A. glacialis, A. australis*
Uruguay: North east coast	Baysse *et al.* 1989	*A. glacialis*
South Africa: 12 beaches between Cape Town & Cintsa Bay	Campbell & Bate 1991	*A. australis*
South Africa: Port Elizabeth	Du Preez *et al.* 1990	*A. glacialis*
Gopalpur: India	Choudhury & Panigrahy 1989	*A. glacialis*
Argentina: Peuhen Co Beach	Gayoso & Muglia 1991	*A. armata*
Australia: Waratha Bay	Campbell 1996	*A. australis*
New Zealand: Waiuku Beach	Campbell 1996	*A. armata, A. kittonii, A. australis,*
		Asterionella cf. *formosa, A. glacialis*
Venezuela	Gianuca, pers. comm.	Unknown
Brazil	Garcia, pers. comm.	*A. australis*
Australia: Queensland	Hewson *et al.* 2001	*A. australis*

to be blooms, but rather semipermanent features of high-energy surf zones where the diatom cells divide at a constant rate. In some cases, dominance can change with seasons. They have been recorded from most continents (Table 4.1, Figure 4.5) and are characteristic of beaches with broad dissipative surf zones exposed to strong wave action. These accumulations seem to be more common in the Southern Hemisphere, where *Anaulus australis* is endemic. Most other species occur in both hemispheres and some are cosmopolitan. Distribution limits seem to be about 46° N and S and most species only occur in surf zones. Further, they are typical of extensive beaches, not being found along short stretches of sandy coastline or pocket beaches.

These diatoms accumulate at the water surface during the day, often associating with bubbles (possibly due to the action of mucilaginous coats) or even forming a semistable foam. Being concentrated at the water surface, they are transported and further concentrated by waves and currents. Because of surface advection by bores, they are moved shoreward, concentrating in the inner surf zone and sometimes being stranded on the beach as deposits of foam. Longshore concentrations also occur at certain points, often adjacent to rip currents (Figure 4.6). It is thought that the opposing forces of incoming waves and outgoing rips create a bottleneck or eddy effect, where the diatom foam accumulates. In fully dissipative beaches without rip currents, such patches are not as

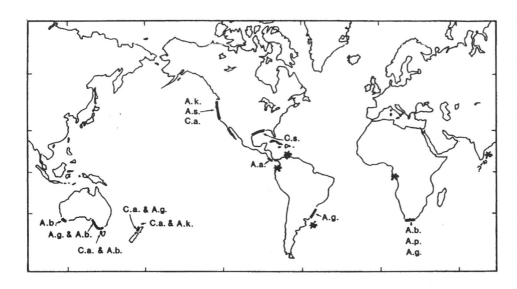

Figure 4.5. Locations of the most persistently recorded patches of surf diatoms (after Campbell 1996). All species recorded are listed, although diatom accumulations generally consist of only one, or at most two, species at any one time. On the northwest U.S. coast, for example, four species are listed but usually *Chaetoceros armatum* is the only species present. A.b. = *Anaulus birostratus*, A.s. = *Asterionella socialis*, A.g. = *Asterionella glacialis*, C.a. = *Attheya armatum*, and A.k. = *Aulacodiscus kittoni*. Isolated locations are indicated as *.

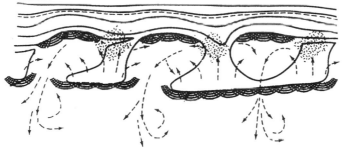

MODAL LOW ENERGY STATE – TRANSVERSE BAR – RIP

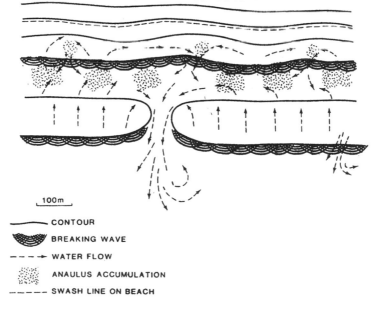

MODAL HIGH ENERGY STATE – LONGSHORE BAR – TROUGH

100m

——— CONTOUR

🌊 BREAKING WAVE

– – – → WATER FLOW

⁙ ANAULUS ACCUMULATION

– – – – – SWASH LINE ON BEACH

Figure 4.6. Patterns of surf diatom accumulations on a beach with well-developed rip currents (after McLachlan and Bate 1984).

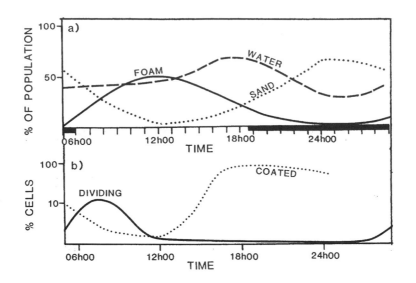

Figure 4.7. Changes over the diel cycle in the percentage of the cells of the surf diatom *Anaulus* (a) in the sand, water column, or surface foam and (b) dividing or coated with mucus (after Talbot 1986).

discrete as on intermediate beaches with clear rip currents. The diatoms not only occur in the foam but are dispersed throughout the water column and in the sediments, moving between these phases. They are not found off reflective beaches or where there is little surf-zone development.

It is important to distinguish between these diatom patches or accumulations and true diatom blooms. Blooms are seasonal phenomena related to water chemistry and temperature, whereas the diatom patches discussed here occur throughout the year and are physically controlled accumulations.

Surf diatom patches exhibit distinct diurnal periodicities (Figure 4.7). For *Anaulus*, a typical pattern is as follows. During the early morning, the diatoms leave the sand and enter the water column, from where they move into the foam by attaching to bubbles. By mid morning, many of the diatoms are in the foam or in the water column and few remain in the surface sand, although a significant part of the population may be buried deeper. During the afternoon, the diatoms lose buoyancy and begin to drop out of the foam into the water column. From here they slowly enter the sand, so that during the night most of the cells are in the substratum. This vertical migration is reminiscent of the short vertical movements of benthic diatoms.

Coupled with these vertical migrations between sand and foam are diel changes in cell division and in the production of mucus (Figure 4.7). There is a clear period of cell division in the early morning, so that most cells are dividing by the time they enter the surface foam. During the afternoon, cell division is complete and the diatoms produce a mucus coat that may allow them to adhere to sand grains and enter the sediment at night. In the early morning, when the cells start dividing, this coat is lost and it is then that they may be eluted from the sediment. The mucus cell coatings of surf-zone diatoms vary between species, from a thick mucilaginous sheath in *Attheya* to a thin layer of mucus that continuously dissolves in *Anaulus* (Lewin and Schaeffer 1983, du Preez and Campbell 1996a). As cell division occurs most actively when the cells are accumulated at the surface, greatest population growth occurs during the development of extensive accumulations (Talbot *et al*. 1990). These diatoms can spend long periods in the sand in a resting state during calms. They are adapted to rapidly fluctuating light intensities in

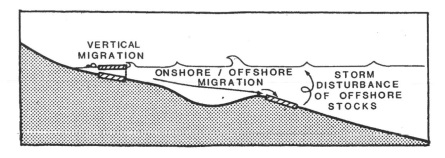

Figure 4.8.
Diagrammatic
representation of the
three migration cycles
that explain the
dynamics of surf diatom
accumulations.

the surf, alternating between intense light (even photoinhibition) when at the surface and much reduced light when submerged under a wave bore. Under good light the photosynthetic rate in *Anaulus* is extremely fast and much surplus production is exuded as mucus polysaccharides (du Preez and Campbell 1996b).

It is thus suggested that three cycles explain the dynamics of surf-zone diatom patch formation and decay (Figure 4.8). First, there is the diurnal vertical movement between the foam during the day and the sediment at night. Mostly, in fact, the bulk of the diatom population remains buried in the sediment, only a portion of it entering the water column each day. Second, there is an on/offshore migration, diatom cells that rise toward the surface during the day being advected shoreward by wave bores. In the afternoon, when the cells start sinking out of the foam, they may be transported beyond the surf zone by rip currents and deposited on the bottom outside the breakers. There they become available for reentry into the surf zone the next morning if they are stirred into the water column. This therefore constitutes a horizontal on/offshore migration pattern superimposed on the daily vertical migration. Third, there is the storm/calm cycle of events, irregular and largely unpredictable. A large proportion of the diatom community is housed within the sediment at any one time. As increased wave action during storms causes a greater disturbance and turnover of the sediment and expands the surf zone seaward, a greater proportion of the sedimentary diatom population enter the water column. Consequently, the richest diatom accumulations occur during and immediately after conditions of high wave energy. During calm weather, much of the diatom population may even be removed from the surf zone and accumulate in the sediment outside the breakers. Onshore winds, which increase wave energy and concentrate diatoms against the shore, are thus an important factor affecting surf-zone diatom accumulation and productivity.

As a consequence of the accumulation dynamics of surf diatom populations and of surf circulation patterns, there may be strong on/offshore gradients in chlorophyll *a* and primary production profiles in the waters of surf zones that tend toward the dissipative extreme (Figure 4.9). Diatoms and chlorophyll show a marked concentration toward the inner surf zone during the day, densities dropping both across the breaker line and across the outer limit of the surf circulation cells. Thus, both surf circulation patterns and the adaptations evolved by the diatoms are geared to retaining cells within the surf zone.

A coupling between the presence of surf diatom accumulations and the seepage of groundwater, rich in nitrogen, from coastal aquifers is a possibility that still needs elucidation. Gunter (1979) described *Chaetoceros*-dominated blooms in the form of patches along the beaches of the U.S. Gulf Coast. These were apparently caused by

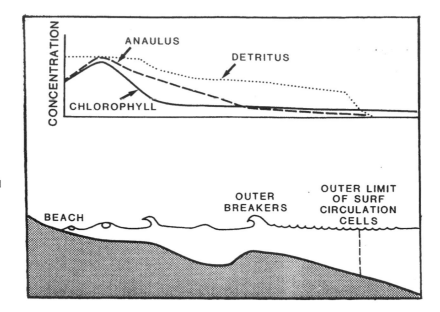

Figure 4.9. Generalized scheme of the concentrations of surf diatoms (*Anaulus*), detritus, and chlorophyll across a beach surf zone in South Africa (after Talbot 1986, McLachlan 1987).

the leaching of nutrients from the land during heavy rains and their retention next to the beach by calm seas. The blooms took the form of strips 5 to 6 m wide hugging the shore for many kilometers.

Surf-zone phytoplankton accumulations have been shown to be relatively consistent on beaches where they occur. They have occurred for many decades on Washington beaches, although the dominant form had changed from *Aulacodiscus* before the 1950s to *Attheya* subsequently (Lewin and Schaeffer 1983). These accumulations may, however, be highly variable in the short term, with large changes in abundance and primary productivity of surf diatoms both from year to year and on a shorter time scale associated with storm/calm cycles.

Diatoms are not the only phytoplanktonic forms to be found in the surf. Autotrophic flagellates and cyanobacteria can occur at times in significant numbers. Another surf-zone primary producer is the free-living alga *Pilayella*. This ball-like macrophyte occurs on the bottom in the surf off beaches in Nahant Bay, Massachusetts (Wilce and Quinlan 1983). It is particularly abundant in summer, when it causes the fouling of beaches, which is a problem in this area.

4.4   Seagrasses

Seagrasses are not normal components of open sandy beaches. However, they are common components of the shallow sublittoral of sheltered sandy shores in tropical, subtropical, and temperate regions. Further, when cast ashore as beach wrack they can contribute substantially to sand beach energetics, so they warrant mention here. Seagrasses are true flowering plants that have adapted to marine conditions and propagate vegetatively and by seeds. Only *Thalassia*, turtlegrass, produces flowers. Pollen is water borne and a sticky slime helps it adhere to the stigma. Seeds may float to disperse, and some have bristles to stick to plants. The plants consist of blades or fronds, rhizomes or horizontal stems, and roots. Rhizomes and roots serve for anchorage and absorption of

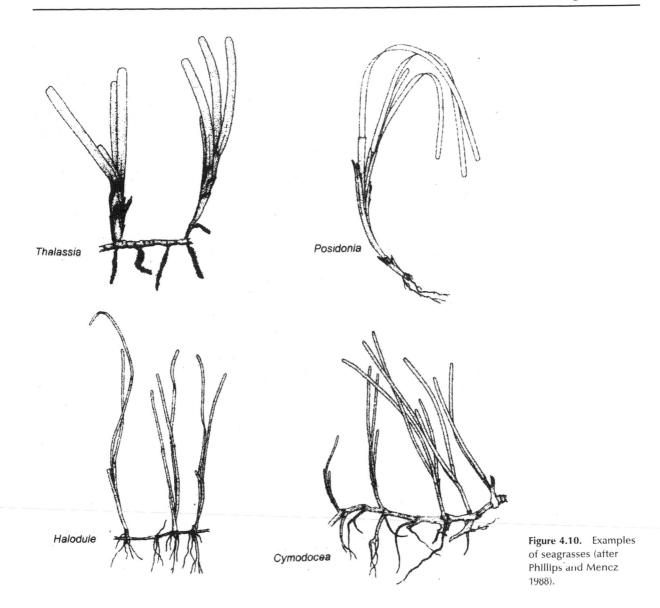

Thalassia

Posidonia

Halodule

Cymodocea

Figure 4.10. Examples of seagrasses (after Phillips and Menez 1988).

nutrients. Most genera have thin blade-like leaves with large surface areas. Twelve genera with about 50 species are recognized. Seven genera are characteristic of tropical and subtropical seas: *Halodule, Cymodocea, Syringodinus, Thalassodendron, Enhalus, Thalassia*, and *Halophila*. Five genera are confined to temperate seas: *Zostera, Phyllospadix, Heterozoster, Posidonia*, and *Amphibolis* (Figure 4.10).

Seagrasses grow in shallow water, less than 10 m deep, often in dense meadows where they may display depth zonation. In some cases, under sheltered conditions distribution may extend upward into the lower intertidal of sandy beaches. The growth of seagrasses creates a unique habitat. It stabilizes the bottom and traps sediment, clarifies the water, and provides habitat for plants and animals. Epiphytes include diatoms and algae, up to 100 species being recorded. Some algae glue themselves to seagrasses and some entwining epiphytes can have negative effects on seagrasses by reducing light reaching the grasses and adding mass that makes the seagrasses more likely to dislodge during storms.

Subhabitats created by seagrasses are leaf surfaces available for colonization by epiphytes; stem/rhizome habitats colonized by macrobenthos such as polychaetes, amphipods, and bivalves; fish, prawns, and cephalopods living among the leaves; and finally the benthos living in the sediment between plants. These habitats can be especially important for fish and prawns, particularly commercially important species, which may use them as nursery areas. Seahorses, pipefishes, and gobies are very typical of seagrass habitats. Most fish feed on invertebrates in the seagrass, especially crustaceans.

Seagrass meadows can be extremely productive. In dense stands there can be > 20,000 stalks per square meter and biomass up to 10 kg dry mass m^{-2}. Rhizomes make up 60 to 80% of biomass. Primary production is boosted by the presence of epiphytes on the fronds and can reach 1 kg C·m^{-2}·y^{-1} in tropical seas and half that in temperate areas. Tropical and subtropical lagoons are most productive. Seagrasses can rapidly lose dissolved organics, but the plants decompose slowly. Less than 10% of biomass is grazed directly, about 5% is exuded as DOM, and the rest enters the decomposer food chain as detritus. Those species that are adapted to graze directly on seagrasses include green turtles (*Chelonia mydas*), the dugong (*Dugong*), three species of manatees (*Ticherus*), and some fishes, crustaceans, and urchins. Whereas turtles and fishes crop the leaves, the sirenians plough up whole plants, including the rhizomes. In many cases, up to 50% of production may be exported.

For sandy beaches, seagrasses can be important when large quantities are torn loose by storms and drift in the surf zone or are cast ashore as wrack to subsidize beach food chains and modify the beach habitat. Threats to seagrass meadows include turbidity, sedimentation, eutrophication, dredging, and erosion.

4.5   Conclusions

Many beach systems have resident primary producers in the form of benthic microflora or surf phytoplankton. Among the species specially adapted to life on sandy beaches, diatoms exhibiting vertical migrations (either within the sediment or between the sediment and the water column) are perhaps most typical. In the case of planktonic forms, these migrations lead to the cells concentrating as semistable foam in the inner surf zone, where their productivity can be exceptionally high. Surf-zone primary production and its impact on the sandy-beach fauna and on food chains are examined in Chapter 12. It clearly presents an important resident food source available to a variety of organisms. Much research remains to be undertaken on these micro-plants, including their adaptations and productivity. Seagrasses are typical of the sublittoral fringe of sheltered beaches and are usually not present on exposed beaches. Where stranded in large quantities, often mixed with algae, seagrasses can significantly add to beach food chains.

Sandy-beach Invertebrates

5

Chapter Outline

5.1   Introduction

Most invertebrate phyla are represented on sandy beaches, either as interstitial forms or as members of the macrofauna, or both. The macrofaunal forms are by far the better known and special attention is given to them here. Some of the macrofaunal genera are typical of intertidal sands and their surf zones, whereas others are more characteristic of sheltered sandbanks, sandy muds, or estuaries and are less common on open beaches of pure sand. We are concerned here chiefly with the fauna of open oceanic beaches, although brief mention is made of genera favoring sheltered sands, where this is considered appropriate. It may be noted that although the macrofauna and meiofauna are ecologically distinct they are not separated on phylogenetic criteria. In some cases, a single family may have both macrofaunal and meiofaunal representatives, and even within a single species larval or subadult forms may be present in the meiofauna (whereas the adults are macrofaunal). For the purposes of this chapter, they are included together, although meiofauna is also covered in Chapter 9.

5.2   Important Groups

Phylum: Porifera

Sponges do not occur in or on sandy beaches but are mentioned because of the occurrence of *Clione* and possibly of related genera of boring sponges in the surface layers of the shells of some sandy-beach mollusks, including the gastropod *Bullia*. Heavy infestation weakens the shell, making the mollusk more vulnerable to predation.

Phylum: Cnidaria

A few interstitial genera of Hydrozoa (such as *Psammohydra*, *Halammohydra*, and *Otohydra*) have been found in sandy beaches for which the meiofauna has been thoroughly investigated. The animals move slowly from grain to grain of sand. Some are adapted to the interstitial environment by being markedly vermiform. Anthozoans, on the other hand, are encountered as members of the macrofauna, on sheltered sands only and usually below low-water mark. Genera include *Peachia*, *Bunadosoma*, and *Bunodactis*.

Phylum: Platyhelminthes

Class: Turbellaria. Free-living flatworms form a most important component of the interstitial fauna in many sandy beaches. The best-known example of the order Acoela is probably the green flatworm, *Convoluta*. This genus is not typical of the interstitial flatworms as a whole, however, because its migratory behavior is dictated by the photic requirements of its symbiotic algae and its occurrence is linked to marine intertidal sands under the influence of freshwater seeping from the land. More typical genera include *Thlacorhynchus* and *Diascorhynchus* (family Schizorhynchidae), *Acanthomacrostomum* (Macrostomidae), and *Coelogynopora* (Manocelididae). Kalyptorhynch Turbellaria such as *Philosyrtis* are common on some beaches, whereas otoplanids are sometimes so abundant in the sublittoral that the term *Otoplana* zone has been applied to the region they inhabit. The order Gnathostomulidae differs quite considerably from other turbellarian orders in that its representatives have a polygonal epidermis reminiscent of sponges and coelenterates, as well as pharyngeal cuticular jaws like annelids or rotifers. In fact, they are sometimes placed in a separate phylum. Genera include *Gnathostomula* and *Gnathostomaria*. The gnathostomulids are characteristic of the reduced layers. All Turbellaria (including the gnathostomulids) are carnivorous and mainly predatory.

Class: Trematoda. Flukes are the most common parasites infesting sandy-beach mollusks, both gastropods and bivalves. Some groups of Crustacea also tend to be infected, including decapods such as *Emerita*. Although adult flukes may be found, most are encysted larvae, the mollusk or crustacean acting as intermediate host and the final host commonly a fish or bird.

Class: Cestoda. Larval cestodes are far less common than digenean trematodes but may be found in both mollusks and crustaceans. Again, a fish or bird is the most likely final host.

Phylum: Nemertea

Nemertines (or ribbon worms) are typically marine and largely intertidal. They are not among the dominant sandy-beach animals but are present on most sandy shores. Nearly all are carnivorous and use the elongate proboscis to capture prey, although particles of dead material may also be eaten. The genus *Cerebratulus* (Order Heteronemertini) appears to be cosmopolitan but is somewhat more abundant on sheltered than on exposed shores. Other genera commonly encountered in sheltered tropical and subtropical sands are *Baseodiscus* and *Zygonemertes*. *Carcinonemertes* and related forms are obligate egg predators that live in and consume the egg masses of decapod crustaceans such as *Blepharipoda*.

Phylum: Nematoda

Marine nematode worms are probably the most ubiquitous of sandy-beach forms, with numerous genera often distinguishable only by an expert. They are nearly always extremely common and often make up the bulk of the meiofauna, especially in fine sands. They occur in all zones of the beach and extend for several meters into the sand, many species not being limited by oxygen availability. Many of the free-living forms frequently encountered — both above and below the water table intertidally and in the surf zone — belong to the family Leptosomatida, whose genera include *Epacanthion*, *Mesacanthion*, and *Platycoma*. Other genera often found include *Bathlaimus*, *Metoncholaimus*, *Nannolaimus*, and *Trileptium*. Chapter 9 provides more information on interstitial nematodes. Nematodes also occur as parasites in the guts of mollusks and crustaceans.

Phylum: Acanthocephala

Thorny-head worms are parasites that occur in the body cavities and tissues of decapod crustaceans as encysted forms called cystocanths. The life cycle includes sharks, skates, and rays that may feed on hippid and albuneid crabs.

Phylum: Rotifera

Seisonoid rotifers (adapted to an epizoic existence) are found on sheltered and even some moderately exposed beaches, attached to sand grains and sometimes to the hard parts of the bodies of macrofaunal animals. Frequently encountered genera include *Encentrum* and *Proales*.

Phylum: Gastrotricha

Gastrotrichs are common members of the sandy-beach meiofauna, their tiny worm-like bodies ideally suiting them to an interstitial environment. Moreover, they are protected from abrasion by the possession of scales and spines. Nearly all marine sandy-beach forms belong to the order Macrodasyoidea. Typical genera are *Turbanella* and

Xenotrichula. Other common genera are *Urodasys, Diplodasys, Thanumastoderma,* and *Pseudostomella.* Gastrotricha are also covered in Chapter 9.

Phylum: Kinorhyncha

The kinorhynchs are minute cylindrical animals related to the rotifers and gastrotrichs, and typical of mud and muddy sand rather than of sandy beaches. They are, however, sometimes found in some numbers in sheltered sands, becoming rarer as wave action or the size of the interstitial pore spaces increases. They are absent from exposed oceanic beaches.

Phylum: Loricifera

Another group of pseudocoelomate animals, this phylum was described only in 1986. Its members appear to be exclusively interstitial, and they are found in a wide variety of sediments ranging from intertidal sands to abyssal environments.

Phylum: Annelida

Class: Archiannelida. Archiannelids occur on most sandy beaches as part of the meiofauna but are not always common, although they may be locally abundant, especially in coarse sands. Three families are generally recognized: the Nerillidae, including the genera *Nerilla, Nerillidium,* and *Troglochaetus;* the family Polygordiidae, with the genera *Polygordius, Protodrilus, Protannelis,* and *Saccocirrus;* and the family Dinophilidae, containing the genera *Diurodrilus* and *Trilobodrilus.*

Class: Polychaeta. Polychaete worms are abundant as members of both the meiofauna and the macrofauna. On relatively sheltered sands they may dominate the macrofauna in the lower part of the beach and are represented by more genera than any other group of animals, except possibly the Nematoda. They become less common as wave action increases and sands become coarser, but some species hold their own even on exposed shores. The geographical distribution of many genera is imperfectly known and ranges are bound to be extended as surf zones become better sampled. This is particularly so in that genera that occur intertidally in some parts of the world are found subtidally elsewhere.

Family: Aphroditidae (including the Polynoidae). These errant polychaetes are not generally important members of sandy-beach communities, although some (such as *Antinoe*) favor sandy habitats. They are for the most part slow-moving predators, taking small prey despite their powerful jaws. The genus *Sthenelais* is common on the shores of tropical South America. The genera *Harmothoe, Malmgrenia, Pholoe, Lepidasthenia,* and *Sigalion* are found in muddy sand and less commonly in pure sand on sheltered sandbanks. However, *Thalenessa* (a genus found in tropical and subtropical southern Africa) is more typically a sandy-beach animal, invading moderately exposed situations and displaying a burrowing behavior unlike that of other polychaetes and clearly an adaptation to wave action.

Family: Phyllodocidae. These primitive highly active polychaetes with uniramous parapodia are found in sheltered beaches and sand flats all over the world and occasion-

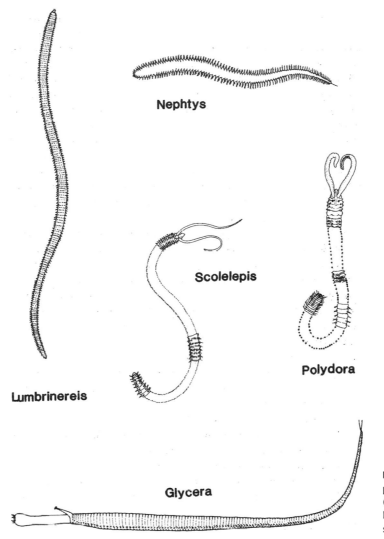

Figure 5.1. Some polychaete worms (about 5 to 15 cm in length) typical of sandy shores.

ally on more exposed beaches. Common genera include *Phyllodoce, Notophyllum,* and *Eteone.*

Family: Glyceridae. Glycerids are elongate predatory worms, rounded in section and tapered at both ends. They burrow vigorously and rapidly into particulate substrata, using a long eversible proboscis to do so. The proboscis is armed with at least four jaws and it is assumed that the animals are mainly predators, or at least carnivores. The family includes a number of forms typical of sandy beaches, such as *Glycinde, Hemipodus* (a worm common on the coasts of California, Mexico, Costa Rica, and Columbia, and on both sides of the Isthmus of Panama), and *Goniada,* a widespread genus with only a few species. However, by far the most common genus in sandy beaches is *Glycera.*

Genus: Glycera. Glycera (Figure 5.1) is a virtually cosmopolitan genus found in sandy beaches from Norway and Scotland in the north, through the tropics to southern Chile, the Cape of Good Hope, and the island of Tristan da Cunha. Mostly of medium size, some members of the genus can reach 20 cm in length. They are most common on the lower shore, but may also be found throughout the surf zone and out into the nearshore region. The larger species are often known as "blood worms" and are used as bait. When extracted from the sand in which they live, they tend to perform vigorous figure-of-eight movements rather than displaying a more appropriate escape response. There are some 15 species within the genus.

Family: Nephtyidae. These are small to medium-size polychaetes with a large muscular eversible proboscis armed with a pair of internal jaws. As in the Glyceridae, the proboscis is used in rapid burrowing through mud or sand. In addition, the animals are active swimmers.

Genus: Nephtys. This is a truly cosmopolitan genus, found even in Antarctica. Species of *Nephtys* (Figure 5.1) are found in all grades of mud and sand. Some, such as *N. capensis*, are quite common on moderately exposed surf beaches. Some can reach large sizes (20 cm). Like *Glycera* and many other polychaetes, they tend to occupy the lower zones of the shore and their distribution may extend subtidally. They are thought to be highly selective omnivores, although some may be active predators.

Family: Syllidae. These are small worms with uniramous appendages, often the best represented polychaete family among the sandy-beach meiofauna. The genus *Exogene* is widespread and is often common in relatively sheltered intertidal sands, swarming at night to release gametes. Other genera are *Syllis*, *Eurysyllis*, *Syllides*, *Sphaerosyllis*, *Streptosyllis*, and *Brania*.

Family: Hesionidae. Members of this family are not typical of sandy beaches but occasional representatives of the genera *Hesione*, *Hesionella*, and *Hesionides* are encountered in sheltered sands.

Family: Pisionidae. Pisionid worms are typical of the sandy-beach meiofauna in many parts of the world, and the family as a whole has a cosmopolitan distribution. Some species have great adhesive ability, attaching themselves to sand grains by means of modified parapodia. Genera include *Pisione* and *Pisionidens*.

Family: Nereidae. These are chiefly fairly robust omnivorous polychaetes, living mainly on rocky shores. The genera *Nereis* (Figure 5.2), *Perinereis* and *Ceratonereis* are, however, widely distributed — with species found in a variety of marine habitats, including sandy beaches. *Ceratonereis*, in particular, is often common in tropical to warm temperate sheltered sands.

Family: Paraonidae. The Paraonidae are small thread-like worms, up to 40 or 50 mm in length, widespread but seldom abundant. The most common genera are *Paranonis*, *Paraonides*, and *Paradoneis*.

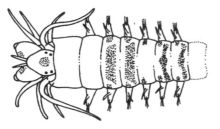

Figure 5.2. Anterior portion of *Nereis*, showing characteristic features.

Nereis

Family: Eunicidae. This is a family of elongate worms occupying diverse habitats, with many tubiculous forms. They are common in muddy sand but also found in sheltered intertidal sands, some species being restricted to this habitat. *Eunice* is a cosmopolitan genus, often found in some numbers in sheltered sandbanks or subtidally on more exposed beaches. *Marphysa* has 11 species, some of which are limited to sheltered intertidal sands. The genus is circumglobal in the tropics and subtropics and south to the Antarctic. *Ophryotrocha* is another common genus with a wide distribution. The subfamily Onuphinae includes *Diopatra* (another genus circumglobal in tropical and subtropical waters) and *Onuphis*, which is cosmopolitan and frequently found in sheltered sand near the water's edge or subtidally. The subfamily Lumbrinereidae contains some of the largest and most robust members of the family, including *Ninoe* (from tropical West Africa) *Lumbriconereis* (very common on Indian shores), and *Lumbrinereis* (Figure 5.1), which is circumglobal in both tropical and temperate waters and extends south to Antarctica.

Family: Orbiniidae. This family of sedentary polychaetes is well represented on sheltered sandy beaches, typical genera including *Orbinia* (Figure 5.3) and *Scoloplos*. Although belonging to the Sedentaria, these worms do not make permanent tubes but burrow through the sand by means of the pointed prostomium, rather than employing their unarmed proboscides. Most ingest particulate organic matter, not very selectively, so that the gut is commonly found to be packed with sand grains. *Leitoscoloplos* feeds largely on benthic diatoms, the production of which may be the major factor controlling the density of the worm population.

Family: Psammodrilidae. Representatives of this family display marked adaptations to interstitial life. The semi-sessile *Psammodrilus* has a specialized pharyngeal apparatus that functions as a pump, with the aid of which the worm sucks in diatoms, bacteria, and protozoans.

Family: Spionidae. Unlike the Orbiniidae, spionids are typically tube dwellers. Sandy-beach genera include *Polydora* (Figure 5.1), *Pseudopolydora*, *Spio*, *Dispio*, *Prionospio*, *Nerine*, and *Scolelepis*. They range in size from small worms that may be considered members of the meiofauna to robust species 10 cm or more in length. Many are deposit feeders, but some forms (such as *Dispio*) are voracious predators. *Spio* often forms dense colonies on sheltered sandbanks. When the current washes away loose sand from between the tubes, these protrude as small chimneys. Whereas most spionids favor sheltered shores, *Dispio* is common on some fairly exposed beaches in southern Africa. The family may be well represented in sandy-beach surf zones.

Genus: Scolelepis. *Scolelepis* (Figure 5.1), a genus boasting some half-dozen described species, is like *Spio* and many other spionids a deposit feeder. It lives in burrows lined with a fragile mucous secretion and is probably the most commonly

Orbinia

Figure 5.3. Anterior end of a typical member of the genus *Orbinia*.

encountered member of the family on open sandy beaches. It is particularly characteristic of the upper midshore. The genus is extremely widespread, being recorded from British and European coasts as far north as Denmark, both coasts of North and Central America, the Caribbean islands, North Africa and the Red Sea, Mozambique, southern Africa, Senegal, India, and Australia.

Family: Chaetopteridae. These are tube-dwelling polychaetes found on sheltered beaches and sand flats. The common genera include *Chaetopterus*, *Phyllochaetopterus*, and *Mesochaetopterus*.

Family: Capitellidae. These sedentary worms are found in a variety of habitats. They are highly opportunistic and capable of rapid colonization of substrata subjected to bioturbation or organic enrichment. They may be found in small numbers in clean sheltered sands but are more common in unstable coarse sediments and in muddy sands. *Capitella* is a cosmopolitan genus, which has been used as an organic pollution indicator, although this interpretation requires great caution. The only other genus occasionally found on sheltered sandy beaches is *Dasybranchus*.

Family: Arenicolidae. These are the lugworms, one of the largest representatives of the polychaetes to be found on sandy beaches. *Arenicola marina* is one of the most intensively studied of sandy-beach animals. Other genera include *Abarenicola* and *Arenicolides*. Members of the family typically construct U-shaped burrows, a funnel-like depression indicating the head end and worm castings the tail end of the burrow. Water is circulated actively through the burrow, not only ventilating the worm but oxygenating the sand around the burrow and making it habitable for small obligate aerobes. As the worm feeds by ingesting sand from the head shaft, such microorganisms may form an important part of its diet. The circulation of water through the burrow also results in suspended organic particles being trapped in the sand of the head shaft. These, too, form a major part of the worm's diet. When in dense populations, the Arenicolidae perform an important ecological function in turning over the sand and mixing its surface layers. Where the burrows penetrate into the reduced layers, each burrow is surrounded by a layer of oxygenated sand, thus encouraging the development of aerobic meiofauna and microbes. The worms are limited to beaches stable enough to support their semipermanent burrows.

Genus: Arenicola. These are large robust worms ranging in length from 10 to 40 cm. Like other members of the family, *Arenicola* (Figure 5.4) is limited to fairly sheltered stable beaches and tends to occur low down the shore, mainly in saturated sand. The worms are frequently used as bait by anglers, and in some parts of the world their populations are protected by limiting the number of individuals any bait collector may remove on any one day. The genus is cosmopolitan.

Genus: Abarenicola. Superficially resembling *Arenicola*, these worms (Figure 5.4) may be distinguished by the fact that their neuropodial ridges are short and well separated in the branchial region — in contrast to the long neuropodial ridges of *Arenicola*, which almost meet ventrally in the branchial region. They are also generally smaller than worms of the genus *Arenicola*, although there is some overlap in size range.

Family: Opheliidae. All representatives of this family are burrowers in sand and mud. Like other Sedentaria, they are much more common intertidally in sheltered than in exposed sands but may also be found offshore where wave action precludes their inter-

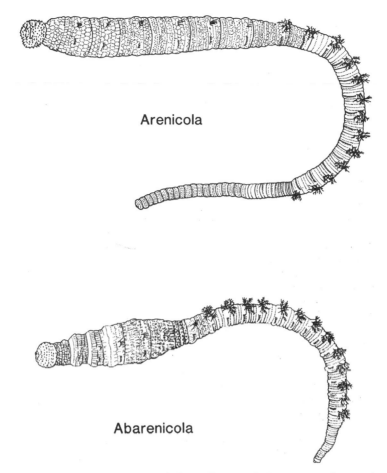

Arenicola

Abarenicola

Figure 5.4. The sedentary tube-dwelling polychaetes *Arenicola* and *Abarenicola* (typically 10 to 20 cm in length).

tidal occurrence. Common genera include *Ophelia*, *Thoracophelia*, *Armandia*, and *Travisia*.

Class: Oligochaeta. The vast majority of marine forms belongs to the families Tubificidae and Enchytraeidae, there being only a few scattered representatives of other families. The Enchytraeidae — including the genera *Enchytraeus*, *Lumbricillus*, and *Marionina* (Figure 9.3) — tend to concentrate toward the top of the shore, whereas the Tubificidae (with the genera *Tubifex*, *Tubificoides*, and *Clitellio*) are more truly marine, being found intertidally and below the tidal limit. All of these genera occur in Europe and North America, but their distributions in other parts of the world are too poorly known to allow detailed zoogeographical coverage, although *Enchytraeus* appears to be virtually cosmopolitan. They can be important members of the meiofauna (Chapter 9).

Phylum: Echiurida

This phylum is not represented on open oceanic beaches but an occasional individual may be found in more sheltered sands, near the water's edge just below the tidal limit.

Phylum: Sipunculoidea

Sipunculids are closely related to the Echiurida, which they resemble in superficial appearance, in size, and in habit. They, too, are restricted to sheltered shores intertidally, where they may be quite abundant in some areas. They seldom occur in more exposed situations. The most common genus is *Sipunculus*.

Phylum: Brachiopoda

Brachiopods do not occur on open oceanic sandy beaches. However, *Lingula* may occur in some numbers below the low-water mark on sheltered sandbanks and flats in the tropics and subtropics. *Glottidia* is reported from Panamanian beaches, and brachiopods occur on cobble beaches in New Zealand.

Phylum: Mollusca

Class Gastropoda, subclass Prosobranchia, order Neogastropoda. Most of the proso-branch gastropods found on sandy beaches belong to the Neogastropoda. Families include the Nassariidae — with the genera *Bullia* (Figures 5.5 and 5.6), *Dorsanum*, *Buccinanops*, and *Nassarius* (Figure 5.7) — the latter perhaps more typical of muddy substrata but

Figure 5.5. The whelk *Bullia digitalis* (Dillwyn). (Photo: P. F. Newell.)

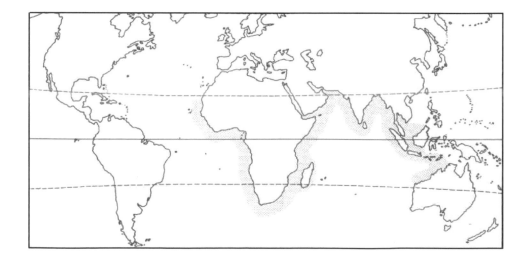

Figure 5.6. Map showing the probable geographical range of whelks of the genus *Bullia*.

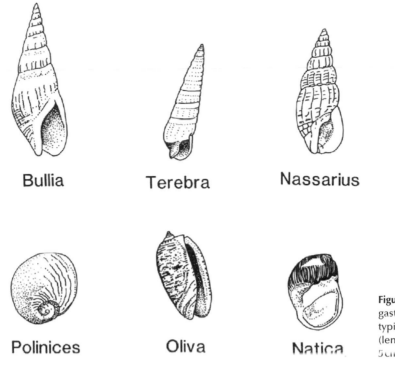

Figure 5.7. Some gastropod mollusks typical of sandy beaches (lengths typically 1 to 5 cm).

often occurring off sheltered sandy beaches. Families also include Naticidae, whose most common sandy-beach genera are *Natica* and *Polinices*; the Terebridae, notably *Terebra* itself; and the Olivacea, including the genera *Oliva*, *Olivella*, and *Olivancillaria*. These are all essentially carnivorous whelks, either predators or scavengers, although *Bullia* is both a scavenger and a predator and grazes on algae growing on its shell.

Genus: Bullia. Bullia (Figures 5.5 and 5.7) are small to medium-size whelks, with thin widely expanded feet terminating posteriorly in a pair of cirri. Eyes are absent. They are confined to sandy substrata in the intertidal zone and shallow water. The truly intertidal species surf by expanding the foot as an underwater sail, and by this means migrate up and down the shore with the tides. Distribution covers both coasts of Africa, from the northern subtropical west coast, south around the Cape of Good Hope, and along the east coast at least as far as Somalia, Arabia, and both east and west coasts of India (Figure 5.6). The genus has also been recorded from Australia.

Genus: Terebra. A circumglobal, tropical genus, *Terebra* (Figure 5.7) is a voracious predator armed with harpoon-like radular teeth and a poison gland with a terminal bulb. The prey, often a polychaete worm, is subdued by the injection of venom before it is swallowed.

Genus: Oliva. Oliva burrows intertidally and in shallow water subtidally. It is typically found buried, with only the tip of its siphon showing. Unlike *Bullia* and *Terebra*, during locomotion it normally ploughs along just beneath the sand surface. This whelk feeds on carrion, and preys on small mollusks and crustaceans. It can turn deposit feeder and even suspension feeder, and rasps algae where these are available.

Genus: Olivella. Similar to *Oliva* in its habits, *Olivella* burrows in both clean and muddy sand, mainly from extreme low-water mark to the edge of the continental shelf. It is common on some North American Pacific beaches, where it is sometimes the numerically dominant gastropod. It also occurs on both Pacific and Atlantic coasts of Costa Rica and Columbia, as well as in Uruguay.

Genus: Natica. Natica (Figure 5.7) is a predatory prosobranch that attacks other mollusks, notably bivalves, although gastropods are not immune. It bores through the shell of its prey and feeds on the soft tissues within. It may be found intertidally on sheltered beaches and sandbanks and subtidally on more exposed shores. It is the only common gastropod on British sandy beaches and occurs throughout Europe and on both coasts of North and South America. It is recorded from many areas on the African coast, including Mozambique and the full length of the South African coastline.

Genus: Polinices. Another predatory naticid, *Polinices* (Figure 5.7) appears to be somewhat more restricted in its zoogeographical distribution, being recorded from both coasts of North America from Alaska to Mexico, from Australia and along the warm-temperate and subtropical east coast of southern Africa, as well as from the Red Sea.

Subclass Opisthobranchia, order Acochlidiacea. These are small opisthobranch gastropods found mainly in coarse sand and occasionally on open beaches. Genera include *Hedylopsis*, *Microhedyle*, and *Parahedyle*.

Order: Acoela. Nudibranchs would appear to be unlikely inhabitants of sandy beaches. They are, in fact, never abundant but can occur as members of the interstitial fauna in both sheltered and exposed beaches in both hemispheres.

Class Bivalvia, order Eulamellibranchia. Typical intertidal sandy-beach bivalves include members of the superfamily Tellinacea, *Donax* being typical of surf-swept beaches in many parts of the world. *Tellina* favors more sheltered shores, although both may occur in macrotidal low-energy sands. The wedge mussel *Donacilla* is also found on sheltered sands in tropical and subtropical Africa and Australia. The surf clam *Spisula* is common on some beaches in North and South America, in Australia, and on some British and European beaches, whereas *Schizodesma* occurs in the surf at the southern tip of Africa. Also worthy of mention is the Pacific razor clam *Siliqua* (Figure 8.12) — common on North American West Coast beaches as well as subtidally in subtropical Africa — and *Sanguinolaria*, abundant on sheltered shores in California and neighboring states. Other bivalves that may sometimes be encountered on sheltered beaches, but are more typical of sand flats and estuaries, include *Mercenaria, Macoma, Mya, Prothothaca, Spisula, Tivela* (Figure 8.12), and *Mactra* (Figure 5.8). These genera may sometimes be found in the surf zones of beaches too exposed to permit their intertidal occurrence. *Paphies (Amphidesma)*, (Figure 8.12) is a common bivalve in New Zealand, with three species on ocean beaches (see Chapter 8). *Strigilla* is abundant on the Pacific coasts of Costa Rica and Colombia. *Mesodesma* (Figure 8.12) is an important genus on beaches and in surf zones in South America (see Chapter 8). The tiny commensal clam *Mysella* lives on the gills of decapod crustaceans on the Pacific coast of the USA. All bivalves are filter feeders and typically occupy the mid- and lower shore, or occur below the tidal limit. Several of the beach clams mentioned previously can form substantial populations that support commercial and artisanal fisheries (see Chapter 8).

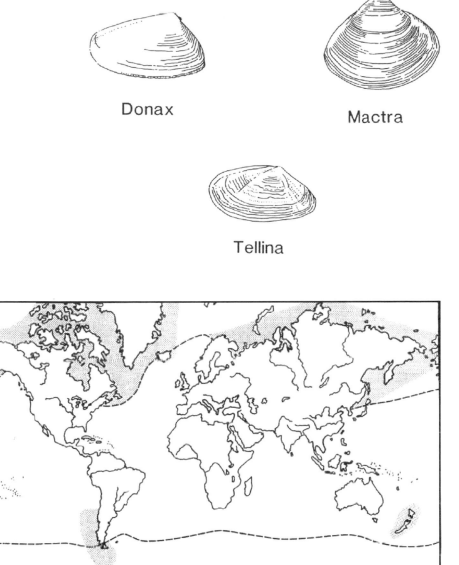

Figure 5.8. Some common sandy-beach bivalves (lengths typically 1 to 7 cm).

Figure 5.9. Map showing the 5° C isotherm and the areas (shaded) where *Donax* does not occur.

Genus: Donax. Donax (Figure 5.8) is a dominant genus on sandy beaches experiencing medium to strong water currents, virtually worldwide. Of the 64 species that have been described, about 75% are to be found in tropical waters, 22% in warm temperate regions, and some 5% in cold temperate areas. The global distribution of the genus is confined only by the 5° C minimum sea-surface isotherm (Figure 5.9), although it has apparently failed to colonize the beaches of Chile and New Zealand. The local species diversity of *Donax* decreases from the tropics to cold temperate regions. The most widely distributed species would appear to be *D. faba*, but this taxon may in reality represent more than one species. The animals are both intertidal and subtidal in areas of strongly circulating water. The intertidal species typically surf up and down the shore with the tides. The shell is wedge-shaped, of small to medium size, the valves equal and not gaping, and the sculpture weak. The umbo is a third of the distance from the hind end of the shell, the ligament is external, and the hinge has two cardinal teeth per valve. Its members are typically suspension feeders.

Genus: Tellina. In contrast to *Donax*, *Tellina* (Figure 5.8) favors sandy substrates not subjected to strong currents. It is thus more typical of sheltered bays, lagoons, and estuaries than of open exposed beaches, and there is little overlap between the two genera. Although the genus represents a wide range of forms, *Tellina* is easily distinguished from *Donax* in that the shell is ovate, with slightly unequal valves that gape narrowly behind. The hinge has one or two lateral teeth. *Tellina* does not surf, unlike most species of *Donax*. The genus is cosmopolitan but is most abundant in tropical waters, where its members tend to be highly colored — in contrast to the delicate shades of pink encountered in temperate waters. *Tellina* is usually a deposit feeder.

Class: Scaphopoda. Tusk shells, such as *Dentalium*, may occur in the subtidal off sheltered beaches in various regions — for example, the Arabian Sea and southwestern India.

Phylum: Tardigrada

A group of minute animals closely related to the Arthropoda, the Tardigrada are well suited to interstitial life, and although seldom abundant are usually present on beaches worldwide. In interstitial forms, the four pairs of unjointed legs end in adhesive discs, instead of the claws of other tardigrades. By means of these discs, the animal moves slowly from grain to grain. Some species are able to withstand desiccation for long periods. A typical interstitial genus is *Batillipes* (Figure 5.10).

Phylum: Arthropoda

Class: Merostomata, subclass Xiphosura. Xiphosurans (or horseshoe crabs) are not permanent members of the fauna of any sandy beach but are mentioned because they come ashore to breed in the intertidal zone, sometimes in very large numbers. The female digs a shallow burrow intertidally, depositing into it several hundred externally fertilized eggs that take a few months to hatch into trilobite larvae. *Limulus* occurs along both coasts of North America, whereas *Tachypleus* and *Carcinoscorpius* are found in Southeast Asia.

Batillipes
(0.4mm)

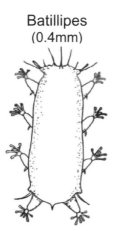

Figure 5.10. A typical interstitial tardigrade.

Class: Myriapoda. Centipedes can be beach predators in wrack deposited on the upper shore in California, for example.

Class: Arachnida. Spiders, solifugids, and scorpions live in supratidal and even intertidal sands, and pseudoscorpions are sometimes associated with wrack at the top of the shore.

Order: Acarina. The Acarina (mites) are also important permanent residents of the sandy-beach ecosystem. These are largely interstitial forms found both above and below the water table, with others found in wrack. Their systematics, distribution, and ecology are poorly known, although on some beaches they may be extremely abundant. Mites also occur as parasites of kelp insects and marine vertebrates and invertebrates. The interstitial forms are small, many being only 0.1 mm or so in length, and scavengers, herbivores, and predators are all represented. The most common free-living sandy-beach forms belong to the family Halacaridae, some species of which extend well into the subtidal.

Class: Pycnogonida. Adult pycnogonids have not been recorded from intertidal sandy beaches, although they may occur on sandy substrata offshore. However, the young stages of some pycnogonids are found as parasites in bivalve mollusks, including *Donax* on South African beaches.

Class Crustacea, subclass Ostracoda. Ostracods are fairly common members of the meiofauna in sheltered sands, less so on high-energy beaches. A number of genera are represented, some being cosmopolitan.

Subclass: Mystacocarida. Mystacocarids are typical members of the sandy-beach meiofauna and are found in no other habitat. They are especially common in fine to medium sands. The only genus is *Derocheilocaris*, and it appears to be cosmopolitan in its distribution (see Figure 9.3). Only a few species have been described.

Subclass: Copepoda. Copepods are prominent members of the interstitial meiofauna, some species being considerably modified for life between the sand grains. Although cyclopoid copepods are found in the interstitial environment, harpacticoids are by far the most common and best known. Genera include *Arenosetella, Hastigerella, Leptastacus, Cylindropsyllis, Arenopontia,* and *Psammastacus*. However, these truly interstitial genera are really characteristic of sands coarser than 180 to 200 µm. In finer sands, they are replaced by burrowers such as *Asellopsis*. A large variety of copepods are also to be found in the plankton of the surf zone (see Chapter 10).

Subclass Malacostraca, order Mysidacea. Sandy-beach mysids are in fact extremely common inhabitants of surf zones throughout the world. *Gastrosaccus* (Figure 5.11) is

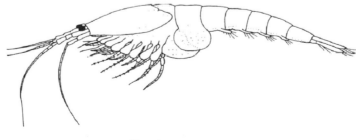

Gastrosaccus

Figure 5.11. The surf-zone benthoplanktonic mysid *Gastrosaccus* (about 1 cm).

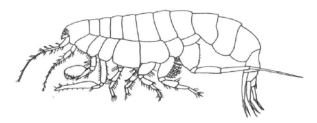

Synchelidium

perhaps the most widespread genus, although a number of others are encountered regularly, including *Mysidopsis*, *Metamysidopsis*, and *Mesopodopsis*. *Bowmaniella*, with a number of species, is common along South American coasts. Two groups of mysids may be distinguished ecologically: those that are benthoplanktonic, spending some time in the sand (e.g., *Gastrosaccus*), and those that are fully planktonic in the surf zone (e.g., *Mesopodopsis*).

Order: Cumacea. The cumacean family Bodatriidae is well represented in the sandy-beach fauna, both intertidally and below the tidal limit. Genera include *Cyclaspis*, *Leptocuma*, and *Pseudocuma* and can be meiofaunal or macrofaunal in size.

Order: Amphipoda. Talitrid amphipods are characteristic of the upper zones of sandy beaches in temperate latitudes. Genera include *Talitrus*, *Orchestia*, *Orchestoidea*, and *Talorchestia*. They are herbivorous or omnivorous and are generally associated with wrack and kelp (feeding on this plant material and burying themselves in the sand beneath it), although they also migrate up and down the shore in search of food. The talitrid *Talorchestes* is also associated with wrack, but occurs on plant material not only stranded on the beach but floating in the water.

In addition to talitrids, truly aquatic species of amphipods may be found on the beach. These include *Synchelidium* (Figure 5.12), an oedicerotid amphipod with a boreal distribution; the tropical *Atylus*, *Paraphoxus*, (Figure 5.13), and a number of haustoriid amphipods such as *Haustorius* (Figure 5.14), *Acanthohaustorius*, *Eohaustorius*, *Neohaustorius*, *Parahaustorius*, *Protohaustorius*, and *Amphiporeia*; the cold-temperate *Bathyporeia*; and the virtually cosmopolitan *Urothoe* (Figure 5.14). On some shores (e.g., Gulf and Atlantic coasts of the United States), haustoriid amphipods may dominate the aquatic biota. Gammarid forms may also occur, notably *Marinogammarus*. Mention should also be made of lysianassid amphipods such as *Psammonyx*, common on the shores of New England. *Pontharpinia* is abundant on some sandy beaches in tropical West Africa.

Genus Talitrus. This is a large genus of semiterrestrial amphipods, with many species in 21 subgenera. *Talitrus* (Figure 5.15) occurs sparsely in the tropics and subtropics as well as on temperate beaches, where it reaches maximum abundance. It is not, however,

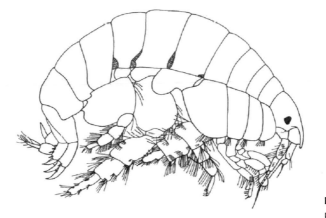

Paraphoxus

Figure 5.13. The phoxocephalid amphipod *Paraphoxus* (about 1 cm).

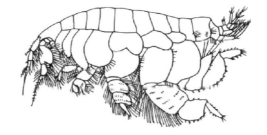

Haustorius

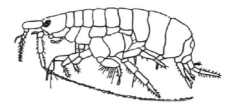

Bathyporeia

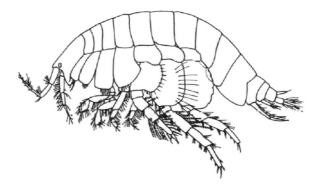

Urothoe

Figure 5.14. Three haustoriid amphipods of sandy beaches (about 1 cm).

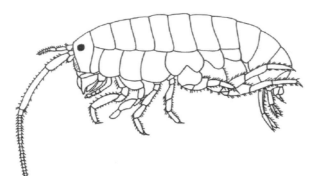

Talitrus

found in Antarctica and is apparently absent from Western Australia. It is found exclusively at the top of the shore in the supralittoral zone. The behavior of *Talitrus saltator* has been extremely well studied (see Chapter 6).

Genus Talorchestia. Talorchestia has fewer species and is somewhat more limited in its distribution than is *Talitrus*, although its mode of life is very similar. It is restricted to the supralittoral. It, too, reaches maximum abundance on temperate sandy beaches around the world, although it is found on some tropical shores (e.g., Mozambique).

Genus Bathyporeia. Bathyporeia (Figure 5.14) is a typical haustoriid amphipod in having no carapace covering the thorax, and having feeble gnathopods and numerous spines on the peraeopods and elsewhere, which aid in burrowing. The body is slender and fusiform. Burrowing takes place head first, and the animal literally swims through the sand, using a hydraulic tunneling action. Distribution of the genus is circumglobal in temperate and cold waters as far north as the Arctic. Although it occurs intertidally, maximum densities are frequently encountered in the breaker zone.

Order: Tanaidacea. Tanaids may be found in saturated sand on sheltered beaches and sandbanks in many parts of the world. Genera include *Tanais*, *Apseudes*, *Apseudomorpha*, and *Leptochelia*.

Order: Isopoda. As in the case of amphipods, isopods occur near the top of the shore on many beaches, as well as on the intertidal slope and in the surf zone. The almost cosmopolitan oniscid (family Oniscidae) isopod, *Tylos*, is characteristic of the upper zones of sandy shores with adequate back-beaches. *Ligia*, more typical of rocky shores, may invade the beach where piles of wrack are available.

Characteristic of the mid-tide zone are cirolanid (family Cirolanidae) isopods such as *Eurydice* (a widespread genus), *Cirolana*, *Neocirolana* (Australia), *Excirolana*, *Pseudolana* (Australia, Chile), *Pseudaega* (New Zealand), and *Eurylana* (New Zealand). *Excirolana* is perhaps the most ubiquitous genus of macrofauna on sandy beaches worldwide. *Pontogeloides*, thus far recorded only from Africa, occurs from the outer limit of the surf zone to the top of the intertidal slope, as well as in brackish waters. *Ancinus* is recorded from the coasts of Costa Rica and Columbia. Aquatic isopods belonging to other families may also be encountered: *Exosphaeroma* may sometimes be found on-relatively sheltered sands, whereas still other sphaeromid genera occur locally (e.g., *Sphaeromop-*

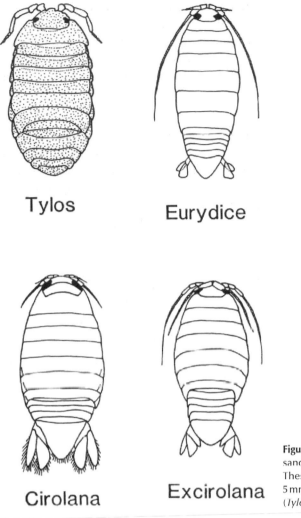

Figure 5.16. Common sandy-beach isopods. These are typically 5 mm (*Eurydice*) to 2 cm (*Tylos*).

sis on the coast of Kenya). Isopods such as *Chiridotea* may totally replace cirolanids on some shores.

Genus Tylos. Tylos (Figure 5.16) is a large semiterrestrial oniscid isopod. Like the talitrid amphipods mentioned previously, it displays well-marked tidal rhythms of activity and is essentially nocturnal in habit. The adults are primarily herbivorous, feeding on cast-up seaweed, whereas the juveniles have catholic tastes but tend toward a carnivorous diet. The genus is circumglobal from the tropics to cold-temperate latitudes but is absent from the Arctic and Antarctic. The adults are restricted to the supralittoral, but juveniles may commonly be found lower down the beach and even surfing in the swash. The biology of the genus *Tylos* has recently been reviewed (Brown and Odendaal 1994).

Genus Eurydice. Eurydice (Figure 5.16) is a fully aquatic cirolanid isopod, often characteristic of the upper midshore. Like other intertidal cirolanids, it displays tidal rhythms of swimming activity (see Chapter 6). Its zonation on the beach differs somewhat from species to species, and with prevailing conditions. Some species are found only subtidally, in the surf zone or beyond. It occurs from Norwegian and British coasts through the tropics to cold-temperate shores as far south as New Zealand.

Genus Cirolana. The behavior of intertidal sandy-beach species of *Cirolana* (Figure 5.16) is very similar to that of *Eurydice*. The genus is widely distributed on open sandy beaches, however, and some species are found only on very sheltered sands or in estuaries (whereas others favor hard substrata). The genus is recorded from both coasts of Africa; from Australia and New Zealand; from the tropical Pacific and Atlantic coasts of Panama, Costa Rica, and Columbia; from California; and from Mexico.

Genus Excirolana. Excirolana, unlike *Cirolana,* appears to be mainly confined to sandy beaches (where many species inhabit a zone around the upper midshore), although some (such as *E. natalensis* and *E. brasiliensis*) occupy positions higher up the shore around the drift line. It is also more widespread geographically than *Cirolana,* being found around the world from the tropics to warm-temperate regions. It is certainly the most ubiquitous intertidal invertebrate on low-latitude temperate, subtropical, and tropical sandy beaches of the New World.

Order Decapoda, suborder Macrura. Swimming prawns are characteristic of well-developed sandy surf zones. A common genus in South Africa is *Macropetasma,* the juveniles of which exploit the surf zone as a nursery area. Also common in some parts of the world, particularly in sheltered bays, is the snapper shrimp *Crangon* (Figure 5.17). *Lissocrangon* occurs buried in the sand of the lower shore on some North American beaches (Oregon). Other genera may also be present.

Order Decapoda, suborder Anomura. A number of anomurans are permanent inhabitants of intertidal sandy beaches. A few (such as *Callianassa*) live on relatively sheltered shores, where the sand is stable enough to allow semipermanent burrows to exist. They may also be abundant in the outer turbulent zones of beaches too unstable to support them intertidally. The more mobile sand crabs include *Lepidopa, Blepharipoda, Hippa,*

Figure 5.17. *Crangon,* a typical surf-zone swimming prawn (about 5 cm).

Crangon

and the tropical mole crab *Emerita*. *Emerita* is more characteristic of moderately exposed surf-swept beaches. It migrates up and down the slope with the tides. *Hippa* behaves in a similar manner and is able to inhabit beaches subject to heavy wave action. Hermit crabs (such as *Diogenes* and *Clibanarius*) may invade very gently sloping dissipative beaches in considerable numbers, and under these conditions are sometimes the most conspicuous and abundant members of the macrofauna. The land hermit crab *Coenobita* is sometimes common in the tropics, at the top of the shore.

Genus Callianassa. This burrowing prawn or ghost shrimp, found in muddy sand and in salt marshes as well as on sheltered shores of pure sand in the lower shore and in the outer turbulent zones of more exposed beaches, is a much-studied animal. Like the polychaete *Arenicola*, it lives in a burrow and plays an important role in turning over and oxygenating the sand, promoting the establishment of aerobic meiofauna and microbiota. It is also commercially important as a bait animal. The prawn feeds by cleaning the sand grains of organic particles, both living and dead, and its alimentary canal is often found to be full of diatoms and/or dinoflagellates. The genus is virtually cosmopolitan.

Genus Emerita. Unlike *Callianassa*, the mole crab *Emerita* (family Hippidae) is a highly mobile animal and favors more exposed beaches. Its barrel-shaped body and tough exoskeleton (Figure 5.18) ideally suit it to be rolled up and down the shore by the waves, chiefly in the lower midshore. Its geographical distribution is, however, more limited than that of *Callianassa*. It is found mainly in the tropics and subtropics, in a few places also invading temperate regions. It is a suspension feeder, using modified antennae to filter the water. A remarkable feature is that these antennae, which when extended are almost the length of the body, can be rolled up under the mouth parts when the animal is surfing in the swash.

Genus Hippa. Similar in its general appearance and surfing behavior to *Emerita*, but dorsoventrally compressed, *Hippa* (Figure 5.18) is commonly referred to as a "sea louse." The most distinctive difference, however, is that the antennae of *Hippa* are short. Its geographical distribution includes the tropical east and west coasts of Africa, the Red Sea, the Atlantic coasts of the USA, Costa Rica and Columbia, and Australia.

Order Decapoda, suborder Brachyura. In the tropics and subtropics, ocypodid crabs are conspicuous members of the macrofauna, occupying permanent burrows at the top

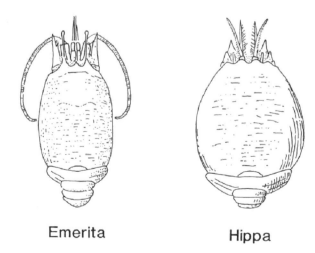

Emerita   **Hippa**

Figure 5.18. Two surfing anomurans from exposed shores: the mole crab *Emerita* and the sea louse *Hippa* (typically 2 to 4 cm).

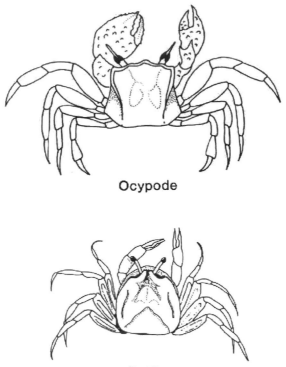

Ocypode

Dotilla

Figure 5.19. The semiterrestrial crabs *Ocypode* (5 cm) and *Dotilla* (1 cm).

of the shore. The most common genus is *Ocypode* (the ghost crab, Figure 5.19), but in some areas this genus is replaced by *Dotilla*. It was thought that the geographical distributions of the ocypodid crabs (tropical) and the talitrid amphipods (mainly temperate) showed virtually no overlap, but this has not been substantiated by subsequent work. The fiddler crab *Uca* is more typical of sandy and muddy areas associated with salt marshes and mangroves, but it sometimes invades open sandy beaches from these areas.

Truly aquatic brachyurans are also common on sandy beaches. Portunid swimming crabs (such as *Ovalipes*) are found on temperate beaches in both hemispheres, whereas *Lupa* and *Thalamita* are sometimes common on sheltered beaches in the tropics, as is the burrow-inhabiting crab *Macrophthalmus*. *Matuta* (Figure 5.20) is tropical and subtropical and is common on both the west and east coasts of Africa and adjacent tropical islands. Other common brachyuran genera include *Arenaeus* and *Callinectes*. Crabs such as the European *Carcinus* prefer muddy sand to pure sand but are sometimes to be found on sheltered sand flats.

Genus Ocypode. *Ocypode* (Figure 5.19), the most widespread of the Ocypodidae, is a much-studied animal and of all sandy-beach invertebrates has the most sophisticated behavior patterns. It is territorial and lives in semipermanent burrows near the top of the shore, which it defends with ritual display. Its distribution is circumglobal in the tropics and subtropics, with marginal invasion into warm temperate regions — including the eastern part of the Mediterranean and the Red Sea. The crab is not only a scavenger and a formidable predator but a deposit sorter. Its burrows, in some species marked by mounds, are typical of warm shores.

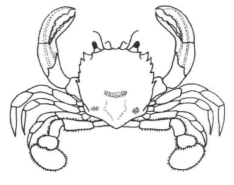

Ovalipes

Matuta

Figure 5.20. Aquatic sandy-shore crabs *Ovalipes* and *Matuta* (about 5 to 10 cm).

Genus Dotilla. This ocypodid (Figure 5.19) is also essentially tropical and subtropical in its distribution. Its mouth parts are highly specialized for deposit sorting, and it shows very high efficiency in extracting organic material from organically poor sands. Unlike *Ocypode*, it tends to be active during daylight low tides, remaining buried at night. It leaves a distinctive pattern of balls of sorted sand radiating from its burrows.

Genus Ovalipes. Ovalipes (Figure 5.20) is a voracious aquatic carnivore, preying largely on sandy-beach mollusks. It is extremely well adapted to life on sand, being a powerful swimmer and a rapid burrower. It is also well protected from higher predators by its robust chelipeds, which it holds aloft while facing a potential threat, burrowing backward at the same time. It is an inhabitant of subtropical and temperate waters around the world, occurring as far south as New Zealand. Although essentially subtidal, it may follow the rising tide up the slope and some species may actually come out of the water for short periods during low water at night.

Class: Insecta. Although insects have in general failed to establish themselves in marine environments, a number of insects are common obligate inhabitants on sandy beaches — both associated with wrack and stranded kelp at the top of the intertidal slope and between tide marks. There is, in addition, a wealth of insect species in the dunes, and these may sometimes be found on the beach either as active invaders or blown there by the wind.

Order: Collembola. These minute primitive insects, sometimes known as springtails (Figure 5.21), occur in soils of all types, all over the world. Only the suborder Arthopleona has marine representatives, however. By far the best known species on sandy beaches is *Anurida maritima*, although this genus in fact has about a dozen species on

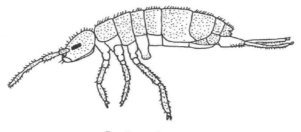

Collembola

Figure 5.21. A typical collembolan or springtail (about 1 mm).

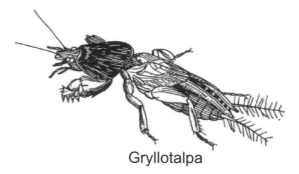

Gryllotalpa

Figure 5.22. The mole cricket *Gryllotalpa* (about 1 to 2 cm).

the shore. *Onychiurus* is found on European beaches, whereas *Hypogastrura* is cosmopolitan in wrack at the top of the shore, as is *Xenylla*.

Order: Orthoptera. These insects can be quite important on sandy shores, and on some beaches on the east coast of southern Africa pycno-mole crickets (Tridactylidae) are abundant members of the macrofauna and are tolerant of being submerged. *Tridactylus* appears to be the commonest genus, occurring most abundantly in subtropical and warm-temperate latitudes. In addition to these small orthopterans, much larger robust mole crickets — such as *Gryllotalpa* (Figure 5.22) — may be common above the high-water mark and invade the intertidal beach while the tide is out, burrowing just below the surface of the sand.

Order: Diptera. Wrack flies (or seaweed flies), most of which belong to the family Coelopidae, are found on sandy beaches all over the world. They are dependent on wrack and kelp for both food and shelter. Their surfaces are so hydrophobic that most species can crawl through densely packed mucilaginous weed without sticking, whereas some emerge completely dry after being submerged by the swash. Indeed, the larvae of some species can tolerate immersion in seawater for days. *Coelopa* (Figure 5.23), with a large number of species, is cosmopolitan. Other genera of Coelopidae include *Orygma*, *Malacomyia*, and *Oedoparea*. Other dipteran families, such as blowflies and houseflies, are also found on sandy beaches, sometimes in great numbers. Many of these are visitors, attracted for one reason or another or simply blown off course from the dunes or the land behind the beach. Among those with stronger sandy-beach affinities are *Helcomyza*, *Fucellia*, and *Thoracochaeta* — all associated with wrack.

Order: Coleoptera. Sandy beaches support numerous and diverse species of beetles, the most obvious frequently being cicindelids (tiger beetles) and staphylinids (rove

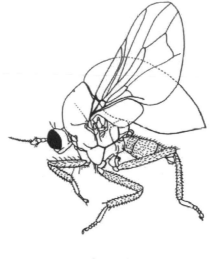

Coelopa

Figure 5.23. The wrack or seaweed fly *Coelopa* (about 3 mm).

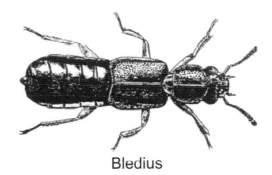

Bledius

Figure 5.24. *Bledius,* a staphylinid beetle characteristic of sandy shores (about 1 cm).

beetles) — although many others are represented, mostly around the drift line and commonly associated with decaying wrack, on which they feed. These include representatives of the families Anthicidae, Melyridae, Lathridiidae, Oedemeridae, and Ptiliidae — beetles seldom subject to submersion. Others probably often find themselves underwater, at least for short periods, including some Carabidae (e.g., *Cillanus, Dyschirius,* and *Halocomyza*), Tenebrionidae (e.g., *Phaleria* and *Epiantius*), and others. Hydrophilids of the genus *Cercyon* are common in most parts of the world, except for the Arctic and Antarctic. These are associated exclusively with piles of seaweed, although some other beetles may be found buried in sand.

Staphylinid beetles are extremely common on sandy beaches worldwide. Of the more than 30 subfamilies, nine are represented by permanent inhabitants of the seashore. Among the 60 or so genera, those most commonly referred to in the sandy-beach literature are *Bledius* (Figure 5.24), *Psammathobledius, Cafius, Omalium,* and *Philonthus*. Some species of Staphylinidae are submarine, in that they not only tolerate submergence in seawater but can continue their activities at reduced levels of metabolism while underwater. Others, including some forms associated with wrack (e.g., *Cafius*), are truly littoral — migrating upshore as soon as the sand or wrack is wetted by the swash.

Phylum: Ectoprocta

Ectoproct bryozoans might be considered the least likely animals to occur on sandy beaches. They are, indeed, rare but are worth keeping a lookout for in the interstitial spaces of sheltered sands. *Monobryozoon* is the best-known genus. It is solitary and free moving. It has outgrowths on the lower part of the body that not only act as adhesive organs but have reproductive capacity, budding off new individuals.

Phylum: Echinodermata

Echinoderms are not typical of sandy beaches, although on sheltered shores some (such as the echinoids *Echinodiscus* and *Dendraster*) may move shoreward to the bottom of the intertidal slope. Exceptions are the sand dollars *Encope* and *Mellita*, which occur intertidally as well as subtidally. The echinoderm more commonly associated with sandy beaches is *Echinocardium*, which is confined to intertidal and shallow-water sands, usually in sheltered areas but occasionally in moderately exposed situations. It excavates a semipermanent burrow for itself so that like other such invertebrates (*Arenicola*, *Callianassa*) its intertidal distribution is limited to beaches stable enough to support its burrows. The larvae of *Echinocardium* settle offshore and migrate toward the intertidal as they mature, the adults burrowing either in the immediate subtidal or intertidally in the zone of saturation. *Enchinocardium* is cosmopolitan in its distribution. The surf zones of sandy beaches, and particularly the outer turbulent zone, may be rich in echinoderms — including not only echinoids but holothurians; ophiuroids such as *Ophionema*, *Ophiothrix*, and *Paracrocnida*; and asteroids such as *Asterias*.

Phylum: Hemichordata

The acorn worm *Balanoglossus* may occur in some numbers on sheltered sand flats in the tropics. It is not recorded from open surf-swept beaches.

5.3   Conclusions

This chapter has introduced the main groups of invertebrates likely to be encountered on sandy beaches around the world. Although individual sandy beaches do not usually have more than 20 or 30 macrofauna species present, and often fewer (see Chapter 7), the range of beach types and geographical areas available to be colonized means that a very wide range of invertebrates have been recorded. Among the macrofauna, crustaceans, polychaetes, and mollusks are usually dominant, whereas among the meiofauna nematodes and harpacticoid copepods are usually most common on sandy beaches. The meiofauna also includes many unique groups (see Chapter 9). These invertebrates have been the main focus of study by most beach ecologists. Their adaptations to these harsh environments are covered in Chapter 6; their ecology in Chapters 7, 8, and 9; and their trophic relations and energetics in Chapter 12.

Adaptations to Sandy-beach Life

6

Chapter Outline

6.1   Introduction

Many of the adaptations that distinguish sandy-beach animals from those of other marine habitats result from instability of the substratum coupled with heavy wave action. Thus, burrowing behavior (which is displayed by animals inhabiting all types of soft substrata) must on high-energy sandy beaches be both rapid and powerful if the animal is not to be swept away by incoming waves and swash. It is also essential that the fauna displays a high degree of mobility and ability to deal with the swash climate. This in turn implies the necessity for mechanisms enabling the animals to maintain their positions on the shore and to regain those positions once lost. Responses such as rheotaxis and burial at an appropriate point in the tidal cycle aid in such maintenance of position, but sophisticated sensory mechanisms are equally important. The ability to orientate to environmental situations is developed to a high degree, not only among the aquatic faunas but particularly among the air-breathing amphipods, isopods, and brachyurans inhabiting the upper part of the shore. Although these semiterrestrial forms can tolerate submersion in seawater, they cannot risk being caught by the surf and swept out to sea. Neither can they allow themselves to wander too far inland. The combination of a high degree of mobility and complex responses to environmental cues has led to typical sandy-beach animals developing night-day and tidal rhythms of migration, which maximize food resources and possibly attenuate predation. In many such species, endogenous clocks (often of a complicated nature) may be demonstrated.

Heavy surf brings other problems, some of which are concerned with feeding. Prey are more difficult to locate and capture under turbulent conditions, and carrion is moved around in the surf and swash until it is deposited at the edge of the tide, where it may be unavailable to many of the aquatic scavengers. This, coupled with the highly erratic supply of carrion to the beach, makes it essential for scavengers to be able to locate and reach their food rapidly and to ensure that they are not parted from it until they have consumed as much of it as they are able. High rates of ingestion, efficient digestion and assimilation, and subsequent energy conservation are called for if the animal is to be sure of surviving until the next unpredictable meal.

These and other adaptations are much more marked in the animals of high-energy sandy beaches than in those on more sheltered shores. Other adaptations show the reverse trend — a trend consistent with differences in environmental conditions. Thus, on sheltered sandy beaches (and particularly in muddy sands) respiratory adaptations are necessary to tolerate poor or declining oxygen tensions, especially during ebb tide. On high-energy sands, oxygen tensions are normally high to a depth of a meter or more below the sand surface (see Chapter 3), so that no consideration of hypoxia arises.

The present chapter treats only the adaptations of the invertebrate macrofauna in any detail. Meiofaunal adaptations have been far less studied but are clearly different in a number of respects, the most obvious requirement of an interstitial animal being the ability to move between the sand grains. Morphological adaptations are thus in evidence in some groups — a thin elongate shape being advantageous. As in the case of the macrofauna, endogenous tidal rhythms are much in evidence, but unlike many members of the macrofauna the meiofaunas tend to migrate up and down through the sand column and do not habitually leave the shelter of the substratum *en masse*, so that adaptations related to locomotion and position finding above the sand surface are not well developed.

6.2   Locomotion

There are no truly sessile animals on intertidal sandy beaches, and locomotion plays a much larger part in the lives of the animals than it does on other types of shore. A number of different forms of locomotion are in evidence — the one type all sandy-beach animals have in common being burrowing. In addition, swimming is common among the Crustacea, whereas crawling, leaping, running, and surfing also occur in one group or another.

Burrowing

Mechanisms of burrowing into soft substrata have received much attention. In the majority of cases, burrowing involves two distinct phases: penetration of the substratum, followed by burrowing movements. Penetration is frequently facilitated by the thixotropic

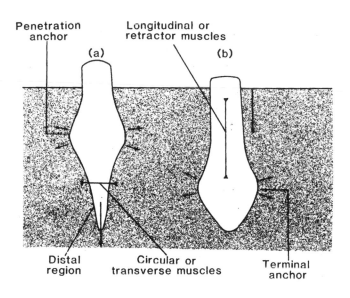

Penetration anchor

Longitudinal or retractor muscles

(a)   (b)

Distal region

Circular or transverse muscles

Terminal anchor

Figure 6.1. The two principal stages in the burrowing process of a generalized soft-bodied animal (after Trueman and Ansell 1969). (a) Formation of a penetration anchor that holds the animal while the distal region is elongated by contraction of circular or transverse muscles. (b) Dilation of the distal region to form a terminal anchor, which allows contraction of longitudinal or retractor muscles to pull the animal into the substratum. Maximum pressures are developed in the fluid system at this stage.

properties of particulate substrata, so that the sand is liquefied by repeated probing movements of the head (in a worm) or of the foot (in a mollusk). In this way, the sand may be penetrated without the application of large forces. Initial penetration can be relatively slow on very sheltered beaches, where waves and swash will not easily dislodge the animal. It may take anywhere from 20 seconds in the polychaete worm *Arenicola marina* up to 15 minutes in the burrowing anemone *Peachia* (Ansell and Trueman 1968, Trueman and Ansell 1969). It has to be far more rapid on surf-swept beaches, where the danger of being swept away is always present, so that penetration in only a few seconds is mandatory and larger forces may be applied.

The burrowing movements that follow penetration, once the animal has gained adequate purchase in the sand, are essentially similar in virtually all soft-bodied forms — being based on the alternate application of two types of anchorage, each of which allows another part of the body to move into the sand (Figure 6.1). This results in a stepping motion, each step being termed a digging cycle. The first, or penetration, anchor is formed by the dilation of part of the posterior region of the body and prevents the animal from being pushed out of the sand as it thrusts downward. The second, or terminal, anchor is formed by the swelling of the anterior or leading region (the head in a worm, the distal portion of the foot in a bivalve mollusk) so as to allow the posterior to be drawn down. To permit such changes in shape, soft-bodied animals have characteristically developed large fluid-filled cavities that effect both changes in shape and the transfer of muscular forces. The various stages of the digging cycle are thus accompanied by pressure changes in the coelom or haemocoel (Trueman and Brown 1976, 1985).

Although the burrowing movements of nearly all soft-bodied forms are based on the same principles, the details differ from group to group. Each digging cycle in *Arenicola* consists of a powerful retraction into the sand, followed by a pause during which the animal elongates and everts its proboscis — its flanges and chaetae functioning as the

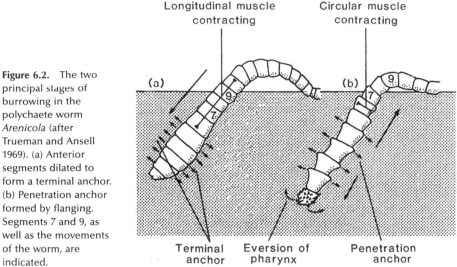

Figure 6.2. The two principal stages of burrowing in the polychaete worm *Arenicola* (after Trueman and Ansell 1969). (a) Anterior segments dilated to form a terminal anchor. (b) Penetration anchor formed by flanging. Segments 7 and 9, as well as the movements of the worm, are indicated.

penetration anchor (Figure 6.2). Increasing time intervals as the worm burrows deeper are associated with the development of a flange-proboscis sequence in which the two structures are combined to form a digging tool. The flange-proboscis sequence begins with the shortening of the *Arenicola*, the trunk moving into the sand while the anterior end is pulled back with the flanges erect. Eversion of the proboscis scrapes sand away from the end of the burrow, the sand moving outward and backward. When the worm has shortened, the proboscis is suddenly retracted — an event that draws water through the sand into the cavity formed in front of the animal. A recovery phase then ensues, the trunk being elongated and the head penetrating the water space — forcing the water into the sand ahead, which liquefies it. The flange-proboscis sequence is thus important in penetration of the sand, in the removal of sand from the burrow, in softening the sand ahead of the worm to facilitate further burrowing, and in the drawing of the trunk into the sand. These events, together with the associated forces involved and pressures recorded from the coelom, are summarized in Figure 6.3. The actual rate of burrowing, or the time taken for the animal to bury itself by a series of digging cycles, is closely related to the penetrability of the sand.

Bivalve mollusks are primitively infaunal, with a laterally compressed and blade-like foot with which to thrust downward into mud or sand. Like the burrowing polychaetes, they too show a preliminary penetration phase followed by a series of digging cycles in which the alternate formation of penetration and terminal anchors is in evidence. In the initial penetration phase, the animal lies on its side on the sand, while the foot probes downward until sufficient purchase is obtained to allow the shell to be drawn erect. Digging cycles then commence. Each cycle involves anchorage of the shell (penetration anchor), while the foot thrusts downward by contraction of the pedal transverse muscles in antagonism to the distal retractor fibers. The shell is then drawn downward, while dilation of the distal part of the foot forms a terminal anchor (Figure 6.4). The downward movement of the shell is facilitated by adduction of the valves. Adduction also generates a pulse of high pressure in the haemocoel, which is directly responsible for the distal pedal dilation. During shell adduction, the siphons are closed so that water can only escape from the mantle cavity downward, through the pedal gape. This jet of water

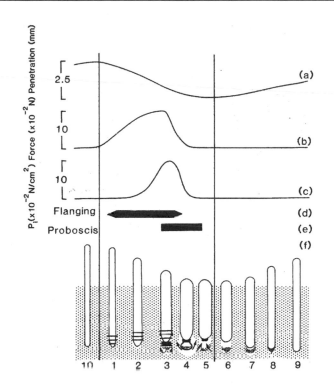

Figure 6.3. The principal events comprising a single flange-proboscis sequence in *Arenicola* (after Seymour 1971). (a) Movement of posterior trunk into the sand, from an isotonic myograph record. (b) The downward force exerted by the worm. (c) Haemocoelic pressure. (d, e) Timing of flange erection/depression and proboscis eversion/retraction, respectively. (f) Successive diagrams of the worm at different stages of the flange-proboscis sequence. The large stippled area represents sand. Coarse stippling in diagrams 5 to 8 represents the volume of sand liquefied when water is drawn through it in the direction of the arrows by retraction of the proboscis. The degree of thickening and shortening of the worm is exaggerated for clarity.

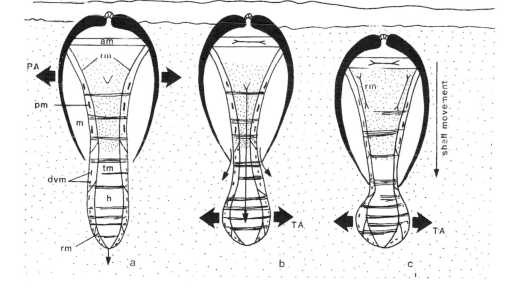

Figure 6.4. Sections of a bivalve mollusk showing the main movements used in burrowing (after Brown *et al.* 1989). (a) Valves pressed outward against the sand by the opening thrust of the ligament, to provide a penetration anchor, while the foot probes downward (arrow). The adductor muscle is relaxed and the foot extends and retracts by muscle antagonism distally about the pedal haemocoel. (b) The valves partially adduct (arrows indicate tension in the adductors), causing blood flow and dilation of the foot distally, forming a terminal anchor and causing water to be ejected from the mantle cavity to facilitate the movement of the shell through the sand. (c) Contraction of the retractor muscles (arrows rm) draws the valves down toward the anchored foot (am, adductor muscle; dvm, dorso-ventral muscle; PA, penetration anchor; pm, protractor muscle; m, mantle cavity; TA, terminal anchor; tm, tranverse muscle; and v, viscera).

liquefies and excavates sand from beneath the shell, making its passage easier through the substratum.

All species of *Donax* are rapid and powerful burrowers, achieving complete burial in only a few digging cycles. The time required for burial in tropical species is only five or six seconds, and even less. Temperate species take longer, some 20 seconds in *D. serra*, but still burrow very rapidly compared with *Arenicola* and other forms from sheltered shores (Brown *et al.* 1989).

Burrowing in the whelk *Bullia* is essentially a continuation of surface crawling, the propodium being inserted into the sand as a mobile freely progressing wedge. Burrowing takes place at an angle of no more than 10 to 15 degrees to the sand surface and ceases once the shell is covered with sand, the short siphon protruding into the water above. This contrasts with the vertical deeper burrowing of *Donax*. The force required to penetrate the sand is greatest for vertical burrowing (as in bivalves) and decreases as the angle of burrowing becomes shallower, being greatly reduced in *Bullia* (Brown and Trueman 1991). Thus, the initial probing or penetration phase — which is such an obvious feature in burrowing bivalves and polychaete worms — is not recognizable as a separate activity in *Bullia*. As in bivalves, however, the whelk burrows in a stepwise manner. These slow steps may be viewed as a modification of the locomotory waves employed by typical gastropods over hard substrata (Trueman and Brown 1987). Each step or digging cycle consists of propodial extension (in a movement resembling the breast stroke of a swimmer) and anchorage posteriorly (penetration anchor) by the expanded metapodium (and when buried also by the shell), followed by propodial dilation to form an anterior terminal anchor that allows the metapodium and shell to be drawn forward. The efficiency of this terminal anchor is increased by a flap of tissue on the anterior dorsal surface of the foot, which is angled backward to resist the anterior end being drawn back toward the shell during stepping (Hodgson and Trueman 1985).

Ejection of water from the mantle cavity during burrowing, a feature common to all infaunal bivalves, is a convergent adaptation in *Bullia*. It is also possible that the aquiferous spaces of the metapodium of *Bullia*, which open to the ventral surface, serve to lubricate this surface as the metapodium is pulled forward during crawling and burrowing (Brown *et al.* 1989).

There are several published figures for burrowing performances in species of *Donax* and *Bullia* (e.g., Trueman and Brown 1989, McLachlan and Young 1982). Burial on a beach over which waves pass at about 10-second intervals is probably restricted to small individuals, as only these can bury themselves rapidly enough to avoid being washed out of the sand by the succeeding wave. Larger animals, which take longer to burrow, would be restricted either to calmer waters subtidally or to the water-saturated sand of the swash zone. The period of swashes crossing 50% of this zone on a dissipative beach typically ranges from 20 to 50 seconds, and once an animal is two-thirds buried swashes do not dislodge it. The effects of sand particle size and swash on burrowing are covered in Section 6.5.

Other gastropods, such as *Oliva* and *Natica*, burrow in a manner similar to that of *Bullia* — although they appear to be not as highly adapted. *Bullia* is probably the most powerful and efficient gastropod burrower so far studied.

To live in sand exposed to wave action, it is necessary that the animals avoid being buried too deeply by sand deposition. Both *Donax* and *Bullia* achieve this, the latter by normal stepping motion in an upward direction and *Donax* by reversal of direction by means of powerful downward thrusts of the foot. It would appear that when confronted with heavy surf both animals constantly adjust their positions in the sand, taking upward or downward steps in response to wave crash.

The types of digging cycles described previously are only applicable to soft-bodied animals, such as worms and mollusks, in which the body can elongate and regions of it dilate. In sand-burrowing arthropods, on the other hand, the body has a fixed shape imposed upon it by the exoskeleton — and a burrow must be excavated by the jointed appendages. This is by no means a disadvantage, and in general the aquatic Crustacea burrow more rapidly than any of the soft-bodied animals (the mole crab *Emerita* displaying a total digging period of less than 2 seconds). The use of jointed appendages also allows many semiterrestrial Crustacea and insects to burrow rapidly into dry or moist sand, whereas soft-bodied invertebrates can usually only burrow into water-saturated sand. A variety of digging methods are employed by the Arthropoda, including the backward burrowing of forms such as *Ovalipes* and *Emerita*; the sideways digging of the ocypodid crabs, the head-first burrowing of *Tylos*, *Eurydice*, and talitrids, and burrowing while keeping the body virtually horizontal, as in *Gastrosaccus* and the isopod *Exosphaeroma*. Despite this variety of burrowing methods, it may be said that in general some limbs grip the sand and others excavate. The same gradual process is apparent in burrowing echinoderms, such as *Echinocardium*.

Although most sandy-beach arthropods (including all of the truly aquatic species) burrow only shallowly into the sand, a few semiterrestrial forms excavate deep burrows, expending much energy in doing so. In the case of ocypodid crabs, these burrows are permanent. However, in forms such as *Tylos* and *Talitrus* the animal does not return to its old burrow after a feeding excursion. It does, however, tend to seek out an abandoned burrow and will use it even if it has collapsed, for the energy then used is much reduced. The animal also commonly starts a new burrow where the sand surface has been disturbed, often in a human footprint, presumably also with a beneficial saving in energy.

A few soft-bodied animals do not burrow in the manner outlined previously, dependent on the successive application of distinct anchors. For example, a notable exception is the polychaete worm *Thalanessa*, which instead of digging head first, lies flat on the sand, rotating its parapodia to scoop out the sand below the animal until the entire body is covered. This method is much faster than more conventional polychaete burrowers can hope to achieve, and is clearly an adaptation to life on relatively high-energy beaches.

Finally, it must be stressed that not only can animals from relatively sheltered shores afford to burrow less rapidly than those on exposed shores, in general, they also need to bury themselves less frequently. Thus, it is doubtful whether an animal such as *Sipunculus* ever needs to reburrow under normal circumstances, although it requires that potential, as do all sandy-beach animals.

Surfing and Coping with Swash

Surfing is a common form of locomotion among the resident invertebrates of exposed sandy beaches, although it is much less in evidence on relatively sheltered shores. It has been observed in *Donax*, in *Emerita*, in the young of some species of *Tylos*, in sandy-beach whelks, and in other forms, and has been studied in detail in the whelk *Bullia*. Surfing is normally associated with tidal migrations up and down the beach face, and at first sight appears to be a cheap form of locomotion in terms of energy expenditure. This is not always the case, however, and surfing in *Bullia* is about as costly per unit time as is burrowing — and far more costly than crawling, although it is relatively cheap in terms of distance covered (Brown 1982a). The rapidity of this form of locomotion may, in fact, be more important to tidally migrating animals than a saving of energy. Odendaal *et al.* (1992) showed how efficiently *Bullia* uses swash riding to track prey through the surf.

The cost of surfing in *Bullia* relates chiefly to contraction of the dorso-ventral pedal muscles, maintaining maximum turgidity of the foot and presenting maximum surface area to the water so that the foot functions efficiently as an underwater sail (Brown *et al.* 1989). Efficiency is enhanced by the fact that the fully turgid foot is not quite flat but is concave in dorsal aspect, thus offering the greatest possible resistance to the water. Only the intertidal species of *Bullia* surf, and these do not all surf at the same rate under the same conditions. For example, *B. rhodostoma* is carried faster and farther in a given current than is *B. digitalis*. This is related to the surface area of the foot and the total specific gravity of the animal. Differences in surfing ability appear to account in large part for the fact that where more than one species of *Bullia* occur on the same beach they tend to occupy different areas that show little or no overlap.

The bivalve *Donax* surfs with both its foot and siphons extended. The cost of this activity in bivalves may well be less than in *Bullia*. Most species of *Donax* surf on every tide, as do the surfing species of *Bullia*. However, the largest species (*Donax serra*), shows a decreasing tendency to surf with increasing size, the biggest individuals probably never leaving the sand. Smaller individuals surf intermittently, chiefly on spring tides, with some migration at neaps — daily tidal changes being largely ignored by the adults (Brown *et al.* 1989).

Even where a species surfs on every tide, this does not imply that every individual of the population does so. The percentage of *Bullia digitalis* surfing on a particular tide depends on the nutritional state of the animals, the availability of carrion, beach slope, wave action, and other factors. It may be as low as 12% or it may involve most of the population. Females carrying eggs do not surf under any circumstances, however, but remain buried in the sand below low-water mark (Brown *et al.* 1989). Crustaceans, being more mobile, regularly swim and surf in the swash — some entering the plankton at night.

All sandy-beach animals must be able to cope with swash. They must move and orient and burrow in it, even if they are not active migrators or surface forms. Thus, the ability to move and burrow in the swash is essential, and because swash climate becomes harsher toward reflective beaches (see Section 2.11) the ability to handle the swash is a key adaptation enabling colonization of different beach types. Ellers (1995a, 1995b, 1995c) has shown how well-adapted *Donax variabilis* is to swash on Florida beaches. This clam jumps out of the sand in response to the acoustic shock of waves to ride swashes up and down the beach during tidal migrations. A small streamlined shape and high density provide stability and orientation in turbulent swash and enhance the clam's ability to gain a foothold and dig in after a swash ride. A comparison of burrowing abilities and swash

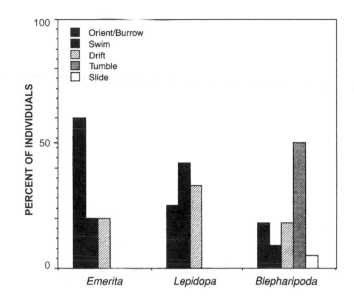

Figure 6.5. Behavior of three hippid crabs following release in the swash zone (after Dugan *et al.* 2000). *Emerita* displays the most effective ability to orient and burrow. *Lepidopa* spends more time swimming and drifting and *Blepharipoda* is tumbled around much of the time.

behavior of three hippid crabs showed that *Emerita analoga* was best able to cope with the swash and was most successful in the intertidal zone, whereas two other species (*Blepharipoda* and *Lepidopa*) that were less able to handle the swash tended to be more subtidal and were less successful in establishing intertidal populations (Figure 6.5, Dugan *et al.* 2000). It was suggested that retreat to the subtidal during rough conditions may be the strategy for species less able to cope with harsh swash conditions in the intertidal zone. Behavioral responses to swash climate are therefore important. Thus, *Hippa* (which can inhabit even harsh reflective beaches) has the ability to swim and orient effectively in swash and to burrow into a wide range of particle sizes (Lastra *et al.* 2002). Isopods of the genus *Excirolana,* also successful over a wide range of beach types, display a distribution across the shore related to their ability to deal with swash (Yannicelli *et al.* 2002).

6.3   Rhythms of Activity

Tidal rhythms of activity are by no means limited to sandy-beach animals. Indeed, all animals subjected to tidal rise and fall on all substrata must of necessity suffer discontinuities of behavior imposed upon them by the tidal cycle. The implications of tidal ebb and flow on a soft substratum are, however, different from those on a rocky shore. In particular, the danger of desiccation is not an overriding concern, as the animals can retreat below the surface of the substratum or even below the water table. However, intertidal filter-feeders cannot feed while the tide is out. Carrion beached by the previous rising tide lies too far up the slope to be available to aquatic scavengers, whereas other activities (such as reproductive behavior, which may depend on the animals being submerged) are at a minimum.

Interest in tidal rhythms of activity was first stimulated by the work of Gamble and Keeble (1904) on an endogenous rhythm of vertical migration in the turbellarian

Convoluta. Since then, vertical tidal migrations through the sand column have been demonstrated for many members of the meiofauna, including other turbellarians. Mass migration of a red turbellarian occurs on Hout Bay Harbor Beach near Cape Town, such that the brilliant white sand surface quite suddenly turns bright red — the color only beginning to fade after about two hours, as the animals retreat below the surface once more.

Macrofaunal arthropods and mollusks, on the other hand, tend to leave the substratum and to show excursions up and down the beach face with the tides, particularly on exposed beaches showing a moderate tidal range. Among Crustacea, such tidal migrations are displayed (among others) by the isopods *Eurydice*, *Excirolana*, *Pseudaga*, and *Tylos*: the amphipods *Bathyporeia*, *Synchelidium*, *Marinogammarus*, *Talitrus*, *Orchestia*, *Orchestoidea*, and *Talorchestia*: the mysid *Gastrosaccus*; and decapods such as *Emerita*, *Ovalipes*, *Ocypode*, and *Dotilla*. Tidally migrant mollusks include the intertidal species of *Bullia* and *Donax* (see Brown and McLachlan 1990 for a list of references).

In many of these animals, an endogenous rhythm has been either demonstrated or assumed, triggered by factors related to the rise and fall of the tides. In most of the Crustacea that have been studied, tidal activity rhythms continue to manifest themselves in the laboratory (in the absence of tidal cues, which is good evidence for the operation of an internal clock). Sandy-beach mollusks, on the other hand, have not been shown to possess internal clocks, and their activity rhythms are not continued under constant conditions. The tidal migratory responses of *Donax*, for example, appear to be dictated entirely by changing physical conditions during ebb and flow (Ansell 1983). Even semicircalunar migrations in *Donax serra* are responses to changes on the beach face over spring-neap cycles and not endogenous rhythms (Donn *et al.* 1986). In fact, emergence by clams from the substrate at the bottom of the shore is probably a response to increasing liquefaction of the sand as the tide rises, whereas emergence from their upper positions is triggered by increasing periods of nonsaturation between waves. Ellers (1995a, 1995b) studied the migratory behavior of the wedge clam *Donax variabilis* in the laboratory, showing that these clams actively jump out of the sand to ride swashes — selecting the strongest swashes for this by responding to low-frequency sounds as cues of breaking waves and swashes. They even appeared to display some endogenous rhythmicity, with jumping behavior being strong during high-tide periods and least during low tide. These factors may be important in triggering the migrations of other psammophiles as well.

More complex than the behavior of sandy-beach mollusks is that of the cirolanid isopods exemplified by *Eurydice*, a genus that has been studied extensively (see Naylor and Rejeki 1996). The distribution of *Eurydice pulchra* varies according to an endogenous semilunar rhythm. If sand was provided in a laboratory tank, the animals swam in the overlying water to a maximum extent at the times of high water a few days after new and full moon (in other words, on waning spring tides). In the field, such behavior ensures that the isopods remain in the water long enough to be carried down the slope, where they will be covered by subsequent neap tides. In contrast, very little swimming takes place during neap tides, the animals mostly remaining buried in the sand or leaving it only for short periods — although a circatidal swimming rhythm is seen at both spring and neap tides if no sand is provided (Figures 6.6 and 6.7).

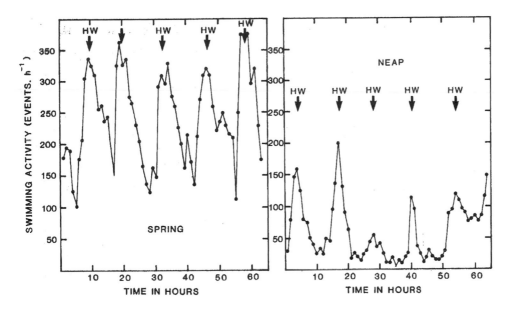

Figure 6.6. Spontaneous swimming activity of the isopod *Eurydice pulchra* collected at spring and neap tides and kept without sand (after Alheit and Naylor 1976). Arrows indicate the time of high water in the natural environment. There is a circatidal rhythm of swimming activity synchronized with high water both at spring and neap tides, although the amplitude is lower at neaps. This suggests the presence of a circatidal swimming rhythm independent of emergence, in that the latter is inhibited during neap tides.

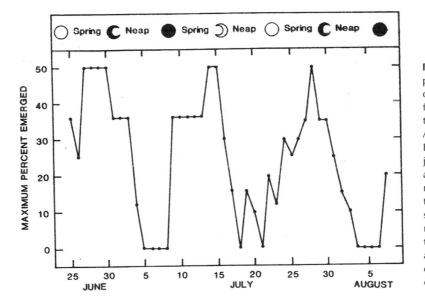

Figure 6.7. Maximum percentage emergence of *Eurydice pulchra* from aquarium sand in the laboratory (after Alheit and Naylor 1976). Emergence is greatest just after spring tides and lowest during neaps. This indicates the presence of a semicircalunar 14-day rhythm of emergence that allows the animals to be carried downshore prior to the onset of neap tides.

Swimming activity is also more marked at night than during the day. This diurnal swimming rhythm and the circa-semilunar rhythm of emergence from the sand are probably distinct from one another, being controlled by separate internal clocks — a suggestion that highlights the complexity of the responses of sandy-beach invertebrates. Similar responses have been described for the isopod *Excirolana* (Enright 1972), as well as for the amphipod *Marinogammarus* (Fincham 1971), which has a particularly marked diurnal rhythm expressed as a photonegative response during the day.

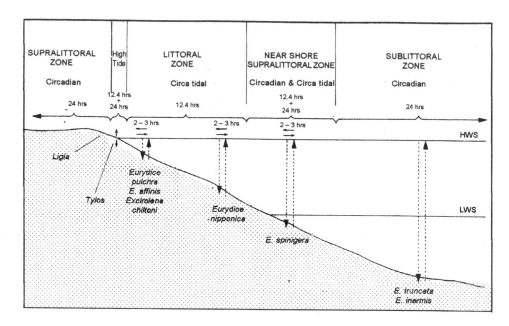

Figure 6.8. The zonation and activity rhythms of some sandy-beach isopods with distributions ranging from the supralittoral to offshore (after Jones and Hobbins 1985).

Mechanical disturbance of the sand can act as a trigger for the emergence of *Eurydice*, *Excirolana*, *Synchelidium*, *Emerita*, *Donax*, and *Bullia*. The difference in this respect between animals from exposed and sheltered sandy beaches is that water currents of sufficient strength to disturb the sand may provoke emergence in species from high-energy beaches but cause burrowing in sheltered-shore species (Brown 1973).

Jones and Hobbins (1985) examined the roles of swimming rhythms in cirolanid isopods, particularly with regard to maintenance of zonation patterns. The zonation of terrestrial and offshore species is largely unaffected by the tides, and these forms tend to display only circadian rhythms. On the other hand, intertidal forms and those immediately adjacent to the intertidal slope employ both circadian and circatidal rhythms (Figure 6.8). This is largely based on studies of species in areas subject to large tidal ranges and moderate to weak wave action. However, species in Mediterranean beaches with small tidal ranges display only a circadian rhythm. In situations where tides are smaller but wave energy higher, tidal rhythms are less clear. Intertidal cirolanids from high-energy microtidal beaches in South Africa, for example, have less pronounced and more variable rhythms than reported from more sheltered macro-mesotidal situations. *Eurydice*, which occurs lowest on the shore, exhibits the strongest circatidal rhythm. *Excirolana*, which occurs highest on the shore, displays some semilunar modulation of the circatidal rhythm. These two species, as well as the species occupying the middle zone (*Pontogeloides*), display some circadian modulation of the circatidal rhythm — the latter having circatidal and circadian rhythms about equally strong (de Ruyck *et al.* 1991). Activity differs between adults and juveniles in these three species, but is mostly greatest during nocturnal high tides. The dynamic conditions on these high-energy South African beaches seem to have resulted in less distinct rhythms and more plasticity in behavior. Indeed, plasticity may be a key survival feature of sandy-beach macrofauna.

The rhythmic behavior outlined previously is not restricted to aquatic forms but is also found in semiterrestrial species, such as the isopod *Tylos* and even coleopterans (Jaramillo

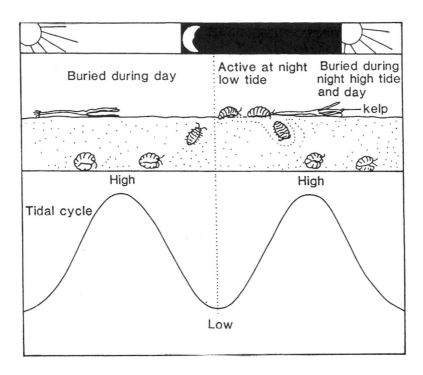

Figure 6.9. The semiterrestrial isopod *Tylos* displays endogenous rhythms of activity and is only active during low tide at night, when it emerges from its burrow to roam the shore in search of kelp, on which it feeds (after Kensley 1974, Branch and Branch 1984).

et al. 2000). Most species of *Tylos* are strictly nocturnal, emerging from their burrows above the high-water mark on the ebb tide, feeding on stranded macrophytes and returning to the burrowing area before the tide can reach them. *T. granulatus* and *T. capensis* demonstrate a 24.8-hour diurnal rhythm of activity, the isopods thus emerging later each night. After some 15 nights, when maintenance of the cycle would result in it growing light while the animals were still on the surface, a tide is skipped so that nocturnal foraging is maintained (Kensley 1974) (Figures 6.9 and 6.10). These animals thus resemble many of the aquatic Crustacea in displaying a nocturnal semicircatidal rhythm upon which is imposed a semicircalunar rhythm. Similar rhythmic behavior patterns are to be found in talitrid amphipods.

The strength of the endogenous tidal clock in talitrid amphipods appears to be correlated with the extent of their migrations down the shore and with their zonation. Talitrid amphipods living on the upper shore generally display circadian rhythms with no circatidal component and are mainly night active. This endogenous modulation of surface spontaneous activity permits them to avoid exposure to direct sunlight and dry atmospheric conditions. Studies on *Talitrus saltator* from the British Isles using groups of individuals revealed clear endogenous circadian rhythmicity that lasted for 30 to 40 days in constant conditions (Naylor 1988), whereas recording single animals indicated an unexpected variation in rhythms between individuals. Nardi *et al.* (2003) compared endogenous activity rhythms in two populations of *Talitrus saltator* from Mediterranean beaches subject to different impacts: one beach was strongly eroded and the second was in dynamic equilibrium between erosion and accretion. The population from the eroded beach showed more variation between individuals than the population from the more stable beach. Moreover, the mean circadian period differed seasonally in both populations, being closer to 24 hours in spring and summer and longer than 24 hours in autumn and winter. These findings all point to some plasticity in rhythmic behavior. Ontogenetic

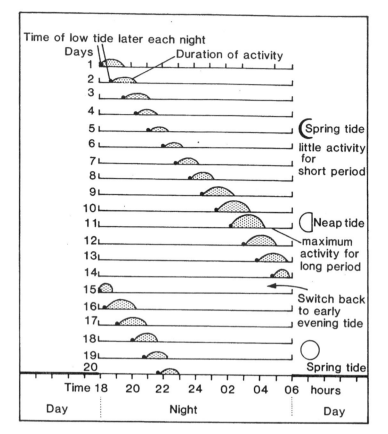

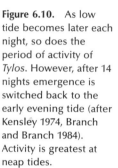

Figure 6.10. As low tide becomes later each night, so does the period of activity of *Tylos*. However, after 14 nights emergence is switched back to the early evening tide (after Kensley 1974, Branch and Branch 1984). Activity is greatest at neap tides.

differences in the expression of circadian rhythms were found in a population of *Orchestoidea tuberculata* from south-central Chile (Kennedy *et al.* 2000). Adults showed a peak of activity after midnight, whereas juveniles were active before and after the adults. The authors interpreted the difference as a strategy to avoid competition for food and/or cannibalism.

In ocypodid crabs, there may be a semicircalunar rhythm synchronized with a 14-day spring/neap cycle, as well as with a diurnal sequence and with the ebb and flow of the tides. However, endogenous rhythms may not be strong, emergence being largely related to the amount of water in the burrow (Jones 1972). The strength of the endogenous tidal clock in talitrid amphipods appears to be correlated with the extent of their migrations down the shore (Bregazzi and Naylor 1972).

The essential role of environmental cycles in phasing and maintaining endogenous rhythms has been demonstrated on numerous occasions and with many different species by keeping the animals in the laboratory under constant conditions (Naylor and Rejeki 1996). The gradual loss of rhythmicity when the animals are deprived of environmental cues is thought to be due to a slight imbalance between endogenous factors that tend to advance or delay the phase of the rhythm. Under natural conditions, such opposing tendencies allow the rhythm to adapt rapidly to alterations in environmental phase periods — a mechanism termed *autophasing* (Brown 1972).

Whereas some aspects of rhythmicity have received much attention, there has been relatively little interest in the importance of rhythmic activity to the physiological well-being of the animals. Jones and Hobbins (1985) suggested that rhythmicity plays a central role in coordinating all basic functions in intertidal and nearshore cirolanid isopods, and rhythms may exert significant control over the physiology of crustaceans in general. On the other hand, mollusks (for example, the whelk *Bullia digitalis*) will survive almost indefinitely under constant conditions and the growth rate and survival of young whelks appear to be unaffected (Brown 1982b). It may well be that the absence of an endogenous rhythm in mollusks renders rhythmic behavior relatively unimportant for the animal's well-being.

6.4   Sensory Responses and Orientation

It will be apparent from what has been written previously that endogenous rhythms on their own are quite inadequate for the maintenance of appropriate rhythmic behavior and that responses to changing environmental factors are essential. The interactions of such responses are every bit as complex as the innate rhythms themselves. In general, nondirectional stimuli (such as disturbance of the sand or its liquefaction, changes in temperature, or hydrostatic pressure) act as releasing factors, whereas directional stimuli (such as light, slope of the beach, or water currents) are orientational cues. Both directional and nondirectional stimuli often act together or in sequence to invoke appropriate behavior.

This is well demonstrated by the responses of the amphipod *Synchelidium* as it moves up the beach during the day at the leading edge of the waves. A wave passing over a buried individual causes an increase in hydrostatic pressure, invoking negative geotaxis, and the animal emerges from the sand. Upon emergence, it is strongly positively phototaxic and this leads it to ascend into the wave. However, the response to light is reversed within a few seconds, negative phototaxis resulting in reburial before the animal is swept back down the beach (Forward 1986). It is clear that reports of single factors producing single responses (although they may certainly occur in the laboratory) are for the most part simplistic as far as field realities are concerned. Thus, while liquefaction of the sand is an important factor in triggering the emergence of *Emerita*, the mole crab repeatedly moves to its optimal feeding zone on the beach by reference to a combination of stimuli derived from currents, depth of water, slope of beach, and light intensity (Cubit 1969).

The most complex and sophisticated responses to changing environmental factors are to be found among the semiterrestrial Crustacea at the top of the slope. Isopods of the genus *Tylos* orientate both to the sun and the moon (Pardi 1954, 1955; Hamner *et al.* 1968). Astronomic orientation requires a capacity for compensation for azimuthal changes throughout the solar and lunar days. *Tylos* also responds to the slope of the beach, in the absence of light, even when the slope is only 1.5 degrees. Moreover, the latter response reverses according to whether the sand is wet or dry. Indeed, a tendency to move upshore when on wet sand and downshore on dry sand has been demonstrated in several supralittoral forms, including *Tylos*, tenebrionid beetles, and talitrids (Brown and Odendaal 1994, Avellanal *et al.* 2000, Naylor and Kennedy 2003).

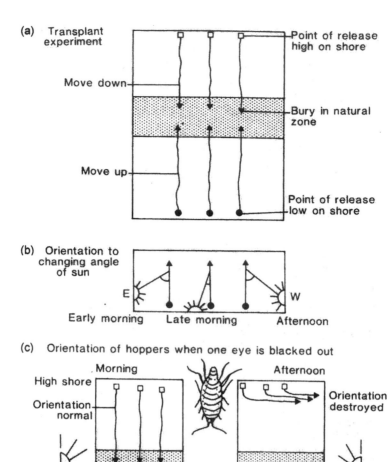

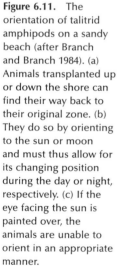

Figure 6.11. The orientation of talitrid amphipods on a sandy beach (after Branch and Branch 1984). (a) Animals transplanted up or down the shore can find their way back to their original zone. (b) They do so by orienting to the sun or moon and must thus allow for its changing position during the day or night, respectively. (c) If the eye facing the sun is painted over, the animals are unable to orient in an appropriate manner.

The orientational responses of sandy-beach talitrid amphipods have fascinated invertebrate biologists for a number of decades, and the experiments in which Pardi and Papi (1952) transferred *Talitrus saltator* from their home beach to one facing the opposite direction — with the result that the animals when released on dry sand moved up the slope (landward) when they should have moved down it (seaward) — are justifiably famous. This experiment has been repeated with other talitrids, with similar results (Figures 6.11 and 6.12). The animals orientate to the sun, to the moon, and to polarized sky light, and compensate for the apparent movement of the sun or moon across the sky. This latter ability indicates the presence of an internal clock distinct from that concerned with the tides. How the clocks (solar, lunar, and tidal) interact is still controversial. Orientation also involves sight of the landscape, slope of the beach, direction of the wind, and other local references — such as the different wavelength composition of the sky over land and sea (Ercolini and Scapini 1974, Ercolini *et al.* 1983, Ugolini *et al.* 1986).

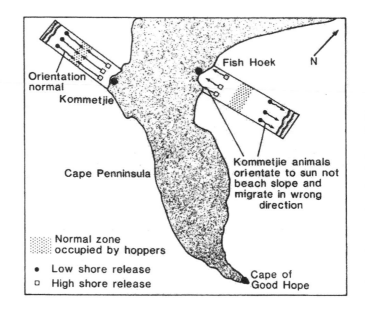

Figure 6.12. The talitrid amphipod *Talorchestia* — transplanted across the Cape Peninsula, South Africa, from Kommetjie to Fish Hoek — continues to orient in the same way it did on its home beach, thus migrating in the wrong direction and never locating its preferred zone (after Branch and Branch 1984).

The amphipods are thus able to return to their preferred zone by the shortest route and at the most appropriate time. In an interesting series of experiments, *Talitrus saltator* females were kept at length in the laboratory without any view of the sky. Their eggs, which had been fertilized by males from the same population, developed into hoppers showing the same orientation as the parents (when they were tested under the sun and sky for the first time). Young derived from crossings between differently oriented parents oriented in an intermediate direction to that of the parents. The capacity for sun orientation is thus inherited (Pardi and Scapini 1983, Scapini *et al.* 1985). In addition to innate responses, there may be a learned component to orientational behavior. Adults can modify their orientation under certain conditions, using local cues, whereas juveniles can modify their responses by recalibrating their sun compass. Such learning ability must clearly improve survival. On a particular beach, the resident population shows the most appropriate response (i.e., adaptation to the prevailing complex of stimuli). Within-population variation is least on relatively stable beaches, whereas orientation behavior on temporally changing beaches appears more scattered (Scapini *et al.* 1995). In the latter situation, scattered orientation can be considered an adaptation to a shoreline changing in an unpredictable way.

Orientational responses by semiterrestrial brachyurans, such as the ocypodid crabs, must in some respect be even more precise — for whereas talitrid amphipods return always to the same zone of the beach, ocypodids return to the same burrow. Thus, orientation toward prominent landmarks becomes just as important as orientation to the sun and moon. The possibility even exists of the actual recognition of some such landmarks and of a sort of kinesthetic memory — a homing ability based on a memory of past movements (Herrnkind 1972).

In the face of such sophisticated responses, one may well wonder how the sandy-beach mollusks, with their lack of endogenous rhythms and the apparent simplicity of their sensory mechanisms, manage to be so successful on surf-swept beaches. It is clear that we are missing something important and that further work is likely to demonstrate orientational responses of which we at present have no knowledge.

6.5 Choice of Habitat

At least once during their lives, a majority of intertidal psammophiles must choose their habitat. Such choice may be critical at the end of larval development or at metamorphosis, or (as is the case with the macrofauna of high-energy beaches) the choice may have to be made repeatedly as their migratory behavior or unpredictable conditions (such as storms) carry the animals into less favorable sites. Once carried out to sea, tidal rhythms and endogenous clocks are of no use in regaining their positions on the shore and they must rely entirely on orientational cues to reach a favorable area. Responses to water currents, to patterns of polarized light, and to changes in hydrostatic pressure may all be important in this regard — although other factors, which have not yet been investigated, may also be involved. *Bullia digitalis* is commonly found to a water depth of at least 25 m off beaches unsuitable for it in terms of slope and wave action. Yet after storms have cut back the beach and temporarily reduced the slope the whelks may be found intertidally within a few days (Brown 1982b). How they achieve this is unknown.

The two most important factors defining the beach habitat for benthic macrofauna are sand texture and swash flow over the beach face (see Section 7.4 for an expanded discussion on this), and sandy beach animals may select for optimal conditions of both — either when settling as larvae or as mobile adults regaining position after displacement. Sand particle size is perhaps the most critical factor and exerts a considerable influence on burrowing. All species tested, crustaceans and mollusks, have shown a clear response to sand particle size in terms of burrowing performance. Burrowing performance can be measured in several ways: burial time, number of digging cycles for burial, and burrowing rate index. Burrowing rate index (BRI, Stanley 1970) is useful and is defined as

$$\mathrm{BRI} = \frac{\sqrt[3]{\text{individual mass (g)}}}{\text{Burrowing time(s)}} \times 10^2$$

All intertidal species of crustaceans and mollusks that have been investigated have shown better burrowing performance in fine to medium sands than in coarser sands (e.g., Alexander *et al.* 1993; Nel 1995, 2001; de la Huz *et al.* 2002; Yannicelli *et al.* 2002), although some (such as *Hippa*) can operate effectively from a wide range of sand particle sizes (Lastra *et al.* 2002). Some typical responses of cirolanid isopod, a whelk and a clam, plotted as BRI (Figure 6.13) illustrate this clearly — showing optimum burrowing (i.e., fastest burial times and highest BRI values) in sands of 125 to 500 mm, with performance slightly decreased in finer sands but greatly decreased in coarser sands. It would seem that for most species characteristic of the intertidal zones of exposed sandy beaches sand particle sizes in the range 125 to 500 mm are optimal and sands coarser than 1 mm severely impair burrowing. It may be expected that settling larvae will tend to select for these grain sizes and there is some evidence for this (Lastra and McLachlan 1996). Water content is also important, as outlined earlier. Finer sands retain more moisture and most species can only burrow into saturated sand.

In supralittoral species, in contrast, the situation may be quite different, as shown by the oniscid isopod *Tylos* (Figure 6.14). Burrowing into dry sand, this species performs better in the coarser the sand. This is probably a reflection of decreasing compactness as dry sand becomes coarser and of the method of excavation. In the intertidal zone, in con-

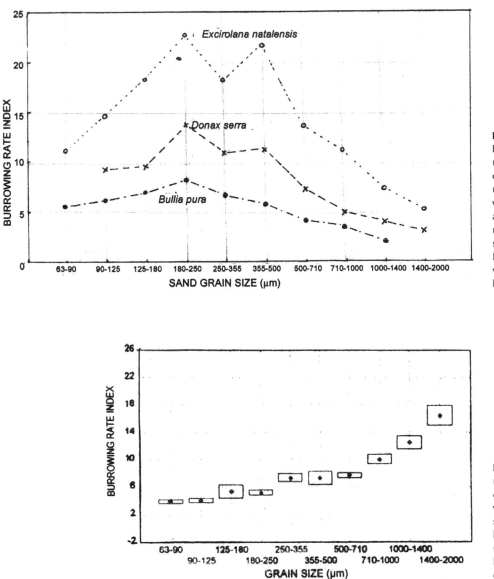

Figure 6.13. Mean burrowing rate index (BRI) values for a cirolanid isopod (*Excirolana natalensis*), a whelk (*Bullia pura*), and a clam *Donax serra* in a range of sand particle sizes (after Nel 1995, Nel *et al.* 2001). Highest values indicate fastest burial.

Figure 6.14. Burial responses of *Tylos capensis* to a range of well-sorted beach sediments in terms of burrowing rate index (BRI) (after Nel 1995). Boxes indicate standard errors.

trast, finer sands have better water-holding capacity and better thixotropy and are more amenable to burrowing.

In most species there are size differences in burrowing, small individuals burying faster than large individuals. There are also temperature effects, invertebrates burrowing faster at higher temperatures within tolerance limits. Thus, sand particle size in the habitat is a critical factor affecting the most important feature of sandy-beach animals: mobility and burrowing ability.

Like macrofaunal psammophiles, the distribution of some species of the meiofauna is governed partly by a choice of particle-size range. However, grain size and factors dependent on it may interact with variables such as oxygen tension and pore space to determine the distribution of many members of the meiofauna. Indeed, the apparent preferences of many species for particular grades of sand may be due mainly to other related factors, such as water circulation and the avaibility of oxygen. In harpacticoid copepods, endopsammic forms choose the coarsest matrices, mesopsammic species the intermediate, and epipsammic the finest (Hockin 1982). In the absence of other abiotic differences, the nonrandom distribution of the species may be attributed directly to differences in interstitial pore diameter (see Chapter 9 for meiofauna).

In those animals with subtidal, often planktonic, young stages the settlement of larvae or spat in suitable areas is critical for the continuance of the species. In the early literature it was assumed that settlement occurred more or less at random, only those individuals settling in appropriate areas having a chance of survival. It has since been shown that a number of sandy-beach species explore the substratum before settling and may even delay metamorphosis until suitable conditions can be found. Most of this work has been done on relatively sheltered shores, including muddy sands.

Various studies have shown the importance of sediment chemistry and microorganisms in substrate selection by polychaetes and archiannelids in sheltered sands (Wilson 1955, Gray 1967). Organic films and bacterial composition are clearly less likely to play a role in attracting settlement on high-energy sandy beaches, but the species inhabiting such shores have not been sufficiently studied to allow generalizations to be made. *Donax* spat might appear to settle at random, often on beaches that are quite unsuitable for adult survival. However, it may be that such settlement is only temporary before resuspension in the water column for transport to areas more suitable for the adults. Some *Donax* spat tend to settle near river mouths, later migrating to other areas. A great deal more work needs to be done in this regard (see review by Defeo and McLachlan 2005).

The swash exclusion hypothesis (McLachlan *et al.* 1993) suggests that in order to establish on a beach a species must be able to burrow between swashes (i.e., burial time must be shorter than swash period) and that a harsh swash climate with short swash periods and greater turbulence (such as encountered on reflective beaches) may exclude many species (see Chapter 7). Bivalves may display morphological and behavioral adaptations to the swash climates of different beach types (McLachlan *et al.* 1995). There is no clear pattern or relationship between burrowing ability and beach type, but there are some differences in shell shape, species from reflective beaches tending to be more wedge shaped. However, there are clear differences in size and density, bivalves from reflective beaches being small and of high density (adaptations to life on beaches with harsh swash climates; see Section 6.2). Thus, although bivalves from dissipative beaches are more variable (often large and of low density and varying in burrowing ability) and although some intermediate forms occur that are typical of intermediate beaches, those able to successfully colonize reflective beaches are generally small, with streamlined shapes and high density. As early as 1967, Wade recognized that *Donax denticulatus* in the Caribbean had a small streamlined shape as an adaptation to swash, and that it responded to the acoustic shock of breaking waves by jumping out of the sand to ride swashes. Similarly, *Donax variabilis* (a small tidal migrant on Florida beaches) shows strong orientation in the swash coupled to small size, high density, and streamlined shape. This makes the clams more stable in the swash, and the stability, drag, and reduced speed of movement enables the clams to gain a foothold and dig in rapidly after swash riding (Ellers 1995c).

Habitat selection may also exhibit temporal changes. On an exposed Mediterranean beach, mobile crustaceans and insects were found to shift their zones on both small (daily) and large (seasonal) temporal scales (Colombini *et al.* 2002). Differences in zonation across the shore and in longshore distribution were found between burrowed adult and juvenile crustaceans in relation to salinity, sand characteristics, and likely food distribution. In spring, crustaceans were active closer to the littoral, whereas coleopterans were active higher up the beach. However, in autumn crustacean zonation shifted landward, whereas some coleopterans moved seaward. Further, orientation performances of talitrids varied seasonally — those distributed more landward exhibiting more scattered orientation as opposed to individuals near the waterline, which consistently oriented seaward (Scapini *et al.* 2002). Within species there may be considerable variability, and the importance of phenotypic plasticity as an adaptation to harsh and changing environments such as sandy beaches is discussed in Section 6.12.

6.6 Nutrition

The absence of attached macrophytes on intertidal sands dictates a predominance of filter-feeders and scavengers among the resident invertebrate macrofauna. The physiological and behavioral adaptations of carnivorous scavengers are largely linked to the highly erratic nature of their food supply and unpredictability as to what type of carrion will be available, whereas the filter-feeders rely on a more constant but poorer diet. There are thus few highly specialized feeders on these beaches, opportunism being the order of the day. The scavengers will not only accept a wide variety of food but will typically turn predator when the opportunity arises. They also tend to have more than one method of feeding. For example, ocypodid crabs are both scavengers and predators, displaying a remarkable ability to locate, dig up, and capture their prey. Yet they will also feed by scooping up the top few millimeters of sand by means of their chelae — this sand being transferred to the mouth parts, which sort it into fine and coarse fractions, the latter emerging as psuedofeces (Robertson and Pfeiffer 1982). This method may be used to ingest meiofauna, microflora, or organic matter in general. *Ocypode quadrata* is said to be able to ingest up to 70% of the available algal cells present in the harvested sand. *Dotilla* also displays a wide range of diet and removes small organisms from sand grains rotated between its mouth parts. It extracts between 10 and 30% of the total organic material available in the sand (Fishelson 1983).

The mysid *Gastrosaccus psammodytes* also has several methods of feeding. It is both a suspension and a deposit feeder, and can scavenge soft animal matter. The amphipod *Haustorius arenarius* has different feeding methods for dealing with large and small particles. The suspension feeder *Donax serra* cannot be said to have more than one method of feeding, although it has been observed to deposit-feed in the laboratory. It will, however, ingest a wide range particle sizes and can make use of algae, particulate plant detritus, and the foam resulting from phytoplankton decay and bacteria, although not all of these are cleared and ingested to the same extent (Brown *et al.* 1989).

Bullia digitalis displays extreme opportunism in feeding, accepting any animal material stranded on the shore (especially cnidarians). It can also absorb dissolved organic matter through the surface of the foot and crop an algal garden growing on its shell (Colclough and Brown 1984, Harris *et al.* 1986). *Bullia* detects the stranded siphonophores and scyphozoans (which form its staple diet) by means of its osphradium, which is particularly sensitive to low concentrations of volatile amines. This detection invokes emergence from the sand, followed by surfing if the animal is buried below the swash zone. After surfing has brought it ashore, the animal rights itself and crawls in the direction of the food, holding its siphon horizontally in the surface film of water (which is always present in the swash zone). A current of water thus continues to flow through the mantle cavity and over the osphradium. Once contact is made with the food, the presence of volatile amines ceases to be of importance and the animal will only feed on the object contacted if amino acids are present (Hodgson and Brown 1985). The animal is thereby enabled to reach its food rapidly by the shortest route available to it and does not waste time or energy on objects that have no food value.

Having reached the food, the animal's problems are by no means over, for the carrion is typically stranded in the swash zone and is consequently moved (often a considerable distance) by the waves. *Bullia* takes every precaution not to be parted from it. If the food is soft, the whelk's proboscis is thrust deeply into it so that it is anchored to the food. If, on the other hand, the food is a bivalve or some other animal with a hard shell *Bullia*'s foot becomes a highly efficient suction cup that renders the animal and its food inseparable by the waves. Last, if the food is very small — a fragment of jellyfish, for example — the whelk drags it below the surface of the sand and consumes it there. A different technique is used when *Bullia* preys on small animals, such as amphipods and isopods. Under these circumstances, the whelk crawls over the prey and folds its foot so that the prey animal is trapped within the fold. At the same time, it turns onto its back and inserts its proboscis into the fold, eating the animal it has caught regardless of being washed about by the waves.

Most usually, however, the preferred food of *Bullia* arrives on the beach in large pieces. As there is no way of predicting when the next meal will arrive, it is appropriate for the whelk to eat as much of it as possible as rapidly as possible, before it is carried too far up the slope. An adult *B. digitalis* can consume up to a third of its own tissue weight in a single 10-minute meal. This casual observation has been quantified for a range of sizes of *B. digitalis* feeding on bivalve gill tissue. They consumed between 138% (for an animal of 16 mg dry tissue weight) and 18% (for one 1,000 mg) in a single meal. This high ingestion efficiency is made possible by well-developed musculature associated with the radula and proboscis, allowing the animal to rasp quickly and effectively and to suck in food powerfully. It also implies storage of food in a much dilated alimentary canal (Brown *et al.* 1989).

Absorption efficiency in *B. digitalis* feeding on bivalve gill is high (at 88%), and this is linked to the long post-feeding phase of between 7 and 10 days. When the daily ingested ration was calculated from the duration of the entire feeding cycle and used in the construction of an energy budget, scope for growth was found to be positive for all sizes of animal — indicating that the whelk can forego feeding for periods longer than 10 days. If the energy balance is recalculated in terms of utilization of absorbed energy for maintenance only, negative scope for growth is reached after 11 days. A single meal might last *Bullia* some 14 days. This does not, of course, take into account energy derived from the algal garden. The whelk is thus well suited to survive an erratic food supply of unpredictable nature.

The nutritional problems faced by filter-feeders, such as the burrowing bivalves of the genus *Donax*, are different — for food particles of varying quality are virtually always present in the water bathing the beach and represent a continuously renewable resource. *Donax* is therefore adapted to making the best use of the highest-quality food, when it occurs, with the least effort. On a diet of detrital foam, assimilation efficiency in *Donax serra* seldom exceeds 50%, but it can rise to over 70% on a diet of cultured algae (Stenton-Dozey 1989).

A special problem is presented to the semiterrestrial Crustacea and the insects inhabiting piles of kelp or other macrophytes stranded at the top of the shore. The animals are literally surrounded by food, which also protects them from desiccation. However, as the kelp ages it dries out, becoming unsuitable both as food and protection. Insect larvae dependent on these piles of kelp have had to adapt their life cycles so that the length of larval life is shorter than the viable life of the habitat, whereas the talitrid amphipods and other semiterrestrial crustaceans have adapted to survive the off-season (when little kelp arrives on the beach) although their numbers are often dramatically reduced.

6.7   Respiration

The respiratory adaptations of aquatic animals inhabiting low-energy sandy beaches are different from those found on surf-swept beaches (or indeed on rocky shores) and resemble the adaptations of animals burrowing into marine mud — being associated mainly with the incidence of low oxygen tensions, particularly during ebb tide. The lugworm *Arenicola* provides a classic example of tidally linked adaptations to reduced oxygen availability. Such adaptations may be behavioral or physiological, or both. For example, *Callianassa* exposed to poorly oxygenated water can either increase its ventilation rate or increase the efficiency with which it extracts oxygen from the water. Alternatively, it may employ both methods simultaneously. On high-energy intertidal sands, where oxygen tensions are high, most of the macrofaunal species are very much more active than those of low-energy shores — this higher activity being especially associated with tidal migrations. Yet the amount of food available may be lower than is found on sheltered shores. It is thus not surprising to find that the accent here appears to be on reduction of metabolic rate and on other ways of conserving energy (Brown 1983a).

In many of the animals displaying endogenous rhythms of activity, coincidental rhythms of oxygen consumption have been demonstrated and it would appear that the amplitude of these rhythms bears some relationship to the strength of the activity rhythm. Examples are *Emerita* and *Tylos*, which show circadian and circatidal rhythms, with some suggestion of a semilunar rhythm in addition (Figure 6.15).

The rate of oxygen consumption of *T. granulatus* during its three-hour period of activity proved to be at least six times that of its 21 hours of quiescence and was in some cases nearly 10 times higher. The lower rates maintained for most of the day represent a significant saving of energy. In some species, this suspension of activity (with its attendant saving in energy) may be very long term — as is the case with *T. punctatus*, which hibernates during the winter at depths of up to 70 cm. During starvation, its rate of oxygen uptake drops to about a sixth of its initial value, and a comparable or greater decline presumably occurs during hibernation (Hayes 1977). Tidal rhythms of oxygen uptake have been noted in *Gastrosaccus*, in some cirolanid isopods, in the crab *Ovalipes*, and in other forms (Brown 1983a).

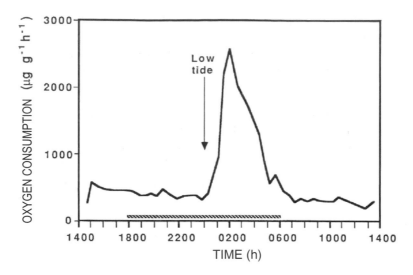

Figure 6.15. Typical respiratory rhythm of an individual of the isopod *Tylos granulatus*, showing a peak following low tide, a marked decline as dawn approaches, and a low inactive rate of oxygen consumption during the day (after Marsh and Branch 1979). The horizontal bar denotes hours of darkness.

Table 6.1. Average energy expenditure by eight nonfeeding *Bullia digitalis* females during a single tidal cycle of activity, referred to a standard-sized animal of 750 mg dry tissue weight. (From Brown 1981.)

Activity	Oxygen uptake per min (μg)	Actual time (min)	Oxygen uptake per period (μg)	Energetic equivalent (cal)
Transport in surf	20.8	14.5	302	1.03
Crawling	11.3	23.0	260	0.88
Burrowing	18.8	7.5	141	0.48
Emerging	18.8	3.8	72	0.25
Buried (observed)	9.3	42.0	391	1.33
Buried (residual)	10.6	653.2	6,924	23.54
Total for tidal cycle		744	8,090	27.51

Those sandy-beach invertebrates that do not possess endogenous rhythms do not display rhythms of oxygen consumption either (Brown 1983a, Ansell 1983). Moreover, the difference in rates of oxygen consumption between active and quiescent states is not as marked as in those animals having endogenous rhythms. The active rate in the whelk *Bullia digitalis* is only a little more than twice the best estimate of the inactive rate. In addition, behavioral and metabolic control mechanisms keep the energy expenditure of such an animal as low as possible. The cost of free existence of an individual of *B. digitalis* of 750 mg dry tissue weight is only about 52 cal over a 24-hour period (Table 6.1). Such low values are not encountered in all species of *Bullia*, however, and weight-for-weight the tropical Indian *B. melanoides* has a routine rate of oxygen uptake an order of magnitude higher than *B. digitalis* at ambient seawater temperatures — implying non-acclimation. The tropical bivalve *Donax incarnatus* also respires much faster at its environmental temperature of 30° C than does the temperate *D. vittatus* at 15° C, possibly representing another case of nonacclimation (McLusky and Stirling 1975).

An important method of conserving energy is to reduce, where possible, the metabolic effects of acute increases in environmental factors such as temperature. This may be achieved behaviorly by avoiding the higher temperatures, as in the case of *Tylos*, which spends the daylight hours deep within its relatively cool burrow. It may also be achieved by a flattening of the metabolic rate: temperature curve. A notable example of the latter

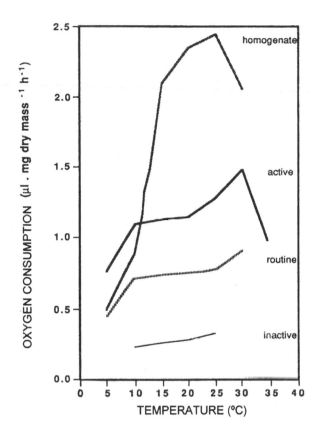

Figure 6.16. Acute rate-temperature curves for adult females of the sandy-beach whelk *Bullia digitalis* (referred to a standard animal of 750 mg dry tissue weight) (after Brown and da Silva 1983). The three rates for animals are *inactive*, suspended in a constant current (*routine* rate), and *active* in a constantly changing current (such as they encounter during surfing). The relative temperature independence displayed at all levels of activity is not shown by whole-tissue homogenates.

strategy occurs in the whelk *Bullia digitalis*, whose rate of metabolism is virtually temperature independent (with a Q_{10} of only 1.1 over the range of temperatures normally experienced by animals in the field). This relative temperature independence is, moreover, in evidence at all levels of activity (Figure 6.16). It is of particular benefit to *B. digitalis* on the west coast of southern Africa, where rapid and marked fluctuations in water temperature occur due to upwelling. *B. rhodostoma* and *B. pura* do not show this metabolic adaptation and this may be a significant factor in their failure to colonize beaches subject to regular upwelling.

It is quite common for the metabolic energy expenditure of small individuals of a species to be less temperature dependent than that of the adults or for the rate of oxygen consumption to be more temperature sensitive in winter than in summer. Most sandy-beach invertebrates show some seasonal temperature acclimation.

The effects of salinity changes on oxygen consumption have been less studied than the effects of changes of temperature. However, in *Bullia digitalis* both increased and decreased salinities result in a marked lowering of oxygen uptake. This is not usual among marine invertebrates, and constitutes a further example of energy conservation under sub-optimal conditions.

In the isopods *Excirolana natalensis* and *Pontogeloides latipes* the slope of the acute rate : temperature curve decreased above 17.5° C, an adaptation to survival in shallowly buried individuals during ebb tide — under which circumstances a temperature of 17.5° C may well be exceeded (Bally 1983). Reduction in respiration will not only conserve energy but reduce water loss, which may be vital under these conditions. On South African west coast beaches, *Excirolana natalensis*, *Pontogeloides latipes*, and *Eurydice longicornis* display differences in the effects of temperature changes, body weight, and activity on metabolism — which may be correlated with the zones of the beach they inhabit.

A common error on the part of marine biologists has been to rely on the measurement of rate of oxygen consumption as an index of total metabolism, thus ignoring the anaerobic component of respiration (which may be considerable in some cases). Among sandy-beach animals one might expect the anaerobic component to be highest in animals from very sheltered shores (where hypoxia is a common condition), and least in those from high-energy shores, where fully oxygen-saturated conditions normally occur. This is only partly borne out by the evidence, although it is certainly true that many sheltered-shore animals are facultative anaerobes. Both *Arenicola* and *Callianassa*, for example, may display a large anaerobic metabolic component and may be dependent on anaerobiosis during ebb tide. At the other end of the scale, the metabolism of *Bullia digitalis* inhabiting fully oxygenated surf-swept beaches is essentially aerobic, with an apparently negligible anaerobic component under field conditions (Brown *et al.* 1989). Remarkably, the surface of the foot acts as an auxiliary respiratory surface, admitting enough oxygen into the pedal sinus to satisfy the entire needs of the animal. It is thus surprising that the animal can survive several days of anaerobiosis in the laboratory and is an oxyregulator (Wynberg and Brown 1986). These capabilities are only likely to be displayed in the field under abnormal circumstances, such as severe organic pollution — when, in addition to oxyregulation, haemocyanin (normally undetectable) appears in the haemolymph (Brown *et al.* 1985). Although it is of course possible that facultative anaerobiosis, oxyregulation, and haemocyanin production have developed in response to such occasional suboptimal conditions, it is more likely that they simply reflect the past evolutionary history of the animal.

This is certainly the case in *Donax serra*, whose burrowing activities are largely anaerobic, despite very high ambient oxygen tensions. Oxygen does not diffuse readily through the pedal wall of *Donax* and very low oxygen tensions are recorded from the pedal sinus, whereas mitochondria occur only near the surface of the foot.

Even semiterrestrial Crustacea (such as ocypodid crabs) may undertake anaerobic respiration (particularly while in their burrows during high tide), although *Dotilla* and some other ocypodids commonly trap an air bubble in the burrow during inundation, in which the animal continues to respire. The ocypodids are in any case not true air breathers, air being circulated through water retained in the branchial cavity or water from the cavity being directed over the carapace, where gaseous exchange occurs (Fishelson 1983). In sandy-beach animals in general, whether from sheltered or exposed shores, control of metabolic processes by behavioral or physiological means (or both) is of vital importance to survival.

6.8   Environmental Tolerances

A number of conclusions concerning the tolerances of intertidal animals in general, and sandy-shore animals in particular, can be made with some confidence — despite the poor quality of many of the early investigations and the impossibility of relating them to real conditions in the field. For example, it is clear that the majority of intertidal animals have tolerance levels of natural variables that exceed those necessary for survival in their particular habitats (Newell 1979). Exceptions to this do occur, of course, on sandy beaches as well as rocky shores — and especially among psammophiles living near the top of the shore. Ocypodid crabs, for instance, may regularly encounter conditions close to their upper lethal limits (Brown 1983a, Fishelson 1983). In such cases, a knowledge of the behavior and physiology of the animal is needed to understand how they manage to survive and flourish. Descent into the burrow is an obvious method of escaping high temperatures. In addition, evaporative cooling occurs in ocypodids, the water in the branchial cavity being replaced from time to time by entering the burrow, by plunging into the sea, or by absorption from the substratum. Evaporative cooling may also take place in some aquatic crustaceans while in air.

Meiofauna living in the upper layers of the beach may also on occasion be near their upper limits of tolerance, although in most cases migration downward through the sand column will relieve the situation. Indeed, many aspects of the distribution of the meiofauna are clearer as a result of ecophysiological studies on the responses, preferences, and tolerances of individual species (Wieser and Schiemer 1977).

Another conclusion reached from simple tolerance studies is that in general intertidal species are more tolerant of environmental extremes than are subtidal species. The essentially intertidal mole crab *Emerita talpoida*, for example, survives salinity changes much better than does the more subtidal *E. analoga*. Bursey (1978) subjected *E. talpoida* to temperature-salinity combinations within the range 5 to 35° C and 15 to 65 ppt salinity. The entire range of salinities was easily tolerated at temperatures below 20° C, whereas the optimal salinity for survival at higher temperatures was slightly above that of normal seawater. Some aquatic species may avoid unusually high or extremely cold temperatures by migrating offshore, and it is significant that many psammophiles in cold climates migrate into deeper water in winter.

Differences in tolerance levels are displayed not only between intertidal and subtidal forms but between animals inhabiting different zones on the shore. The correspondence among zonal level, the relative importance of aerial respiration, and resistance to desiccation, temperature, and salinity changes are now almost as well documented for sandy-beach species as for those on rocky shores. Thermal tolerance in *Bullia* is correlated with the zonation of the various species. Large individuals of *B. rhodostoma* have a higher tolerance than small individuals, which tend to occur lower down the shore. Both *B. rhodostoma* (on the south coast of South Africa) and *B. digitalis* (on the west) also display higher thermal tolerances than do the more deeply buried species of *Donax* inhabiting the same beaches — the overlying sand tending to insulate the bivalves to some extent (Brown *et al.* 1989).

Differences in thermal tolerances among meiofaunal organisms are also to be expected in relation to tidal level and mean depth within the sand column. The upper limits of temperature tolerance are related to environmental temperatures experienced by meiofaunal populations (Giere 1993).

It is hardly surprising that most of the work relating to rates of water loss and resistance to desiccation has been undertaken on rocky shore animals or on semiterrestrial forms. When whelks of the genus *Bullia* come ashore in search of food, they bring with them a considerable amount of water in the free space, the mantle cavity, and the aquiferous spaces in the foot — a potentially most important buffer against desiccation of the tissues. However, resistance to desiccation in *Bullia* correlates well with the zonation of the different species on the shore.

Preserving the body temperature at the cost of water loss in semiterrestrial species, such as ocypodids, is clearly disadvantageous and possibly hazardous if continued for long. Only by a critical timing of activities can this be overcome. Surprisingly, sandy-beach talitrid amphipods are poorly adapted to low humidities. The survival times of talitrids (such as *Talitrus*, *Orchestia*, and *Talorchestia*) are considerable at high relative humidity but brief at low relative humidity at similar temperatures, although survival times increase if the animals are allowed to feed on moist wrack. In an experimental humidity choice chamber, they spend most of their time in the moister air. One of the issues seldom faced squarely in attempting to correlate laboratory tolerance studies with environmental situations is the importance of the period of exposure. In some species, the LT_{50} may change little with duration of exposure, whereas in others the difference may be substantial. It is not surprising to find that in the field meiofaunal species tend to be found midway between their upper and lower lethal extremes, adjusting their positions to achieve this by vertical and horizontal migration. A number of workers have shown that such migration occurs in response to heavy rain, to wave disturbance, and to changes in the moisture and oxygen content of the sand, as well as to changes in temperature (Brown 1983a).

In most of the work undertaken on tolerance limits, only one factor has been varied at a time. Although this accords with the classical view of how experiments should be performed, it is unrealistic not only because such changes seldom occur singly in the field but because the animal's responses to different variables commonly interact with one another. For example, the energy cost of osmoregulation frequently increases with increasing temperature and is compounded by decreasing oxygen tensions. There is thus no doubt that the tolerance limits of intertidal animals are best defined and presented in terms of three-dimensional multivariable response surfaces rather than as relationships to a single varying factor (Newell 1979). However, sandy-beach animals are absent from the list of organisms to which this approach has been applied.

It is also not sufficiently recognized that conditions that may be tolerated in the laboratory may be indirectly lethal in the field. Any factor that upsets the normal functioning of the animal or impairs its physiological well-being may lead to malfunction with regard to migratory excursions, seasonal patterns of movement, orientation, responses to food, or reproductive behavior. Any such disruption will be critical in the harsh, uncompromising environment that constitutes a sandy beach. Fortunately for intertidal psammophiles, biological rhythms appear to be relatively independent of temperature — an obvious advantage, indeed a necessity, under field conditions. Rates of locomotion, on the other hand, are often temperature dependent, so that low temperatures may jeopardize the ability to undertake migratory excursions successfully and prove dangerous to the life of the animal even though the temperature does not approach the lethal lower extreme as determined in the laboratory. Low temperatures significantly decrease the rates of burrowing in sandy-beach bivalves and gastropods, increasing the risk of the animals being swept away by the waves before an adequate purchase has been achieved in the substratum (McLachlan and Young 1982). Low temperatures may also prevent migratory behavior altogether.

Most of what has been said previously relates to the effects of environmental changes on survival. Of even greater importance is the influence of prevailing conditions on reproduction, for any decrease in reproductive potential (whether through physiological or behavioral malfunction) spells disaster for the population in the long term. Specifically, reproductive potential may be impaired through the untimely death of adults, reduction in scope for growth, a decrease in the numbers of gametes produced, reduction in sperm motility, fertilization failure, a failure of fertilized eggs to develop, larval mortality or the production of larval abnormalities, or a reduction in ability to undergo metamorphosis. The concept that it is reproductive potential that is ecologically important has borne fruit in the field of population studies (see Chapter 8).

6.9   Reproduction

Much of the published work on the reproduction of sandy-beach invertebrates records the numbers of eggs produced, the anatomy of the reproductive organs, the morphology of egg cases, times of breeding, mating behavior, or developmental stages. One of the problems facing intertidal animals is when to reproduce. It may be critical to coincide gamete production, copulation, or release of young with a particular phase of the tidal cycle. It is thus not surprising that fortnightly or monthly rhythms of reproductive activity are common. For example, the release of juveniles from the brood pouch of the isopod *Excirolana chiltoni* follows the same pattern as molting, being accomplished at a stage of the lunar cycle when the tides are getting higher on successive days. In this way, stranded juveniles (like newly molted adults) will be covered by the next tide (Klapow 1972).

Other animals, such as the whelk *Bullia*, breed but once a year and it is equally critical that they breed at the right time. Changes in temperature may provide the appropriate cue, although in some psammophiles the presence of blooms of phytoplankton on which the larvae can feed act as a stimulus. The influence of temperature may also be indirect in influencing the animal's energy budget, so that gamete production may only be possible at certain times of year. In *Bullia digitalis*, copulation within a given population is synchronized to a remarkable extent, all sexually active members copulating within 48 hours and not again until the following year. How this synchrony is achieved remains to be discovered.

The minimum breeding age and the number of times breeding should appropriately occur during the life span of the animal are also important. The advantages and disadvantages of being once-breeding (semelparous) as opposed to perennial (iteroparous) have been hotly debated. Clearly the optimal condition is to maximize the sum of present and future surviving offspring, but there is more than one way in which this may be achieved. Most members of the sandy-beach macrofauna are, in fact, iteroparous — the number of offspring produced at any one time being geared to the generally limited food supply. In general, reproductive costs increase with successive breeding events. However, although iteroparous forms are in a majority it cannot be demonstrated that semelparous species are at any disadvantage in the sandy-beach environment. Similarly, both r and k strategists appear to be equally successful on sandy beaches. *Donax*, like most bivalves, is a typical r strategist — producing very large numbers of gametes, the oocytes being small and containing little yolk. Fertilization is external and free-swimming planktotrophic larvae are produced. There is often a virtually unlimited breeding season, although peaks of reproductive activity may be apparent. The sex ratio is 1:1. In contrast, *Bullia* produces relatively few eggs, fertilization is internal, larval stages are suppressed (the young hatching as miniature, crawling whelks), and there is maternal care of the developing eggs (Brown *et al.* 1989).

As far as mode of development is concerned, there is a latitudinal gradient in the proportion of species with pelagic larvae, which declines from the equator toward the poles. This may be correlated with a parallel decline in both food availability and temperature. Planktotrophy may be more efficient than lecitotrophy if there is an abundance of suitable food (so that development can be rapid), and if planktonic predation is low. The retention of developing eggs in a brood pouch, as in amphipods and isopods, is also possible only where temperatures are low enough to permit optimal development. This must be of particular importance among semiterrestrial forms and may be a major reason why talitrid amphipods on temperate shores are replaced by ocypodid crabs as dominant supralittoral forms on tropical and subtropical beaches, these crabs having planktotrophic larvae.

Egg production must clearly be adapted to the food supply. Sandy-beach animals tend to be opportunistic in every respect and to adapt rapidly and appropriately to changing conditions. *Bullia digitalis* is capable of packaging its eggs in two distinct ways, although the reasons for this have not been investigated.

Psammophiles having external fertilization simply release their gametes into the water, although they must do so at an appropriate time. On the other hand, aquatic forms with internal fertilization face the problem of locating and copulating with a suitable mate, often under conditions of heavy wave action. How an animal such as *Bullia* achieves this is not clear, but once the male has found a suitable female he makes sure of not being parted from her by using his foot as a suction pad partly encircling her shell. The couple can then copulate while being carried about by the surf. Elaborate courtship — as is found in ocypodid crabs, for example (Vanini 1980) — is clearly out of the question for the aquatic fauna, although copulation itself may last for a considerable period and include behavioral patterns that stimulate the partner.

6.10   Aggregations and Gregariousness

There are other aspects of sandy-beach ecophysiology about which we as yet know very little. One of these is the apparently gregarious behavior of many mobile psammophiles, in contrast to the typically nonaggregated distribution of tube-dwelling invertebrates on more sheltered shores. Such groupings may certainly come about due to the action of water currents, particularly in those animals that surf, and this may be intensified in the case of carnivorous scavengers that home in on fairly large food masses. Aggregations of mole crabs and clams may be maintained passively by the physical factors acting upon them. However, in some cases peak periods of aggregation may coincide with maximum abundance of reproductive females. Thus, although physical forces may play a major part (at least in some species), other phenomena (such as visual cues or chemical stimuli, including pheromones) may also be important. It may be noted that chemically mediated gregariousness has been reported for several members of the meiofauna. It is also suggested that in both *Bullia* and *Donax* aggregations tend to occur in just those areas where food is plentiful. The implication that the animals actively seek the most suitable areas cannot be ignored, although no direct evidence of this is presently available. Aggregations in relation to physical factors, such as rhythmic shorelines, are also discussed in Chapters 7 and 8.

In contrast to this gregariousness, some semiterrestrial animals avoid forming aggregations and are territorial. Ritualistic defense of territory and complex courtship displays, involving learning as well as simple responses to releasing and orientational factors, are a feature of the biology of ocypodid crabs — the only invertebrates of sandy beaches to display such advanced social behavior.

6.11 Avoidance of Predators

Sandy-beach invertebrates are particularly vulnerable to predation, and it is not surprising to find a wide variety of mechanisms whereby the animals avoid predators. Deep burrowing is clearly one device that renders them unavailable to birds and fish alike, and it is the chief protection of bivalves such as *Donax* and *Tellina*. It does not protect their siphons while the bivalves are feeding, however, and bitten-off siphons form an important part of the diet of some fishes and a few birds. The loss of one or both siphons does not seriously incapacitate the animal for any length of time, as the siphons are rapidly regenerated (Hodgson 1982).

Deeper-than-normal burrowing in the whelk *Bullia* occurs in response to certain chemical substances — including combinations of betaines and urea, to which the animals are quite sensitive. This is in marked contrast to the animal's response to pollutants and may be a reaction to the scent of elasmobranch predators. The tidal migrations of *Bullia* and other invertebrates of exposed sandy beaches, keeping the animals in the swash zone, also reduce predation by making it difficult for birds or fishes to get at them.

Diurnal inactivity on the part of semiterrestrial Crustacea, either in deep burrows or below piles of wrack or kelp, also subtend survival — both through avoidance of high temperatures and desiccation and by being inaccessible to birds. The apparently meaningless behavior of talitrids, scattering in various directions after displacement from their burrows or when removed from wrack, can serve to confuse predators.

Some animals have developed, in addition, escape responses. Some species of *Donax* can perform leaping movements over the sand (Ansell and Trevallion 1969). A sudden leaping movement out of the sand has also been reported for the mysid *Gastrosaccus psammodytes* in response to increased pressure within the sand associated with the approach of a potential predator. Some crabs, such as the swimming crab *Ovalipes punctatus*, rely on an impressive threat display — the chelae being held open and aloft, pointing at the predator, while the crab burrows rapidly backward and disappears into the sand.

6.12 Phenotypic Plasticity

Behavior of organisms in variable environments needs to be flexible to cope with changing conditions. Indeed, behavioral patterns in sandy-beach macrofauna are not inflexible but become modified according to physical as well as biological circumstances, resulting in behavioral differences between populations of the same species in different environments. This suggests that plasticity has been selected for survival and evolution

(Brown 1996). Sandhoppers, for example, display plasticity in orientation behavior, demonstrating learning ability and fine modulation of the orientation response under changing environmental conditions. Thus, at birth a sandhopper has a set of simple behaviors concerning internal clocks, taxes, and orientation, among which certain behaviors are used and fine tuned to the specific environment (Scapini Fasinella 1990, Scapini *et al.* 1988, 1993, 1996).

Soares *et al.* (1999) suggested that sandy-beach species such as *Bullia* may have undergone evolutionary inertia because changing/unstable environments lead to selection for phenotypic plasticity, resulting in a co-adapted genome stable through space and time. They suggested this as a general adaptation to sandy-beach environments. Similarly, low degrees of genetic variability but highly plastic behavior seem also to be typical of *Donax* and *Talitrus*. *Talitrus saltator* populations may differ in intrapopulation genetic variation, and populations from rapidly changing shores may have a lower degree of heterozygosity than populations from stable shores (Scapini *et al.* 1995). Plastic behavior permits colonization and survival in changing environments, whereas inherited behaviors result from fine-tuned adaptations to constant environments (Scapini *et al.* 1988, Scapini 1995). In the latter case, which for sandy beaches may well be an exception (shown, for example, in Mediterranean populations of sandhoppers), genetic assimilation could account for the evolution of inherited adaptations (Scapini 1988, Scapini *et al.* 1988).

Soares *et al.* (1998, 1999) and Laudien *et al.* (2003) showed no large-scale genetic differentiation in the whelk *Bullia* (direct development) or the clam *Donax* (planktonic larval phase) in southern Africa and ascribed this to a high degree of plasticity in their behavior, ecophysiology, and morphology that allowed them to rapidly adapt to environmental changes. Life history traits in sandy-beach populations may thus be highly plastic over latitudinal gradients, with large-scale variations in temperature and concurrent environmental variables leading to an adjustment of the phenotype-environment relationship. This plasticity may lead to separate populations displaying differences in behavior and life history patterns (see Defeo and McLachlan 2005). The sandhopper *Talitrus saltator*, which is widely distributed along the European coasts from the Mediterranean to the northeastern Atlantic, shows differences in population traits across a geographic gradient from Tunisia, Italy, and Portugal to the British Isles (Williams 1978, Marques *et al.* 2003).

6.13   Conclusions

Most of the adaptations characteristic of sandy-beach animals are dictated by cycles of changing environmental conditions that are probably more complex and harsher than those faced by any community in any other marine habitat. Waves are themselves a cyclical phenomenon, and the intensity of wave action also tends to be cyclic. There are tidal cycles, cycles of lunar periodicity, diurnal cycles, seasonal cycles, cycles of erosion and deposition of sand, cycles of water movements through the sand, and cycles of food availability. Through these cycles runs the constant threat of unpredictable and sudden change. Aquatic as well as semiterrestrial sandy-beach animals constitute one of the few communities that can do nothing to modify the environment to their own advantage but must accept it as it stands and adapt fully to it if their populations are to survive. They must keep pace until they die not only with the times but with the tides — and their built-in clocks, neurological tide tables, sun compasses, and various navigational aids count for naught unless they can also adapt smoothly and rapidly to the unpredictable. Viewed in this light, it is surprising not that there are so few sandy-beach species but rather that there are any such species at all, particularly on high-energy shores.

As a consequence of their diverse adaptations to cope with the rigors of the sandy-beach environment, not only cyclical changes but also sand and swash features, these animals display a range of abilities to successfully colonize beaches across the morpho-dynamic spectrum. These differences result in varied communities of benthic macrofauna on different beaches, the subject of the next chapter.

Benthic Macrofauna Communities

Chapter Outline

7.1 Introduction

The invertebrate macrofauna is the most well-studied component of the biota on most sandy beaches and comprises benthic forms too large to move in the interstices between the sand grains. These organisms are generally in the size range 1 mg to 2 g dry tissue weight and may be collected on a sieve of 0.5- to 1.0-mm mesh. Their most pronounced characteristic is a high degree of mobility, including the ability to burrow rapidly (see

Chapter 6). Such organisms make up shifting populations in the intertidal and surf zones of beaches. They include representatives of all major taxa, although polychaete worms, mollusks, and crustaceans usually predominate. Other groups include nemerteans, anthozoans, platyhelminthes, sipunculids, echiurids, insects, and echinoderms (Chapter 5). Some of these groups are confined to sheltered beaches, although insects (often overlooked in accounts of sandy beaches) occur on the backshore under all conditions and may sometimes constitute the most abundant group present.

Different species and taxa, depending on their morphology and burrowing abilities, have probably evolved to take advantage of different types of beach. Generally, crustaceans dominate the sands toward the upper tidal levels and often dominate the most exposed shores (polychaetes the most sheltered) and mollusks shores of intermediate exposure (Dexter 1983). In many cases, however, mollusks dominate the biomass (McLachlan 1983). On very exposed coarse-grained reflective beaches only supralittoral crustaceans (and insects) remain, and on very exposed fine-grained dissipative sandy beaches vast populations (often of bivalves) may develop.

Sandy beach benthic macrofauna have received more attention from ecologists than any other component of the biota. There have been significant advances in our knowledge of macrofauna community and population ecology since the first edition, and aspects of the ecology of sandy-beach macrofauna have been reviewed recently by McLachlan and Jaramillo (1995), McLachlan and Dorvlo (2005), and Defeo and McLachlan (2005). This chapter aims to cover the main patterns, processes, and ideas regarding sandy-beach macrofauna ecology but not to provide an exhaustive survey of the now extensive literature. The reader is referred to these three reviews, the first edition, and McLachlan and Erasmus (1983) for a more complete listing of the published literature.

7.2 Sampling

Sandy-beach macrobenthos is usually collected by excavating quadrats or taking cores in transects across the intertidal and passing the sand through sieves of the finest mesh able to let the sand through but trap the fauna. Collecting sieves as fine as 0.5 mm may be effective in fine-sand beaches, but mesh as coarse as 4 mm may be needed in areas of coarse sand. However, 1 mm has tended to become the standard, for good reasons. Hacking (1997) demonstrated that there was little difference in retention of macrofauna between 0.5- and 1.0-mm mesh but that coarser mesh incurred serious losses, especially of small slender species, such as polychaetes and amphipods. The increased efficiency of the 1.0-mm mesh, in terms of reduced labor needed to sieve large volumes of sediment, compared to the 0.5-mm mesh, supports the adoption of 1.0 mm as the standard. When sieving large volumes of sand, mesh bags can be easier to handle than rigid sieves.

Core and quadrat sizes have varied considerably. In view of the fact that smaller samples are easier to handle and replicate, but that quadrat/core size should be at least 10× the body size of the largest organisms expected, samples with a surface area of 0.02 to 0.2m^2 are generally practical and efficient. These correspond to cores or circular quadrat frames of 15 to 50 cm in diameter. However, 20- to 30-cm-diameter cores are probably optimum in most cases. Sample depth has also varied widely, but considering the deep vertical penetration of some species (especially in more exposed situations) the sand should be sampled to at least 20 cm — and if deeper forms are present preferably to 30 cm.

The foregoing means that beach sampling involves moving large volumes of sediment and is hard work. Sampling, therefore, needs to be well planned and a reasonable-size team should be available to help.

Supralittoral forms, including insects, are sometimes collected by pit trapping. Down-shore, in the shallows, various dredges and sleds have been used to capture active and surface-dwelling forms. In the deeper subtidal, the surf-zone fauna may be sampled by these means or by grab from a boat, but rough conditions can make the latter difficult. Some success has been obtained deploying grabs from a helicopter, but the costs involved make this expensive procedure available to few workers. Diver-operated suction (airlift) samplers are often the most appropriate means of sampling in the surf zone. Some studies have used airlift suction samplers operating off compressed air cylinders or a hookah, whereas others have used a water pump to create suction by the venturi effect.

There has been considerable debate recently about sampling strategies for sandy-beach macrobenthos. This has focused on the question of stratified or regular sampling, use of geostatistics, replication of quadrats and transect lines, sample numbers, and sample intervals (James and Fairweather 1996, Defeo and Rueda 2002, Schoeman *et al.* 2003). Species accumulation curves for sandy beaches (Figure 7.1), based on a range of total sample

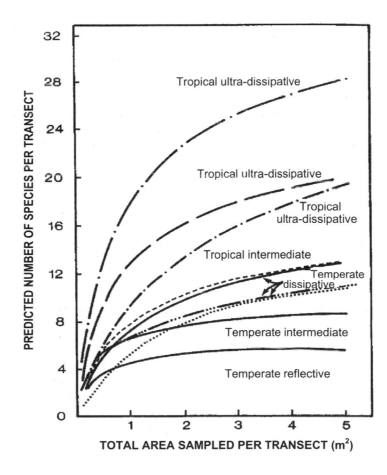

Figure 7.1. Predicted number of species collected in beach transect surveys with increasing sampling area up to a total area of 5 m^2 for a series of beaches from temperate reflective to tropical ultradissipative (after Jaramillo *et al.* 1995).

areas up to $5\,m^2$ per transect survey, show that a large total area (at least $3\,m^2$, and preferably much more) needs to be sampled to capture the majority of species at a site. Indeed, even $5\,m^2$ does not sample all species (Schoeman *et al.* 2003), and the ideal total sample effort will vary in space and time. Most studies have sampled considerably less than this because of practical limitations. Transects for intertidal surveys are usually run from above the drift line or high-water mark (in order to collect supralittoral species, such as talitrids and ocypodids) down to the swash zone during low tides, with at least 10 sampling levels. Either the entire line or the samples at each level should be replicated. In studies of the sublittoral regions, transect lines may extend to some distance offshore.

As most macrofaunal populations (particularly on exposed beaches) are highly mobile and undergo tidal, diurnal, seasonal, storm, and other migrations, population distributions and hence densities per square meter can vary dramatically. For this reason, it is now common practice when sampling intertidal or surf macrobenthos to integrate numbers and/or biomass for meter-wide transects of the entire zone, at the same time stating the width of the zone across the shore or surf zone. In this way, abundance (numbers m^{-1}) or biomass ($g \cdot m^{-1}$) estimates are for whole populations and samples collected during neap tides or storms may be compared directly with those made during spring tides or calms. Finally, a transect survey of a sandy beach should be considered to represent a single point in space and time and not an entire beach or season.

7.3 Taxonomic Composition

The main macrofauna groups were introduced in Chapter 5. In general crustaceans, mollusks, and polychaete worms are most important and usually make up more than 90% of species and biomass on ocean beaches. Crustaceans are common across all beach types and usually dominant toward reflective beaches, whereas polychaetes are sensitive to exposure and coarse sand and are most common toward sheltered and dissipative conditions (see next section). Among crustaceans, peracarids (especially isopods, amphipods, and mysids) and decapods (anomurans and brachyurans) are most important. Mollusks are predominantly bivalves and gastropods, and polychaetes hail from several families, but spionids, ophelids, and nephtyids are perhaps most typical and often very abundant on sandy beaches.

Soares (2003) undertook the most wide-ranging analysis of taxonomic composition of sandy-beach macrofauna. Based on a broad study of more than 50 beaches across the Southern Hemisphere, he showed that tropical regions had larger species pools than temperate regions. Crustaceans, polychaetes, and mollusks dominated the fauna on all beaches. Crustaceans were most diverse on dissipative beaches, mollusks and polychaetes not showing such clear trends. Crustaceans seemed to show the clearest response to changes in beach type, both in abundance and species richness, due largely to the appearance of many small and delicate species (such as amphipods, for example, on dissipative beaches). Polychaetes and mollusks tended to respond more to changes in sediment texture. Mollusks and polychaetes were more abundant on tropical and subtropical beaches, whereas crustaceans had greater biomass on dissipative beaches. Crustaceans and mollusks were larger on temperate than on tropical beaches. Soares' (2003) results on taxonomic composition are shown in Table 7.1, together with those from Hacking's (1997) extensive surveys in Australia. The most widespread taxon on ocean beaches worldwide is the cirolanid isopod genus *Excirolana*, which occurs on the upper shore

Table 7.1. Taxonomic composition of marine macrobenthos on 87 beaches in eight different regions from two extensive studies.

Region	Latitude	# of beaches	Source	Total no. of marine species	Crustaceans %	Polychaetes %	Mollusks %	Others %
Madagascar	Tropical	11	Soares 2003	53	47	26	15	12
North Brazil	Tropical	10	"	42	52	31	12	5
SE Brazil	Subtropical	10	"	37	52	31	12	5
S. Central Chile	Temperate	11	"	28	61	29	3	7
W. Coast South Africa	Temperate	10	"	37	57	30	11	3
South Australia	Temperate	10	Hacking 1997	19	37	42	11	3
East Australia	Warm temperate	10	"	33	48	39	9	3
NE Australia	Tropical	15	"	61	34	34	16	15
			Mean		49	33	11	8

and around the drift line of most beaches. Crustaceans, being most mobile, are best able to cope with harsh swash and may be the only form left on reflective beaches (e.g., cirolanid isopods, talitrids, and ocypodids).

Although Table 7.1 (based on two extensive studies representative of a wide range of beaches) shows that crustaceans typically make up half the species pool on sandy beaches (followed by polychaetes and mollusks), the situation for abundance and biomass can be different. Abundance and biomass data for the 35 beaches studied by Hacking (1997) are not available, but Soares (2003) found that crustaceans dominated abundance on 75% of the beaches, polychaetes on 15%, and mollusks on 5%. However, when it comes to biomass mollusks are often more important, and on most beaches supporting very high biomass (especially those with clam fisheries) bivalves can totally dominate the biomass (see Chapter 8). Of Soares' (2003) 52 beaches, crustaceans dominated biomass on 65%, mollusks on 27%, and polychaetes on 5%.

7.4   Macroscale Patterns

This section considers differences in macrofauna communities between beaches and latitudes. The community indices most commonly used to describe sandy-beach macrobenthos are species richness, abundance, and biomass. They are generally expressed as the number of species recorded per transect, the total abundance per meter-wide transect, and the total biomass per meter-wide transect. The meter-wide strip unit is used, as indicated previously, because it encompasses the entire intertidal population of each species and is therefore relatively constant, whereas density varies as the population expands or contracts in response to changes in beach width caused by changing tide or wave conditions. It has long been recognized that clear trends in these community parameters occur on a global scale. McLachlan and Dorvlo (2005) reviewed this information and brought it together in an analysis of transect studies published in the literature. The beaches were drawn from all regions of the world and only marine species were considered (i.e., insects were excluded). They examined responses in the three previously cited community parameters to a wide range of physical variables: wave height, tide range, sand particle size, beach face slope, and latitude. They also considered a series of indices of beach type, including DFV, RTR, and BI (see Chapter 2). The results of these analyses show clear patterns, despite the fact that the data from the 161 beaches surveyed had considerable noise — due both to the natural variability of beach macrofauna and because of varying sampling techniques used by different workers.

Species Richness

Species richness, the number of species recorded per transect survey, increases in response to decreasing sand particle size, flattening beach face slope, and increasing tide range (Figure 7.2). The number of marine species recorded per beach transect survey ranges between 1 and 30. BI (which incorporates measures of sand particle size, slope, and tide range) therefore shows a very close relationship to species richness per beach (Figure 7.2), explaining 56% of the variability in these data, which are drawn from widely differing beaches all around the world. Species richness showed no clear relationship with wave height and weaker relationships with RFR and DFV than with BI. Multiple regression analysis confirmed that tides, sand, and slope were the primary variables explaining variance in species richness across beaches (McLachlan and Dorvlo 2005). Tide range is clearly very important in fixing the primary dimension of the intertidal habitat, but it has not been considered by many workers because most studies have

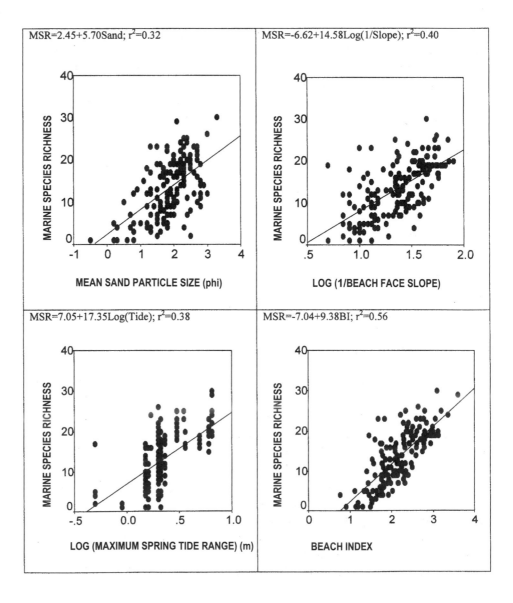

Figure 7.2. Significant relationships (p < 0.01) between marine species richness (total number of species collected per transect survey) and sand particle size, beach face slope, tide range (m), and the beach index for 160 beach surveys (after McLachlan and Dorvlo 2005).

focused on one region where tide range is uniform. Only in large-scale comparative studies does the significant role of tides become clear. This is supported by the impoverished marine faunas associated with sandy beaches in the nearly tideless Mediterranean Sea (Deidun *et al.* 2003) and Polish Baltic coast (Haque *et al.* 1996), as opposed to rich faunas in macrotidal areas (Hacking 1997), and with the tendency for species richness to increase with increasing RTR across a range of intermediate beaches (Rodil *et al.* 2003).

The number of macrobenthic species (intertidal forms and supralittoral crustaceans) able to establish populations on a beach increases linearly from microtidal reflective to macrotidal dissipative conditions. This trend is clear and may now be considered a paradigm in beach ecology. Of course, no transect survey recovers all species present on a beach and most authors have undersampled community species richness quite significantly, especially when working on dissipative beaches (Figure 7.1). Dissipative beaches have higher relative species richness (i.e., a greater proportion of the regional species pool are present). Soares (2003) showed that each transect survey of a dissipative beach in microtidal regions recovered on average 47% of the total species pool of that region, whereas intermediate beaches yielded 36% and reflective beaches 27%. He found that for total macrofauna and for crustaceans and mollusks species richness increased toward dissipative beaches across all regions.

Increasing numbers of species toward dissipative beaches mostly reflect the addition of sublittoral species from the surf zone onto the lower shore. Virtually all sandy-beach macrofauna species are successful on dissipative beaches, but most become less successful moving across the morphodynamic spectrum through intermediate beaches and few can colonize reflective beaches. Those species that are able to operate across the entire morphodynamic spectrum and that can successfully colonize reflective beaches are mostly highly motile crustaceans: the anomuran hippid crabs, *Hippa* and *Emerita*, cirolanid isopods of the genus *Excirolana*, and of course supralittoral species that live above the intertidal and are not subject to the swash climate: ocypodid crabs, talitrid amphipods, and oniscid isopods. However, in extremely reflective situations only one or two supralittoral species may be present.

Abundance, Biomass, and Density

Total abundance and biomass of the macrofauna community on a beach increase exponentially with flatter beach face slopes, finer sand, and larger tides (in the case of abundance) — as well as higher wave energy (in the case of biomass) (Figures 7.3 and 7.4) — and respond more weakly to other factors. Abundance ranges from as low as one animal per meter-wide transect (typically in very reflective situations where only a sparse supralittoral fauna is present) to nearly 1 million macrofauna organisms per meter transect in very rich dissipative systems. However, values between 100 and 10,000 are more typical. Biomass ranges from less than 1 g dry mass per meter transect to nearly 10 kg across the same spectrum, although values between 10 and 1,000 g are more typical.

In general the responses are not as clear as for species richness, particularly in the case of biomass, and there is more scatter in the data. In both cases the response to the BI is significantly positive (i.e., abundance and biomass increase exponentially from microtidal reflective to macrotidal dissipative beaches). These relationships were

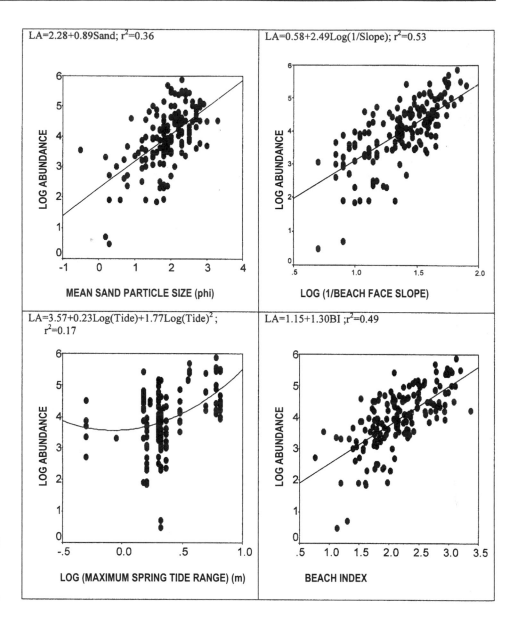

Figure 7.3. Response of total macrofaunal abundance per linear meter transect to physical parameters and the beach index (after McLachlan and Dorvlo 2005). All regressions are significant (p < 0.01).

confirmed by multiple regression analysis and have been found by many workers. This means that beaches of fine sand and flat slopes (i.e., dissipative systems) are able to support high abundance and biomass. Factors beyond the beach must also play a role, and inshore productivity must be important in providing the food resources needed to sustain the macrofauna. However, little information on this is available (see Chapter 12).

Species richness, abundance, and biomass of the macrofauna communities all increase as beach width increases from reflective to dissipative beach types and positive correlations between these variables and beach width have been found in several cases (Nel 2001, McLachlan and Dorvlo 2005). However, it should be noted that these responses in abundance and biomass are not simply a consequence of increasing beach width under-dissipative conditions. Whereas beach width increases by one order of magnitude over the range of beach types, total abundance increases by several orders so that the densi-

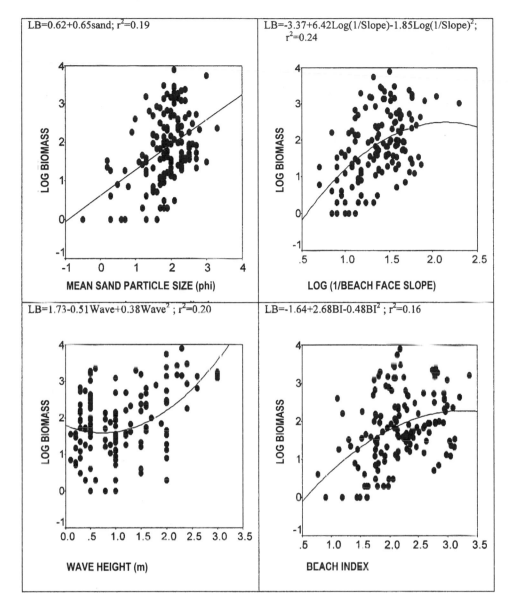

Figure 7.4. Response of total macrofaunal dry biomass (g) per linear meter transect to physical factors and the beach index (after McLachlan and Dorvlo 2005). All regressions are significant (p < 0.01).

ties of macrofauna are greater on dissipative than reflective beaches. Indeed, density increases as abundance increases, sand becomes finer, slope flatter, and beaches more dissipative (Figure 7.5). Density per square meter ranges from less than one individual to in excess of one thousand across the range of beach types.

The foregoing indicate that the changes in sandy-beach macrobenthic community structure across beach types is not simply a response to increasing area of intertidal habitat (beach width) toward dissipative beaches — much more than that. Rather, the nature of the environment changes in a fundamental way, bringing about disproportionate changes in community structure and enabling richer communities to establish on dissipative than reflective beaches. Greater faunal densities also mean a higher probability of encounters between individuals and hence greater potential for biological interactions (such as competition) on dissipative beaches.

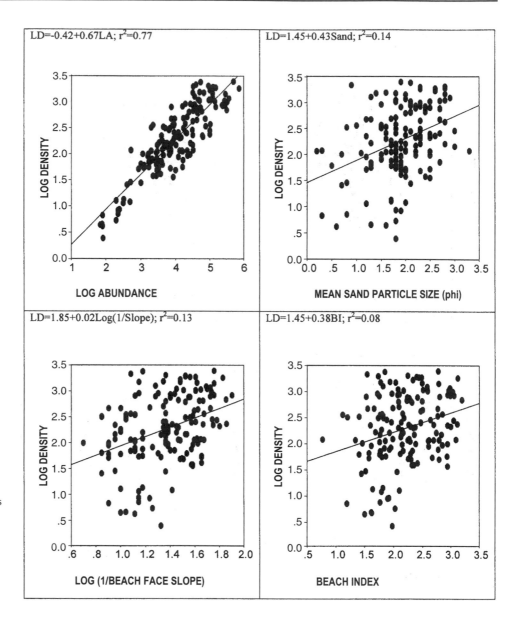

Figure 7.5. The relationship between average density of all macrofauna per square meter on sandy beaches and four other factors (after McLachlan and Dorvlo 2006). Note that density is plotted on a log scale. All regressions are significant (p < 0.05).

Latitude

There is now ample evidence that sandy-beach macrofauna communities display similar responses to latitude as those found in other communities, namely, greater diversity toward the tropics if the same beach types are compared. However, latitudinal influences are weaker than the effects of physical factors (Brazeiro 1999, McLachlan and Dorvlo 2005). Species richness increases toward the tropics (Figure 7.6) because of a greater species pool (Soares 2003, McLachlan and Dorvlo 2005), whereas abundance and biomass tend to drop toward the tropics, probably because of lower inshore productivity.

Biomass usually correlates closely with wave height, suggesting that wave energy (though controlling surf-zone and inshore productivity) influences the standing crop of macrofauna that can be supported on a beach. Wave energy increases with latitude

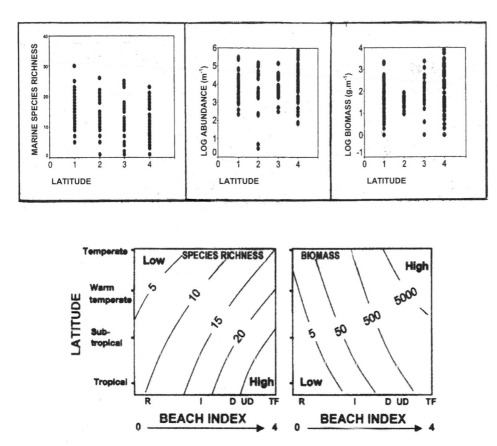

Figure 7.6.
Relationship between
total species richness,
abundance (per meter),
and biomass (per
meter) and latitude for
beach transect surveys
(after McLachlan and
Dorvlo 2005). Latitude is
defined by water
temperature, not
absolute latitude, as
follows: 1 = tropical, 2 =
subtropical, 3 = warm
temperate, and 4 = cold
temperate. Latitudinal
trends are highly
significant in the case of
species richness and
significant in the case of
abundance and
biomass. Variability at
each latitude reflects
the range of beach
types encountered
there.

Figure 7.7.   Conceptual model of latitudinal variations in species richness (number of species per transect survey) and biomass (g·m^{-1}) as a function of beach type, as categorized by the beach index (after Defeo and McLachlan 2005). The number of species increases at low latitudes under conditions of fine sands, flatter slopes, and increasing tide range. Biomass is also highest toward tidal flats and increases from tropical to temperate sandy beaches (R = reflective, I = intermediate, D = dissipative, UD = ultradissipative, and TF = tidal flat).

(Chapter 2). There are latitudinal differences in most physical features and beach types. Wave energy is greater at higher latitudes, and the prevalence of finer river-borne sands results in a greater preponderance of dissipative beaches there than in the tropics. Reflective beaches, which are rare on open coasts at temperate latitudes, are common in the tropics. A typical temperate beach, likely to be more dissipative, may support a richer fauna than a typical tropical beach (which is likely to be more reflective), but beaches of the same morphodynamic type will host more species in the tropics (Soares 2003). Among the three main groups of sandy-beach macrobenthos (crustaceans, mollusks, and polychaetes), the same patterns are found. Temperate beaches also tend to have higher densities per beach type, whereas tropical populations are more sparse and may have narrower niches (Soares 2003). Certain groups may also be more important at certain latitudes. For example, amphipods are generally more prevalent at temperate latitudes.

Thus, the three general community parameters for sandy-beach macrofauna all show very clear large-scale global patterns (which are summarized in Figure 7.7 using a conceptual model). These trends conform to the broader patterns found by Ricciardi and

Bourget (1999), who looked at community changes across a complete spectrum of sediments from mud to exposed sand. They found increases in biomass with finer sand, flatter slopes, and higher waves, but not with increasing exposure. These relationships suggest that the physical environment of the beach — defined as it is by tide, wave, and sediment regimes — exerts an overriding control on large-scale community patterns. Several theories have been proposed to explain this.

Factors Controlling Large-scale Patterns

Beaches are considered physically controlled environments because they lack biogenic structure and are extremely dynamic in space and time. This has led to invocation of the autecological hypothesis (Noy-Meir 1979), which when applied to sandy beaches (McLachlan *et al.* 1993) states that in physically controlled environments communities are structured by the independent response of individual species to the physical environment and that biological interactions are minimal. Thus, the ecology of the community is the sum of the autecologies of the constituent species. How, then, do these species experience the environment?

Sandy-beach macrobenthic species experience their immediate environment as three suites of factors: (1) the sand in which they burrow, (2) swash water movement over the sand, and (3) the intertidal gradient of exposure to air. Because beach sands are generally well sorted, the most important feature of sand is its mean particle size. Particle size affects porosity and permeability, water-holding capacity, and drainage of the sand (Chapters 2 and 3), and strongly influences the burrowing efficiency of macrofauna (Chapter 6). It can also be abrasive, and if coarse can inflict physical damage on organisms. At the extreme, shingle beaches are almost devoid of surface macrofauna. They harbor a macrofauna which remains well below the surface — in a sense behaving like interstitial fauna, distinct from the macrofauna on sandy beaches (Gauci *et al.* 2005).

Water movement, driven by waves and tides, is experienced by macrofauna on the beach face as swash — and the swash climate is closely related to the beach morphodynamic state as reflected in beach face slope (Chapter 2). Beach faunas are dependent on the swash to move (by surfing or active swimming), feed (swash brings their food), distribute their gametes and propagules, and assist in the avoidance of predators (Chapter 6). We have seen (Chapter 2) that swash climates are harsher toward reflective beaches. The intertidal exposure gradient creates opportunities for species with different tolerances of exposure to inhabit different levels on the shore, although this is to some extent mitigated through escape from desiccation by burial and by tidal migrations that enable the more mobile macrofauna to move back and forth across the shore with the tides. How do these three factors (sand, swash, and exposure) tie in with the large-scale patterns?

Species richness on the large scale seems to be controlled by two processes: an ecological process whereby harsh environmental conditions allow fewer species to establish populations on reflective beaches and an evolutionary process that has led to the development of greater species pools in the tropics (Soares 2003). Several hypotheses have been proposed to explain these clear large-scale patterns in sandy-beach macrobenthic communities, all resting to some extent on the autecological hypothesis and its applicability to sandy beaches.

- The *swash exclusion hypothesis* (McLachlan *et al.* 1993) proposed that swash was the controlling factor. It was postulated that, because the swash climate the macrofauna experience on the beach face becomes harsher moving from dissipative to reflective ends of the beach morphodynamic continuum (Chapter 2), it is this increasingly harsh swash that excludes species toward the reflective extreme. According to this hypothesis, all beach species can operate in the benign swash climate experienced under dissipative conditions, but only very mobile and robust forms can tolerate the turbulent short-period swash and rapid drainage of reflective beaches. Indeed, it was suggested that under extremely reflective conditions all intertidal species may be excluded and only supralittoral forms (which live outside the swash climate) remain. Few ecologists have measured swash climate over a range of beach types, but there is mounting evidence of good correlations between community parameters or species presence and swash climate features (Dugan *et al.* 2000, Nel 2001, Soares 2003, Stephenson and McLachlan this text, Chapter 6). Dugan *et al.* (2000) suggested that species less able to cope with harsh swash may also retreat to the subtidal during rough conditions.

- The *multicausal environmental severity hypothesis* suggested that increasing harshness in sediment, swash, and accretion-erosion dynamics may exclude species from reflective systems (Brazeiro 2001). There is good support for the additional role of sand; for example, in the effects of particle size on burrowing (see Chapter 6).

- The *habitat harshness hypothesis* proposed that in reflective beaches the harsh environment forced macrofauna to divert more energy toward maintenance, leaving less for reproduction and causing higher mortality – thus making it more difficult for populations to establish (Defeo *et al.* 2001, 2003). Some population studies have lent support to this hypothesis (Chapter 8), which is more applicable at population level (for which it was derived).

- The *sand and swash exclusion hypothesis* coincides with many of these ideas, suggesting that both sedimentary (coarse sand) and swash (harsh swash climate) factors make reflective beach environments inhospitable and exclude species. Latitude and beach width and length may also play a role (Nel 2001, McLachlan 2001).

- The *hypothesis of macroscale physical control* (McLachlan and Dorvlo 2005) identifies two levels of factors responsible for the clear global patterns in sandy-beach community species richness outlined previously: (1) primary control is by (a) tide range, which defines the dimensions of the intertidal habitat and the number of species/niches that can be accommodated, and (b) latitude, which influences the size of the species pool available to colonize a beach, and (2) secondary control is by (a) harsh swash climates and (b) coarse sand and sediment instability toward reflective beaches excluding species. According to this hypothesis, the primary factors determine the maximum number of species that could occur under ideal conditions on a dissipative beach in a particular region, whereas the secondary factors limit how many of these species are actually able to establish populations across the range of beach types by excluding the less well-adapted species under the harsher conditions that develop toward reflective beach states.

These theories all imply that it is post-settlement processes that prevent less robust/well-adapted species from developing populations on reflective beaches. It does seem clear

Figure 7.8. General conceptual model relating community parameters and beach state (after Defeo and McLachlan 2005). Species richness, abundance, and biomass increase from reflective beaches to tidal flats. Toward the reflective domain, species richness is controlled by the interaction of physical factors such as tides, sand, and swash. At a finer scale and under more dissipative conditions, biological factors start to become more important. Latitudinal influences are not considered here (R = reflective, I = intermediate, D = dissipative, UD = ultradissipative, and TF = tidal flat).

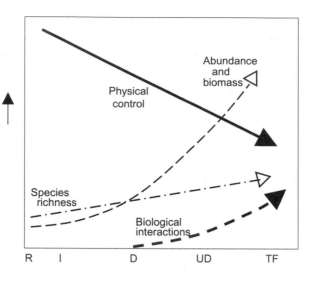

that on the large scale and toward the reflective extreme physical control of the sandy-beach macrofauna communities are overriding, but toward dissipative conditions and on finer scales the more benign environment and greater densities of organisms may allow biological interactions to become more important (Figure 7.8). We will come to biological interactions later. First we need to consider other factors.

Other Trends

Interrelationships among the three community indices (species richness, abundance, and biomass) themselves are clear. Species richness (linear) and biomass (logarithmic) both increase steadily with increasing abundance (logarithmic). However, species richness shows little response to increasing biomass, and biomass increases with abundance up to about 1,000,000 individuals·m^{-1} and then levels off (McLachlan and Dorvlo 2006). The relationship between abundance and distribution of sandy-beach macrofauna follows the rule that abundant species tend to be widely distributed. That is, there is a positive relationship between abundance and extent of occupancy of beaches (Frost *et al.* 2004).

Body Size

There are also clear large-scale trends in the sizes of individual macrofauna on sandy beaches. This has been reviewed by McLachlan and Dorvlo (2006). Mean individual size ranges over five orders of magnitude and shows some response to all environmental variables (Figure 7.9). The mean individual mass for all macrofauna on a beach increases with decreasing species richness and abundance, but also with increasing total biomass, decreasing tide range, and intermediate slopes. It increases weakly with larger waves and coarser sand. It shows some latitudinal variation, animals tending to be largest in the tropics. However, Soares (2003) found that mean individual body sizes of crustaceans, mollusks, and polychaetes were larger on temperate beaches for all microtidal beach types (i.e., wave-dominated beaches). Overall, it can be said that dissipative beaches (with

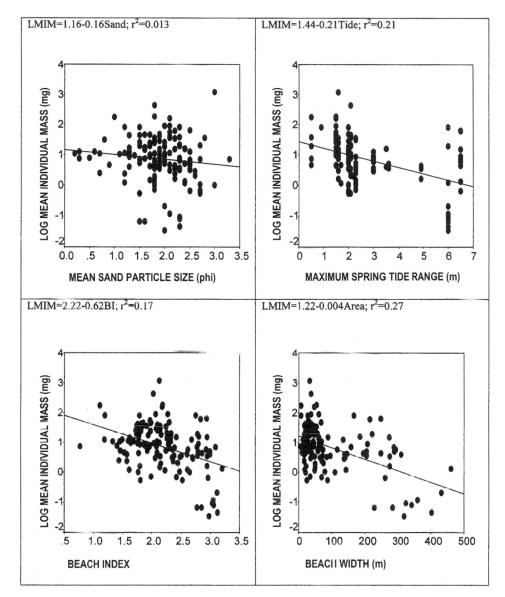

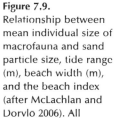

Figure 7.9.
Relationship between mean individual size of macrofauna and sand particle size, tide range (m), beach width (m), and the beach index (after McLachlan and Dorvlo 2006). All regressions are significant (p < 0.01).

greatest densities of macrofauna) tend to contain faunas with smaller mean individual sizes than reflective beaches with low densities. Coarse sand and harsh swash may select against small and delicate forms on reflective beaches, and several species have been found to exhibit larger sizes on reflective beaches.

Most communities show an inverse relationship between body size and density. The more individuals there are per square meter the smaller they tend to be. This has been found for mean individual size versus total community density for a large sample of 140 beaches (Figure 7.10), although many other factors in addition to density must also influence body size. At a different level, Dugan *et al.* (1995) recorded a positive relationship between mean individual mass and population density for communities on some

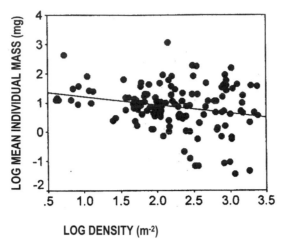

Figure 7.10. Mean individual mass (MIM) as a function of density (D) of the total macrofauna in sandy-beach communities (after McLachlan and Dorvlo 2006). The equation is $\log MIM = 1.49 - 0.28\log D$, $r^2 = 0.06$.

Californian sandy beaches, but the 1.5-mm mesh they used to separate the fauna may have discriminated against small species.

7.5 Mesoscale Patterns

This scale refers to distribution and community structures within a beach and can be divided into two main components: alongshore and across shore. The alongshore dimension is affected by beach length.

Beach Length

Many ecologists have suspected that beach length is a significant factor in beach ecology. We have seen in Chapter 2 how beach length or embaymentization affects beach morphodynamics and surf-zone circulation. For this reason it has long been felt that short pocket beaches (which have a high degree of embaymentization and are often ephemeral in nature, being easily eroded away by storms) should contain poorer communities than longer beaches. Longer beaches develop more regular surf circulation and support large populations of each species, thereby decreasing the risk of local extinction from episodic events. However, only one study has come close to demonstrating a significant effect of beach length. Brazeiro (1999) found that for a range of Chilean beaches of different lengths there was no effect above a beach length of 2 km, but below this communities tended to become more impoverished as beach length shortened (Figure 7.11).

Alongshore Variation

Despite the common presence, or even abundance, of fairly deep-burrowing bivalve mollusks, most macrofaunal species tend to occur at or close to the sand surface. Their distribution may therefore be considered essentially two-dimensional relative to the truly three-dimensional distribution of the interstitial fauna, although it is three-dimensional as compared to faunal distribution on hard substrata. The distribution of macrobenthic animals on sandy beaches exhibits patchiness, zonation, and fluctuations due to tidal and other migrations. Patchiness results chiefly from passive sorting by the waves and swash, from localized food concentrations, from variations in the penetrability of the sand, and

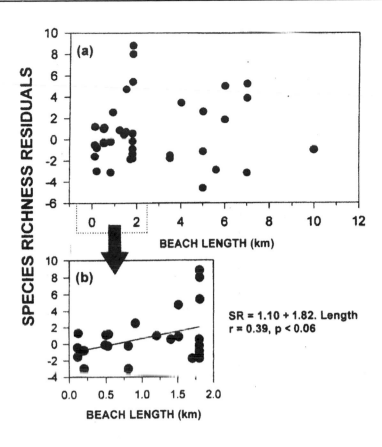

Figure 7.11. The relationship between species richness residuals (variation not explained by beach face slope) and beach length in Chile (a) for a full range of beaches and (b) only for beaches less than 2 km, showing a decrease in the number of species when beach length drops below 2 km (after Brazeiro 1999).

from active biological aggregations (Chapter 6). Some of these factors are interrelated, and swash movement may frequently concentrate the benthos and its food on the same part of the beach. Depending on morphodynamic beach state, the scale of the patches may vary from 10 m on reflective beaches with cusps to > 100 m on high-energy intermediate/dissipative beaches.

Alongshore distribution in sandy-beach macrobenthos shows great variability due to responses to gradients, disturbance, embayment effects, and patchiness on different scales. This is mainly determined by physical factors, but biological factors may also play a role. Factors that can influence longshore patterns include river mouths, local food concentrations, human structures, and activities such as beach nourishment, beach cleaning, and shellfish harvesting. Alongshore variations also occur along curved beaches that change in exposure and other physical conditions along their length. Zetaform or spiral bays display clear gradients in physical factors from shelter at the narrow end to exposure at the open end, with concomitant changes in the fauna. Species richness, abundance, and biomass generally respond to such alongshore changes in wave action, sand particle size, and beach morphodynamics, including swash. There is often clear alongshore structure, including patchiness, with a unimodal bell-shaped distribution.

At a slightly finer scale, the benthic macrofauna may respond to rhythmic features on the shoreline, such as cusps and rip currents. Several intertidal species have been shown to concentrate in cusps, possibly due to passive sorting by the swash and/or active migration (Defeo and McLachlan 2005). Cusp swash circulation may sort the smaller and less mobile forms into cusp bays, whereas more mobile forms may be able to maintain position on promontories (James 1999, Defeo and McLachlan 2005). McLachlan and Hesp (1984) showed that filter-feeding bivalves (*Donax*) concentrated in cusp bays (30 m wavelength) on a reflective Australian shore, whereas a scavenging hippid crab occurred both on the horns and in the bays. As the bivalves were less mobile, the question was raised as to whether their concentration in bays was the result of passive transport by the swash (which flows from horns to bays) or a consequence of active selection for bays, where the gentler gradient allows more uniform swash movement and presumably better conditions for filter feeding.

Across-shore Variation

Across-shore variability is reflected in both changes in general community parameters perpendicular to the shore and zonation of the species within it. Zonation is discussed in the next section. General changes across the shore include clear changes in species richness (Figure 7.12). Species richness is low in the supralittoral, where usually one or two crustaceans and insects predominate. It increases down the intertidal to maximum on the lower shore, where exposure to air is least. Then the dynamic and turbulent conditions in the surf zone reduce species richness, which is most reduced near the break point of the waves, before increasing again offshore as conditions become less turbulent. Abundance and biomass can be more variable but usually tend to track these across-shore trends in species richness, being highest on the lower shore. Benthic macrofauna is therefore usually sparse in the most turbulent part of the surf zone, more abundant in the intertidal zone and in the nearshore region outside the breakers, and sparse in the supralittoral zone, except on beaches with high wrack input.

Where wrack input is significant, an abundant fauna of insects and crustaceans may develop around the drift line (for review see Colombini and Chelazzi 2003). Among supralittoral species, talitrid amphipods occur lowest on the shore and are usually the first colonizers of wrack as it washes up the beach, followed by oniscid isopods and a variety of insects. Flies are the insects quickest to arrive at stranded wrack, followed by many

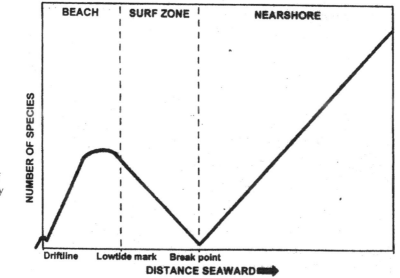

Figure 7.12.
Hypothetical pattern of marine species diversity across a beach/surf-zone system. Insects and terrestrial species, which increase above the drift line, are not considered.

other species, mainly beetles. Ants can also be important around the drift line, where carrion is stranded. A few herbivorous coleopterans of the families Tenebrionidae, Hydrophilidae, Curculonidae, and Scarabaeidae may appear, but carnivorous beetles of the Staphylinidae, Histeridae, and Carabidae (which appear later) are more important and feed on larvae of Diptera and other insects. Species richness, which generally peaks within a week or two and then declines as the wrack dries out, can be high. For example, 30 species of insects and one talitrid were associated with seagrass wrack on Polish Baltic beaches (Jedrzejczak 2002), and 60 species of crustaceans and insects were associated with kelp and seagrass wrack in California (Dugan *et al.* 2003).

Zonation

This topic was reviewed by McLachlan and Jaramillo (1995), who categorized the various conflicting views on zonation of sandy-beach macrobenthos into four sets of ideas (Figure 7.13).

- No clear zonation
- Two zones: air breathers above the drift line and water breathers below (Brown)
- Three zones based on Dahl's (1952) description of the intertidal distribution of characteristic crustaceans
- Four physical zones based on Salvat's (1964) description of changes in sand moisture content across the shore

The weight of evidence from numerous studies summarized by McLachlan and Jaramillo (1995) and more recent work (Defeo and McLachlan 2005) lends support to the recognition of three typical zones along the lines first recognized by Dahl. However, the nature and number of zones is not fixed and varies with changing beach type, creating a complex and dynamic picture.

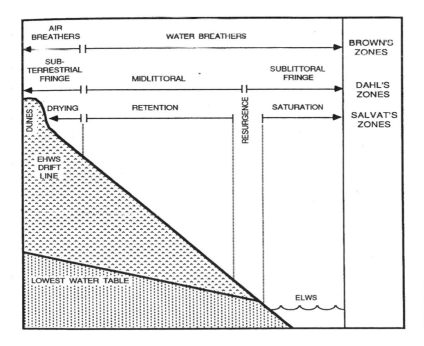

Figure 7.13. Three schemes of macrofauna zonation on sandy beaches.

Many workers have described three zones for beach fauna, and the schemes of Dahl and Salvat generally coincide (only the numbers of recognized boundaries differ). Although it has been vociferously argued that zones on rocky shores should be defined biologically, not physically, there appears to be some correspondence between physical and biological zones on sandy beaches. Only three physical boundaries may be clearly recognized during the low-tide period, and these correspond to Dahl's zones: the drift line (or highest swash line), the low-tide swash zone, and the effluent line (or water table outcrop). There is no relationship between macrofaunal zones and mean tidal levels. Rather, zones adjust each day to the limits of the beach as defined by the excursions of the swash zone. Zones occupied by individual species do not show sharp boundaries. Rather, they exhibit great variability and considerable overlaps with other species. Furthermore, zonation (as studied by ecologists) exists only during the low-tide period. As the tide rises, populations migrate, some species enter the water column, and zones compress.

Zones are clearest and narrowest at the top of the shore and become increasingly blurred and wide moving downshore. This may partly be a simple consequence of the concave slope of most beaches resulting in more gradual horizontal changes associated with the intertidal gradient toward the lower shore. Further, reflective beaches tend to support fewer species and zones than do dissipative beaches. However, none of the previously cited zonation schemes for sandy beaches takes different beach morphodynamic types into account.

We are clearly dealing with a dynamic and variable scenario. Zonation occurs on sandy beaches, but it is not precise, and attempting to delineate it too finely would be hazardous. Rocky shore zones are defined biologically by the upper limits of characteristic species, these making abrupt changes coupled to biological interactions and community-level responses. On the sandy shore, however, each species typically responds independently to the physical gradient (*sensu* the autecological hypothesis) and distributions vary from day to day. Hippid crabs of the genus *Emerita*, for example, may occur anywhere from the low-tide swash zone to the midshore, depending on the conditions on the particular day of sampling. Most shores display three zones, which may be distinguished by the presence of characteristic faunal elements rather than upper or lower distribution limits or moisture levels, but these zones change with changing beach type (Figure 7.14).

- *Supralittoral zone:* A top zone of air-breathing crustaceans, usually situated at and above the drift line in sand dry on the surface. This macrofauna lives outside the swash but may return to it for reproduction and feeding. It corresponds to the top zones of Dahl (1952) and Salvat (1964), and may harbor one or more of the following taxa (other than insects) in any combination: talitrid amphipods, oniscid isopods (*Tylos*), ocypodid crabs, and (infrequently) cirolanid isopods of the genus *Excirolana*. The landward extent of this zone is variable and may be considerable in situations where talitrids, oniscids, or ocypodids wander far into the dunes. This zone and its fauna are strongly influenced by the amount of material cast ashore on the drift line, and insects can be an important component of the fauna.
- *Littoral zone:* Below the supralittoral fauna is a marine intertidal macrofauna corresponding to Dahl's midshore and Salvat's retention and (at most times) resurgence zones. Present as a faunal zone on all shores, except coarse-grained reflective beaches that completely lack intertidal macrofauna, the littoral zone occurs from

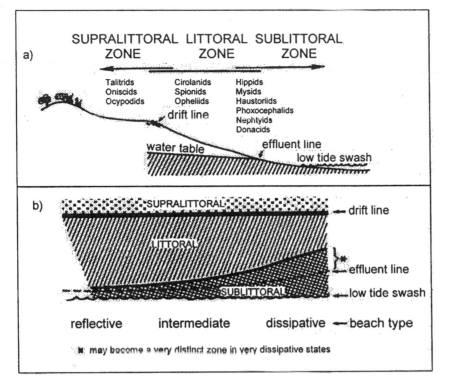

Figure 7.14 General scheme (a) for zonation of macrofauna on sandy beaches and (b) showing effects of changing beach type (after McLachlan and Jaramillo 1995).

the drift line down to damp sand at or just above the effluent line or water table outcrop. Taxa characteristic of this zone are all true intertidal species not normally found in the surf zone and may include cirolanid isopods (including some species of *Excirolana*), other isopods, haustoriid and other amphipods, spionid polychaetes such as *Scolelepis*, and ophelids such as *Euzonus*. This zone extends farthest downshore on well-drained reflective/intermediate beaches, and makes up a smaller portion of the intertidal zone in dissipative beaches where the next zone may penetrate up to the midshore.

- *Sublittoral zone:* This zone is an upward extension of the surf zone and has its upper fringe on the lower or mid-intertidal near the effluent line. It houses hippid crabs, mysids, donacid bivalves or their equivalents, nephtyid and glycerid polychaetes, and idoteid, oedicerotid, and haustoriid amphipods, and it corresponds to Dahl's lower zone and Salvat's zone of saturation. This zone is devoid of macrofauna on most reflective beaches, but these are present as a faunal zone on intermediate and dissipative beach types, where it includes species that occur in saturated sand and extend into the surf zone. They could thus be termed sublittoral species. This zone extends farthest up the shore on dissipative beaches with high-water tables.

On meso/macrotidal dissipative beaches, the last zone may be subdivisible into two zones: a low intertidal zone and a sublittoral zone: However, such a division is not clear and only a few studies on dissipative beaches have shown signs of it as an intertidal resurgence zone (with nephtyid and glycerid polychaetes, haustoriid and phoxocephalid amphipods, hippid crabs, mysids, and donacid bivalves) and as a true sublittoral zone with haustoriid and phoxocephalid amphipods, mysids, portunid crabs, and donacid bivalves. In such cases, the resurgence-zone faunas are intertidal species not normally found in any numbers in the surf zone.

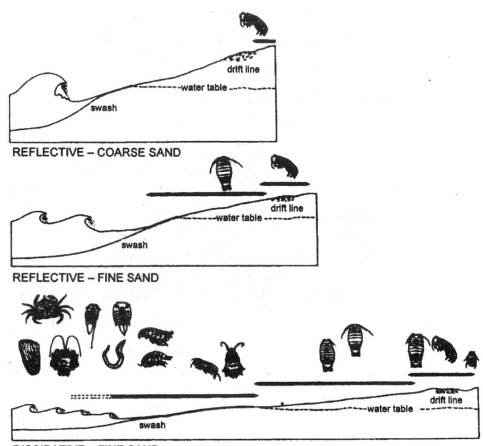

Figure 7.15. Changes in macrofauna zones on different beach types in Chile, showing a single zone (*Orchestoidea* in the supralittoral) on reflective beaches of coarse sand (top), two zones (*Orchestoidea* in the supralittoral and *Excirolana hirsuticauda* in the littoral) on reflective beaches of fine sand (middle), and three zones (additional species in the supralittoral and littoral and many species in the sublittoral) well developed on dissipative beaches (after Jaramillo *et al.* 1993, McLachlan and Jaramillo 1995).

The three zones are broadly similar to those on rocky shores, with the exception that the sublittoral extends farther up the shore on the sandy beach, much as would occur for the fauna of pools on rocky shores. This is because of the elevation of the beach water table above sea level and the escape from desiccation provided by the moist interstitial environment.

Although the moisture zones of Salvat (1964) are certainly valid descriptions of the interstitial environment across open sandy beaches, macrofauna zonation cannot be defined solely in these terms, and the boundary between the littoral and sublittoral zones may lie at or below the resurgence zone in different cases. The high mobility of most species, coupled with the disturbance effects of changing wave energy levels, results in species distributions being quite variable from day to day. Physical boundaries correspond to swash and moisture zones on the shore and not to absolute tide levels. Thus, these zones do not exhibit sharp boundaries and should be defined on the basis of the centers of distribution of characteristic taxa, rather than on sand moisture levels.

It can be concluded that the distribution of macrofauna across sandy beaches is not a smooth continuous response to the intertidal gradient but rather assumes the form of three distinct and universal zones, which can be identified primarily by the distribution of characteristic taxa. However, physical, taxonomic, and physiological differences in the environment and fauna at different beach levels also play a role. The lower two zones may be limited or absent on reflective beaches (Figure 7.15). Recent studies, summarized by

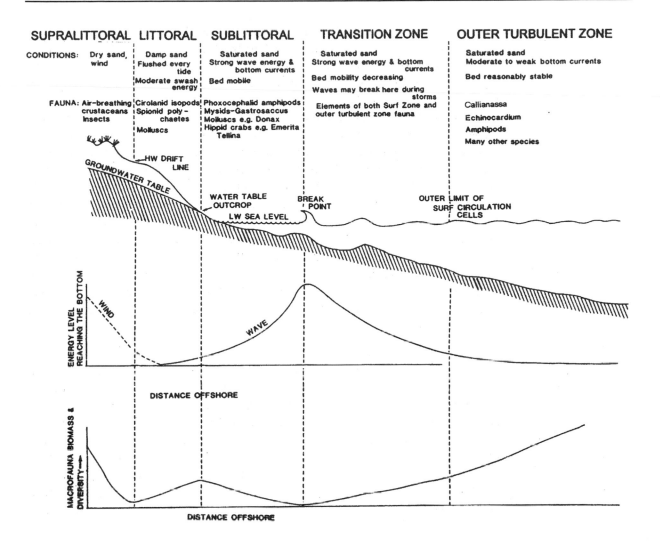

Figure 7.16. Generalized scheme of zonation across sandy shores into the sublittoral.

Defeo and McLachlan (2005), have tended to confirm that at the extremes of beach morphodynamics only the sublittoral zone may be present on very harsh reflective beaches and up to four zones on very dissipative beaches. Temporal variations imposed on these zones include seasonal movements, storm surges, and tidal effects.

The causes and mechanisms for maintenance of intertidal zones are complex. Migrations of the fauna shuffle and recreate zones daily and with each tidal cycle. Zones, as described here for the low-tide period, are thus maintained by the fauna through a complex system of behavioral responses that include endogenous rhythms and responses to cues such as swash processes, sand particle size and moisture content, diel cycles, beach slope, and thixotropy (Chapter 6). The basis of zonation is the sum of the individual responses of different species to physical cues across the intertidal gradient, with sand moisture and swash activity probably being the most important. In a few cases, zonation may be influenced by interspecific competition (see material following) — but other than this it is a result of responses to the physical environment.

This zonation scheme can be extended into the sublittoral to include the surf zone (Figure 7.16) based on quantitative surveys of sublittoral macrobenthos inside and outside the surf zone (McLachlan *et al.* 1984, Janssen and Mulder 2005). The sublittoral or surf zone (also referred to as the inner turbulent zone) includes faunas that extend upward onto the lower shore and seaward to beyond the breakers. Its landward boundary is quite sharp at the water table outcrop, whereas its seaward boundary is blurred outside the breakers, a transitional zone being apparent where it gives way to the outer turbulent zone as the bottom stabilizes. Wave-driven currents (longshore, rip, and so on) dominate the surf zone and the bottom is unstable, bars and troughs migrating in response to changing wave conditions. This fauna, predominantly highly mobile crustaceans and mollusks, is richest around the low-tide mark and depauperate in the most turbulent region around the break point of the waves. Where there are deep and relatively stable troughs between outer bars, a rich fauna may develop. Further offshore, well outside the break point, the outer turbulent zone is the area where the bottom becomes more stable and a rich macrofauna develops — increasing in species richness and biomass in the offshore direction and including forms that construct permanent burrows as the bottom stabilizes. Between these two zones (the relatively stable outer turbulent zone and the dynamic sublittoral surf zone) lies a transition zone that roughly corresponds to the area between the break point and the outer limit of surf circulation cells. This transition zone is turbulent and includes elements of the surf-zone fauna, but outer turbulent zone species may begin to appear as conditions become calmer toward the outside of this zone.

On very sheltered beaches, a surf zone may be practically absent, and in low-energy macrotidal beaches the surf zone may comprise a narrow band of breaking waves that moves back and forth over a much larger beach face. In such cases, there are supralittoral, littoral, and lower zones, the lower zone being equivalent to the surf zone and ending around the level of extreme low water of spring tides. This lower zone of saturation, or surf zone, may harbor species with stable burrows and gives way just below the extreme spring low-water level to a fauna equivalent to that of the outer turbulent zone (including, for example, forms such as *Echinocardium*). This scheme for a macrotidal sheltered Scottish beach is illustrated in Figure 7.17. Similarly, the shallow sublittoral of reflective beaches may harbor more diverse macrofauna than the surf zone of intermediate

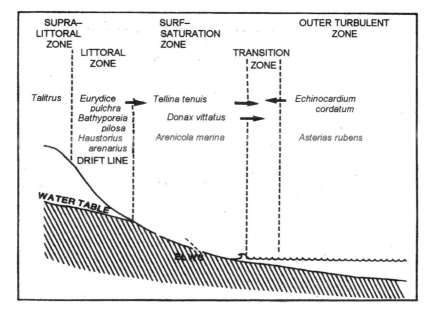

Figure 7.17. Zonation on a macrotidal low-energy sandy shore in Scotland (after Elefheriou and McIntyre 1976).

beaches (Barros *et al.* 2002) because of the location of the outer turbulent zone close inshore.

Temporal Changes in Zonation

For most species on open ocean sandy beaches, zonation (as recorded by the researcher) reflects faunal distribution only at the time of the investigation. These zones are modified in the short term by tidal migrations and in the medium and longer terms by responses to seasons, storms/calm, and accretion/erosion cycles (Brazeiro and Defeo 1996).

Tidal migrations are characteristic of this fauna, typically involving a simple movement up and down the slope with the tides — thus allowing the animals to remain in the swash zone, where conditions for feeding are optimal and access by terrestrial and marine predators is limited. Such migrations may be mediated by endogenous rhythms triggered by changing environmental factors, or (as in those mollusks that have been studied) the animals may rely entirely on physical cues (see Chapter 6). In crustaceans, the rhythms (usually endogenous) are far more complex — being not only circatidal but frequently circadian and circa-semilunar as well, with entry into the plankton at night in addition to movement up and down the beach face. The highest tidal levels are usually avoided, thus preventing the animals from being stranded as the tide falls. Among the few exceptions to this rule are the whelk *Bullia rhodostoma* and the isopod *Pontogeloides latipes*, which may frequently be found buried in sand above the water — although not out of reach of the next high tide. For most of the fauna, however, the extent of the tidal migration is less than the tidal range for that day. *Hippa australis*, for example, minimizes the extent of its tidal migrations, being located near the top of the swash zone at low tide but low down in the swash at high tide (Shepherd *et al.* 1988).

Tidal migratory behavior has several advantages for sandy-beach animals, in addition to maximizing food availability. It keeps them in an area where water coverage is almost continuous but wave action not too severe. The animals are mostly too shallow to be reached by predatory fishes, and it is difficult for birds to get at them between swashes. Migrating animals are also less likely to be stranded as the beach changes with tides, storms, and calms, as they remain in the zone of sediment reworking. Tidal rhythms also enable supralittoral forms to move safely downshore for feeding purposes during low tide. Tidal migrations clearly have marked effects on zonation. All zones compress toward the high-tide period, when most of the invertebrate populations are concentrated into a narrow strip — although this does not apply to beaches sheltered enough to allow the construction of semipermanent burrows.

Longer-term temporal changes in zonation have been recorded both in whole macrofaunal communities and in individual populations. The most dramatic of these are associated with the monsoon in India, where much of the fauna disappears from the beaches during this stormy and erosive period (Ansell *et al.* 1972) (Figure 7.18).

Seasonal changes have been recorded in most communities studied over a period of time, although they are usually not as dramatic as on Indian beaches during the monsoon. Typically, seasonality shows as higher intertidal macrofaunal abundance in summer and reduced numbers during the winter months in temperate regions (e.g., Leber 1982, Degraer *et al.* 1999), whereas in tropical regions it is more related to disturbance due to

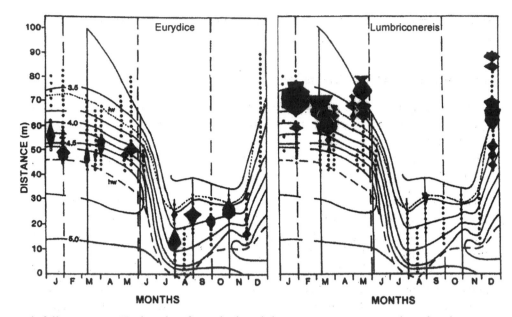

Figure 7.18.
Downshore migrations
of the polychaete
Lumbriconereis and the
isopod *Eurydice* in
India, associated with
the monsoon (after
Ansell *et al.* 1972).
Kites represent the
population and
contours the beach
face.

rainfall or storms. Emigration from the beach in temperate areas may be related to temperature declines and reproductive requirements, but it is more likely that offshore movement is an adaptation to avoid winter storms. Storms are not always associated with offshore migration, however, and some animals simply display deeper burrowing during the storm period.

7.6   Microscale Patterns

This scale refers to the quadrat scale (i.e., distances from millimeters to meters within a beach), which concerns mainly the interactions between individual organisms but also fine-scale environmental gradients (Defeo and McLachlan 2005). It has been pointed out that moving to finer scales and more hospitable dissipative beaches with high macrofauna densities (Figure 7.5) increases the potential for biological interactions (Figure 7.8). However, sandy-beach macrobenthic communities generally lack the obvious level of organization found on rocky shores. Because of the three-dimensional and dynamic nature of the environment and the mobility of all species, space is not limiting and there is not a strong underlying basis for competition to provide biological structure to the communities. Rather, as we have seen, the community is mainly physically controlled and species respond independently to the physical environment. Understanding the autecology of the constituent species should give a reasonable understanding of the community, and deciphering the role of biological interactions is likely only to contribute significantly in the case of benign dissipative systems and on fine scales. There is no reason, from the point of view of sandy beaches, to reject the general idea that predation is most important in benign environments (very sheltered beaches and sand flats), competition in harsher situations (sheltered and dissipative beaches), and physical stress under the harshest conditions (reflective and most intermediate beaches). However, keystone species (or predatory species that may exert an inordinate influence on community structure in other systems) are absent on sandy beaches.

Although the evidence points to competition and predation on sandy beaches both decreasing with increasing exposure, this is difficult to substantiate as there have been no experiments in either regard on high-energy intertidal sands. The highly dynamic nature of such beaches precludes caging or other manipulative experiments, and in their absence conclusions are bound to be largely speculative. In general, however, predation is a much more important factor than competition on sandy beaches (and it always occurs, although it may decrease with increasing exposure). However, whether or not competition or predation occurs is not the issue. The question to be asked is whether and to what extent they affect community structure. These biotic interactions may be significant in dissipative beaches with high species richness, high densities, and relatively stable substrate, whereas populations on reflective beaches are mainly regulated by individual responses to the environment (Defeo *et al.* 2003, McLachlan and Dorvlo 2005).

Competition

Competition may be defined as the struggle of individuals to acquire a resource in short supply, this resulting in the individuals having a negative effect on one another. The resources in question are generally food and space, although what exactly is competed for may be modified by other factors (such as the influence of predators). Competition is always between individuals but may be intraspecific or interspecific, or both at the same time. Two distinct types of competition may be recognized: exploitation (which can be described as a scramble for a common resource) and interference, during which a competitor is denied access to the resource, often by some damaging activity. Interference is usually encountered among individuals belonging to closely related taxa.

One might expect competition on sandy beaches to be less evident than on rocky shores, and this has indeed been generally assumed. First, there appears to be no shortage of space, even bearing in mind that space and suitable space are not necessarily the same thing. Second, the particulate three-dimensional nature of the substratum coupled with the fact that all species can burrow into it must be expected to minimize competition by interference. In particular, the crushing or undercutting of one individual by another (a common phenomenon on rocky shores) is most unlikely to occur. Similarly, the overgrowth of organisms is difficult to envisage in an environment in which they are free to move up and down through the substratum. Furthermore, for sandy-beach deposit feeders and filter-feeders food is by no means as limiting as for rocky-shore grazers. Most sandy-beach animals are opportunistic in their feeding, and this too decreases the likelihood of competition being an important factor.

Therefore, the difficulty in demonstrating competition in sandy-beach macrobenthos is due to (1) the three-dimensional nature of the habitat allowing vertical and horizontal space partitioning, (2) manipulative experiments usually not being possible, and (3) lack of information on the scale dependence of processes structuring populations (Peterson 1991, Defeo and McLachlan 2005). Despite these difficulties, there are some studies at fine scales under dissipative/sheltered conditions where competition has been indicated, if not conclusively demonstrated. However, on exposed shores evidence for competition is mainly circumstantial/correlational, rather than experimental and conclusive. Interference competition is limited because of the three-dimensional nature of the environment and mobility of the fauna. Individuals that interact can usually simply move away from each other. In a few cases, such as dense beds of filter-feeders or populations of scavengers tracking the drift line, exploitation competition (as competition for food resources) may be a factor.

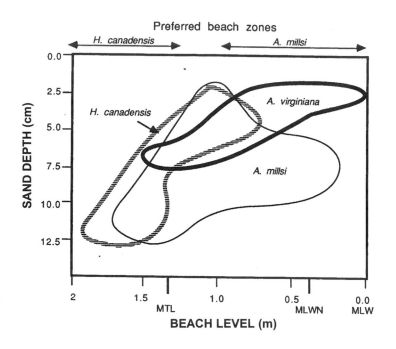

Figure 7.19. Spatial habitat partitioning among species of haustoriid amphipods on a low-energy sandy beach (after Croker 1967). Although there is overlap, species tend to separate horizontally and vertically.

Competition might be expected to be more prevalent among the supratidal animals at the top of the shore than among the completely aquatic forms and here direct interference may play a more significant role. Thus, the ghost crab *Ocypode* is territorial — defending its burrow and the area around it, resulting in the burrows tending to be equidistant from one another. However, the isopod *Tylos* (which also forages at night and returns to supratidal levels during the day) makes a new burrow on each occasion, does not defend it, and does not show a regular dispersion.

In sheltered situations, there are more species and these are often more specialized and less opportunistic than the animals of high-energy shores — so that there is greater potential for competition. Croker (1967) described spatial partitioning of the habitat among five haustoriid amphipods on low-energy beaches (Figure 7.19). Close horizontal and vertical partitioning in the sand and differential survival in coexistence experiments suggested that some competition might occur between these sympatric species (Croker and Hatfield 1980). On a very sheltered shore, Woodin (1977) showed that tube-building polychaetes inhibited deposit feeders. Wilson (1981), also working on a very sheltered shore, demonstrated negative interactions in a dense assemblage of deposit-feeding polychaetes — large burrowers adversely affecting small surface-deposit feeders, which in turn ingested the larvae of the former. Large polychaete tubes have been shown to provide protection from predators and disturbance, thus stimulating an increase in diversity and more opportunity for competition (Woodin 1981). Seagrass beds have a similar effect.

Defeo *et al.* (1997) provided evidence for interactions between two closely related intertidal isopods on South American Atlantic shores in terms of shifts in zonation and substrate selection in sympatry (Figure 7.20). Further, field and laboratory studies showed that abundance and survivorship in these isopods (*Excirolana armata* and *E. braziliensis*) were affected, and both species showed a noticeable reduction in body size in sympatric populations. *E. braziliensis* is most abundant in fine sands, and has a greater niche

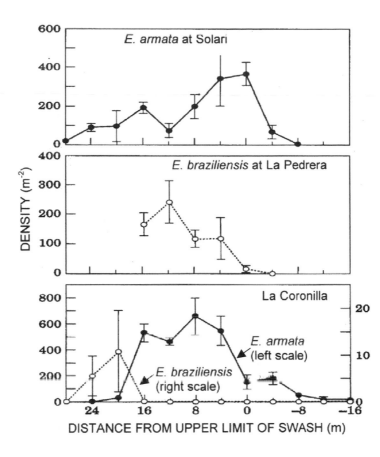

Figure 7.20.
Distribution of two species of intertidal isopods of the genus *Excirolana* in allopatry (top and center) and sympatry (bottom) showing *E. brasiliensis* restricted to upper beach levels under sympatry (after Defeo *et al.* 1997).

breadth through most of its range than that observed in Uruguay — where it mainly inhabits coarse sands and upper beach levels on reflective beaches, being displaced by *E. armata*.

In another example, Jaramillo *et al.* (2003) demonstrated for a guild of supralittoral scavengers that two species (the isopod *Tylos* and the beetle *Phalerisidia*) had activity rhythms and distribution patterns that tended to result in time/space partitioning to separate them from the amphipod *Orchestoidea*, with which they both displayed negative interactions in coexistence experiments. This partitioning was interpreted as allowing coexistence by avoiding competition, implying that competition was important in influencing community structure.

There are several cases of distributional evidence for competition between filter-feeders in the intertidal zone. Cardoso and Veloso (2003) showed a negative relationship between densities of *Donax hanleyanus* and *Emerita brasiliensis* over a two-year study, suggesting competition. Many beaches harbor two species of clams, typically of the genera *Donax* and/or *Mesodesma* — a larger species occupying a fixed zone on the midshore in warm areas and on the lower shore in colder climates and a smaller tidally migrating species. The two are usually vertically separated on the shore, possibly as a consequence of competition.

Human exclusion experiments are useful tools in detecting the role of intra- and inter-specific competition, because humans can strongly impact the dynamics and demography of sandy-beach macrofauna. Fishing pressure on the clam *Mesodesma mactroides* along a 22-km dissipative beach in Uruguay explained temporal variations in recruitment in another species: recruitment density of the sympatric nontargeted bivalve *Donax hanleyanus* was inversely related to the density of juveniles and adults of the yellow clam *Mesodesma mactroides* (de Alava 1993, Defeo 1998). Removal of the former species by a fishery resulted in increases in abundance and shifts in zonation of the latter, a response reversed when the fishery was closed. Similarly, *Donax hanleyanus* and *Emerita brasiliensis* increased dramatically in density for 10 years after the dominant *Mesodesma mactroides* experienced mass mortalities (Defeo 2003, Defeo and de Alava 1995).

Arntz *et al.* (1987) showed fluctuations in the sandy-beach bivalves *Donax peruvianus* and *Mesodesma donacium*, and the mole crab *Emerita analoga* in Peru. Where the dominant *M. donacium* experienced mass mortalities, due to an increase in seawater temperature during an El Niño southern oscillation event, *D. peruvianus* increased its representation within the community from 5 to 60 to 100% and its density from about 20 to 185 m^{-2}. *E. analoga* increased from < 1 to 29%, with densities increasing from 0 to 85 m^{-2}. Further, this was accompanied by an expansion of the distributional range to beaches previously unoccupied by these species. Both of these examples suggest differential responses to climatic events and interspecific interactions because of release of resources by declines in dominant community members. In both of these cases, changes in species composition and abundance occurred after removal of the dominant species, strongly indicating competitive release.

Such instances, where redistribution of animals has been studied following the removal of one of the populations, provide good evidence of competition. However, great caution should be exercised in inferring competition merely from distribution patterns. Whereas correlational/distributional evidence is often convincing, direct experimentation is more difficult.

McLachlan (1998) looked for evidence of interference competition by experimentally assessing effects of the presence of one species buried in the sand on the burrowing ability of another. He found that intra- and interspecific interactions could occur during burrowing at high densities (400 to 600 ind, m^{-2}) of the sympatric clams *Donax serra* and *Donax sordidus*. Asymmetric competition was postulated: the larger *D. serra*, occupying more space at experimental densities, had significant effects on *D. sordidus*, but not vice versa. However, the densities required to produce these effects exceeded those recorded in the field. In a similar study, Dugan *et al.* (2004) investigated potential interference between the hippid crab *Emerita analoga* and the clam *Mesodesma donacium*. Burrowing times of large crabs were significantly longer in all densities of clams than in controls, and large crabs displaced small clams from the sand while burrowing (Figure 7.21).

Nel (2001) undertook a series of innovative experiments to test for biological interactions in exposed sandy beaches. She examined interspecific interactions of three cirolanid isopods in the laboratory by comparing their grain size preferences individually and in combinations. When placed together with other species, all three species displayed shifts in substrate preferences. An example of her results is shown in Figure 7.22. She related this back to field distributions of the three species and concluded that across-shore distribution of the three species where they co-occurred was influenced by interference competition.

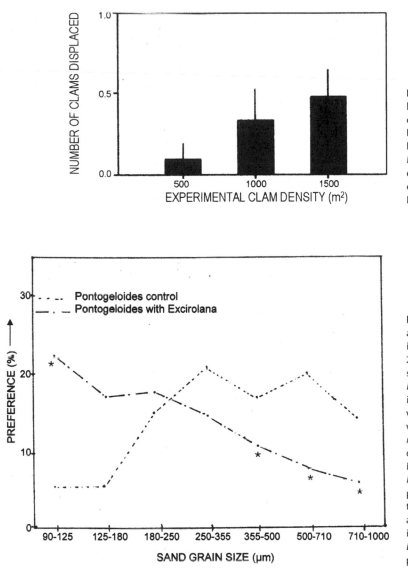

Figure 7.21. Effects of hippid crabs (*Emerita*) on clams (*Mesodesma*) buried at different background densities. Mean numbers of clams displaced by burrowing crabs in each trial (after Dugan *et al.* 2004).

Figure 7.22. Substrate and competition in isopods (after Nel 2001). Changes in substrate selection by *Pontogeloides latipes* in isolation (control) and when placed together with *Excirolana natalensis*. Significant differences are indicated as *. *Pontogeloides* changes preferences to select the finest sand and avoid the coarser sands in the presence of *Excirolana*, which prefers coarser sands.

We are seeing accumulating evidence for interspecific competition in sandy-beach macrofauna, but much of this is inferred rather than demonstrated — and effects are mainly small shifts in distribution rather than significant changes to community structure, such as exclusion or elimination of species. Further, such effects are clearest in more benign dissipative systems with rich macrobenthic faunas. Quite extreme situations need to be created under experimental conditions in the field or in the laboratory to obtain significant effects, and it cannot be concluded that competitive interactions play a major role in structuring sandy-beach macrofauna communities.

Disturbance

Disturbed sediments are commonly invaded by opportunistic species, and this has in the past been considered a result of reduced competition — it being thought that these species are only capable of colonization once the established fauna has been eliminated. The eventual return of the normal species was then assumed to result in the opportunists being outcompeted. A classic example was the colonization of sheltered beaches by capitellid and cirratulid polychaetes following spills of crude oil (Southward 1982). Such a "pollution fauna" may dominate the beach for years before it loses ground to the normal fauna. Although such a sequence of events might be explained in terms of interspecific competition, it is more likely that it reflects a direct response by the opportunists to resources made available by the disturbance.

Other types of disturbance are also exploited by opportunistic species. For example, the feeding pits created by rays may also be used by motile crustaceans, which aggregate in them to feed on the organic debris accumulating there (von Blaricom 1982). This writer was, however, unable to demonstrate competition between these opportunists and resident infaunal organisms.

Sandy-beach macrofaunas have little effect on the sediment and bioturbation is generally unimportant, other than is the case on more sheltered depositing shores. Only in sheltered situations do deposit feeders become important, and only in such benign environments do any species establish semipermanent burrows. Some species may work the sediment (e.g., the crab *Dotilla* sorts the surface in sheltered situations), and species with burrows (e.g., *Callianassa*, *Arenicola*) bring sand to the surface, but all of these effects are localized and limited to sheltered shores.

Disturbance does not occur in the sense of rocky shores, clearing patches of attached organisms and enabling succession to start. On beaches, it is usually localized, and barring major storms (which restructure the entire beach environment and often cause offshore migration into deeper water) disturbance usually has minimal effects, perhaps causing small local accumulations or temporary patchiness. Succession has not been studied on a sandy beach, and in the classical sense probably does not exist. Beach faunas are opportunistic, mobile, and plastic in their behavior, and any defaunated area (created, for example, by artificial beach nourishment) will be rapidly colonized with little interaction between species and both pre- and post-settlement processes playing a role. There is no indication of any sequence of community states on a sandy beach or habitat modification by some species influencing the success of other colonists. Instead, in most cases there is not a structured community, just a number of independent species populations. The colonization of barren sand by microbes and meiofauna may be different. Presettlement processes, recruitment, and removal of larvae can be important determinants of community structure, but these effects are mainly intraspecific (covered in the next chapter).

The example of disturbance in the form of a fishery removing the dominant *Mesodesma* species showed a clear response in other species. Disturbance related to a clam fishery has also been investigated by Schoeman *et al.* (2000), who showed that the beach and its population of donacid clams recovered within a tidal cycle from intense local disturbance (which excavated a substantial area of the intertidal zone and removed all macrofauna). The inherent dynamicism and mobility of this fauna overrode this local disturbance with ease. Lercari and Defeo (2003), and other papers cited by them, examined effects of freshwater discharge over a productive dissipative beach, finding significant spatial effects. Abundance, biomass, species richness, and evenness all decreased toward this source of disturbance, lower shore species being most affected by lower salin-

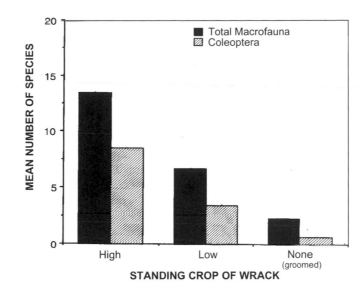

Figure 7.23. Number of wrack-associated macrofauna species collected on beaches with large amounts of wrack, small amounts of wrack, and cleaned of wrack by grooming (after Dugan *et al.* 2003).

ities. Similar local effects of disturbance by freshwater are probably common wherever rivers or canals discharge over sandy beaches.

Disturbance, in the form of beach grooming to remove wrack to make beaches more attractive to tourists in California, has been shown to seriously impact the wrack-associated fauna (Dugan *et al.* 2003) — the number of wrack-associated species dropping from an average of 6 to 13 on beaches with high or moderate amounts of wrack to just 1 to 4 on groomed beaches (Figure 7.23). However, this represents simply a removal of habitat and associated species, not a restructuring of the community.

Effects of monsoonal erosion and disturbance were shown in Figure 7.18. Similarly disruptive effects were caused by an El Niño disturbance on Peruvian beaches (Arntz *et al.* 1987), with dominant species disappearing and others taking over, as outlined previously.

Stranded macrophytes can have a variety of effects on the fauna, not only providing food and habitat around the drift line (and thereby enhancing species richness and abundance) but causing disturbance by macrophytes washing around the intertidal — which may preclude or limit some species, such as the clam *Donax serra* (Soares *et al.* 1996).

Recruitment

Recruitment variability in sandy-beach macrofauna was reviewed by Defeo (1996a). This is more relevant to population-level studies but it does influence ideas on macroscale patterns, specifically the question as to whether exclusion of species from reflective beaches is due to pre- or post-settlement processes. Little is known about this. There may be some biotic interactions involved in settlement success; for example, by filter-feeders removing settling larvae, which could be inter- or intraspecific. This is discussed in the next chapter. How settling larvae select the physical environment in which to settle has hardly been studied on an exposed sandy beach. Lastra and McLachlan (1996) showed a strong tendency for *Donax serra* spat to settle in areas of fine sand in the surf zone, and Dugan and Hubbard (1996) recorded the greatest numbers of *Emerita* megalopa larvae near large adult populations. Density-dependent factors may operate more at population level.

Predation

In all cases where predators have been excluded from soft substrata, diversity has increased. This is in contrast to rocky shores, where predators (sometimes considered keystone species) are considered to hold potentially competing populations below the level at which competition becomes an important factor, thereby increasing diversity (Branch 1984). This fact also supports the contention that competition in sediments is less important than on rocky shores.

Three groups of predators feed on sandy beaches: (1) birds, arachnids, and insects from land, (2) fishes from the sea, and (3) resident invertebrates, notably crabs and gastropods. These forms of predation crop populations and result in significant energy flow, but their effects on population dynamics or community structure are still little known. The invertebrate predators may also include gastropods and asteroids, as well as polychaetes such as *Nephtys*. Most of the work on the effects of these predators has been undertaken on relatively sheltered beaches.

Naticid gastropods are important predators of bivalves in sheltered beaches (Ansell 1982), whereas crabs play a major role over a wide variety of beach types, including very exposed beaches. Crabs are important predators of mollusks, both bivalves and gastropods. Portunids, for example, utilize a variety of different feeding methods to obtain prey and remove whelks and clams from their shells (du Preez 1984). Predation by crabs, birds, and fishes can crop populations, quite substantially in some cases (for example, when large numbers of migrant shorebirds feed on a beach). Reise (1985) studied German sand flats, an environment more estuarine than open marine. By caging, he showed that large carnivorous fish and birds had little effect on the benthos, although they did bring about a reduction in mean size by cropping large individuals. Both birds and fish depressed the numbers of small predators. On the other hand, crabs, shrimps, and juvenile fish (all small epibenthic predators that can significantly crop the fauna on sandflats) had significant effects on the meiofauna and small macrofauna, as well as on juveniles of the larger macrofauna — large population increases taking place in the absence of these predators. He suggested a conceptual hierarchy of biological processes (Figure 7.24) influencing community structure on sand flats, including positive interactions such as promotion. Promotion can take the form of accommodation in burrows, sediment stabilization, and improvement in conditions within the sediment, such as oxygenation. Reise concluded that predation was the most important biological interaction in tidal flats and that it decreased upshore.

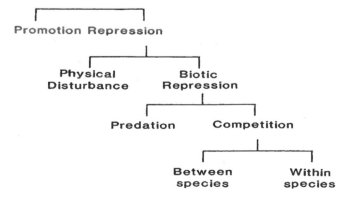

Figure 7.24.
Conceptual hierarchy of biological processes as envisaged by Reise (1985).

However, exclusion experiments are not possible on open beaches, and because of limited (if any) competition predation can only increase evenness by reducing the abundance of dominant species, without changing species richness. There are no keystone species on sandy shores and predation is not important in structuring the community. On open sandy beaches, the effects of predation are certainly less than those found by Reise (1985) on sheltered flats. Indeed, it is generally assumed that the effects of predation are most severe in benign habitats and that they decrease with increasing harshness of the habitat (Connell 1975).

Commensalism and Mutualism

There are several cases of commensalism recorded in sandy-beach macrobenthos. Hydroids attach to the shells of some clams and snails, and small commensals may live in the mantle cavities of mollusks. In sheltered shores some species can construct permanent burrows that can host commensals. For example, burrows of the sand prawn or ghost shrimp *Callianassa* may be inhabited by gobies, polychaete worms, crabs, and even clams.

Positive interspecific interactions within physically stressful habitats such as sandy beaches have been sparsely documented. Manning and Lindquist (2003) demonstrated that the clam *Donax variabilis* facilitated epibiotic occupation by providing a stable substrate for attachment of the hydroid *Lovenella gracilis*. The hydroid in turn defended the clam against fishes by means of its nematocysts. However, it facilitated predation by crabs, because projection of the hydroid above the surface of the sand allowed the crabs to more readily detect clams. Depending on relative predation pressure, the occupation of *D. variabilis* by *L. gracilis* could be characterized as beneficial or detrimental to the host.

7.7 Trophic Relations

Sandy-beach energetics are covered in detail in Chapter 12. Trophic groups among the intertidal macrofaunas include predators, scavengers, filter/suspension feeders, and deposit feeders (the latter only important on relatively sheltered shores). In the turbulent environment of the sandy beach, suspended food is always available, although it may be variable in nature and quantity. Filter/suspension feeders usually dominate the community and consist largely of bivalves (or on tropical shores, mole crabs). Where conditions are optimal on exposed beaches (for example, in dissipative situations where rich surf diatom accumulations develop), suspension feeders can maintain huge populations and biomass. The common tellinacean bivalves of open beaches, *Donax* species, are suspension feeders — unlike most other members of the family, which are typically deposit feeders (Ansell 1983).

In most situations, and certainly on reflective and intermediate beaches, motile feeders and scavenger/predators dominate the sandy-beach macrofauna. Where the fauna is impoverished, such as on steep beaches with coarse sand, supralittoral scavengers such as *Ocypode* may be most important as a consequence of the absence of truly intertidal forms — whereas on most other beaches intertidal scavengers and predators are ubiquitous. Gastropods and crabs are usually the most conspicuous of these.

Deposit feeders are usually a minor component, except on sheltered shores and in the sublittoral, where they may increase in importance. However, on sheltered sands stable enough to allow the construction of semipermanent burrows deposit feeders such as *Arenicola*, *Callianassa*, and *Scolelepis* may be dominant. Interactions between suspension and deposit feeders are therefore usually not an issue on sandy beaches. Ricciardi and Bourget's (1999) analysis of broad community patterns in marine sedimentary communities showed that deposit feeders increase with more sheltered conditions, finer sediments, and flatter slopes (i.e., dissipativeness in the case of sandy beaches) — with carnivores also preferring more sheltered shores.

As there are no macroscopic primary producers intertidally on exposed sandy beaches and limited quantities of epibenthic diatoms only on sheltered sands, the macrofauna is dependent on food imports from the adjacent land and sea. The sea is by far the more important source of food, supplying particulates for the filter-feeders and carrion and plant debris for the scavengers. The size of beach populations is therefore probably closely related to the richness of the inshore waters, particularly in terms of particulate material.

General trophic interactions, as well as predation by birds and fishes, are discussed in Chapter 12. Unlike the birds and fishes, invertebrates represent a source of predation resident in the beach and may have significant impact on intertidal populations. Both resident ghost crabs moving down from the supralittoral and portunid crabs moving up from the surf zone are important in this regard, as are naticid gastropods on sheltered shores. Opportunistic scavengers, such as whelks of the genus *Bullia*, may also turn predator on occasion, but their impact is usually minimal. There are, of course, other major causes of mortality among sandy-beach invertebrates besides predation. Severe storms constitute such a cause of death, as do unseasonal upwelling, poisoning due to red tides, and of course exploitation by man.

Wrack-dominated Shores

A special situation occurs on beaches subject to large wrack inputs. Here, supralittoral scavengers (insects, talitrids, oniscid isopods) dominate biomass, which is concentrated at the top of the shore and birds become the main predators (Colombini and Chelazzi 2003). Midlittoral carnivore/scavenger populations, such as cirolanid isopods and whelks, are found where there is regular stranding of carrion (for example, near seabird colonies). Wrack inputs to sandy beaches may be seagrasses, kelps, and other marine algae. Much of the breakdown of this wrack is accomplished by micro- and meiofauna, and macrofauna usually plays a lesser role quantitatively. Nevertheless, they become very abundant. The supralittoral fauna associated with wrack is typically dominated by insects in terms of species — with beetles and flies most common, but crustaceans, especially talitrid amphipods, are usually the first colonizers and may be numerically dominant and more important in wrack consumption. Talitrids, oniscids, and dipteran larvae are important primary consumers that feed on the wrack and its epiphytes, and assist in breakdown. A variety of macrofauna, including carnivorous beetles and arachnids as well as birds, are important secondary consumers. Although this takes place in most cases in the supralittoral zone, on some beaches wrack mainly washes around in the swash on the lower shore and in the surf zone, and then the associated species are entirely marine.

7.8 Conclusions

Research on sandy-beach macrobenthos has advanced considerably over the past two decades. This chapter on community ecology has demonstrated an emerging field of research being directed by accepted paradigms and new theories under test. Four paradigms may be considered to frame our understanding of large-scale community patterns on exposed sandy beaches.

- The concept of overriding physical control in these harsh environments
- The beach morphodynamic models and associated features, swash, and sand
- Changes in richness/abundance/biomass with morphodynamics
- Latitudinal effects

Beach morphodynamic models have become widely accepted and applied at the community level, with consistent large-scale patterns of increasingly species-rich communities toward macrotidal dissipative beaches and lower latitudes. The distribution and diversity of the invertebrate macrofauna of sandy beaches are largely determined by physical factors — primarily latitude and tide range, as well as wave action and the particle size of the sand — which in turn fix the morphodynamic state of the beach and surf zone. This is ultimately experienced by the benthos as the swash climate and the sand texture and stability. In relatively sheltered situations and on finer scales, predation, disturbance, and competition may influence species distributions and possibly even relative abundance, but not general community structure. Food input to the beach and surf-zone productivity may determine the abundance of populations. Future research should focus on the role of physical and biological factors on finer scales, through population studies and experimental work on ecophysiology and behavior.

The super-parameter controlling macrofauna on exposed beaches is water movement. In the case of ocean sandy beaches, this is vigorous wave-driven water movement experienced as swash on the beach face. It moves the macrofaunas, excavates and buries them, brings them their food, transports their reproductive products, provides most of the energy for their tidal migrations, controls the access of their predators, and influences other physical features such as sediment particle size and stability.

Although there is some evidence for modest biological interactions (such as competition, especially on fine scales and toward dissipative beaches, where conditions are relatively benign and densities are high), there is overwhelming evidence for overriding physical control of sandy-beach communities. In general, factors operating at large scales and community level will be density independent, whereas density-dependent factors may become more important at population level and finer scales. Understanding the community should therefore largely be a case of understanding the autecologies of the constituent species. By deciphering cause-and-effect relationships at the individual and population levels we should begin to understand the processes that underlie the large-scale community patterns outlined in this chapter. This is the topic of the next chapter.

Benthic Macrofauna Populations

8

8.1 Introduction

We have seen in Chapter 7 that macrobenthic communities on sandy beaches exhibit clear macroscale patterns that closely track changes in the physical environment along the gradient of beach types. This implies that to a large extent the ecology of the community is the sum of the independent ecologies of the constituent species (i.e., it is an autecological environment). It is therefore appropriate to examine in some detail the population ecology of sandy-beach benthic macrofauna species in terms of their response to the beach environment and their internal dynamics.

In this chapter we consider population ecology on three scales: macroscale (latitudinal trends and variability between beaches), mesoscale (within beaches), and microscale. We also consider the temporal dimension and examine those invertebrate populations that are sufficiently extensive to support artisanal or commercial fisheries. Much useful work has been done on all of these aspects of beach ecology over the past decade or more. Recent reviews that summarize information on these topics are McLachlan *et al.* (1996a) and Defeo and McLachlan (2005). For a full listing of the literature the reader is referred to these papers and the first edition. Our aim here is to synthesize general trends.

8.2 Macroscale Patterns

Latitude

Most sandy-beach macrofaunas have limited distributions, being restricted to one latitudinal or biogeographic region. However, a few species exhibit more extensive ranges and latitudinal trends have been demonstrated in some of these cases. These trends include variations in reproduction, morphology, genetics, distribution, and sediment preferences. Cardoso and Veloso (2003) compared 10 *Donax* species and found that mortality and P/B ratios increased and longevity decreased toward the tropics, and similar trends probably occur for populations of most species.

In *Emerita analoga* Dugan *et al.* (1994) showed some latitudinal trends in body size in response to changing water temperature along the Californian coast (as well as some effects of beach morphodynamics). Defeo and McLachlan (2005) synthesized such trends for two crustaceans: *Emerita brasiliensis* (which has planktonic larvae) and *Excirolana braziliensis*, which is ovoviviparous, both species occurring from tropical Brazil to temperate Argentina in South America (Figure 8.1). The trends displayed by these two species from temperate to tropical beaches include the following: increases in abundance, growth, and mortality; decreases in the proportion of females, size of ovigerous females, fecundity, length at maturity, mass, and longevity; and a shift from seasonal to continuous reproduction. Further, in *Emerita* body size decreases as population density increases over two orders of magnitude. However, male *Emerita* show some opposite trends toward the tropics: increasing abundance, larger size, and faster growth. These large-scale variations were ascribed to temperature effects associated with latitude in most cases. However, there are a few departures from these patterns where beach type plays a role.

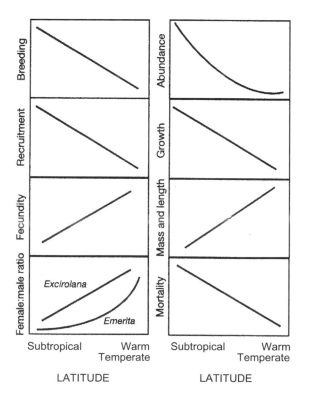

Figure 8.1. Latitudinal patterns in life-history traits of *Emerita brasiliensis* and *Excirolana braziliensis* on South American sandy beaches (after Defeo and Cardoso 2002, 2004; Cardoso and Defeo 2003, 2004; and Defeo and McLachlan 2005). Breeding and recruitment refer, respectively, to the frequency of occurrence of ovigerous females and of juveniles throughout the year. In *Emerita brasiliensis*, most of these patterns are valid only for females.

Latitudinal gradients in population characteristics are therefore caused primarily by temperature but can become asymmetrical due to parallel gradients in beach type (see Chapter 2) (i.e., more reflective beaches toward the tropics and more dissipative beaches toward temperate regions).

We have suggested that adaptations of sandy-beach species may include phenotypic plasticity and genetic stasis in behavior as well as other traits (Chapter 6). This ties in with the previously described latitudinal patterns, which show changes in life history traits over latitudinal temperature gradients. Plasticity may lead to separate populations displaying differences in behavior and life history, so that larval dispersal and connectivity between populations are important at this scale. Sandy-beach species have two chief modes of reproduction and development.

- Internal fertilization and direct development, sometimes with parental care and low fecundity (e.g., whelks and peracarids)
- External fertilization, high fecundity, and planktotrophic larvae (e.g., clams)

Species with larval stages (especially planktotrophic larvae) may be structured as metapopulations, with varying levels of connectivity between these populations. Although the metapopulation model has been considered appropriate for sandy-beach populations (Defeo and McLachlan 2005), it has also been questioned on the basis that wide dispersal and common larval pools result in well-mixed local populations (Hummel 2003, Frost *et al.* 2004). Connectivity, in terms of larval supply and dispersal, will influence latitudinal gradients in population dynamics, life history characteristics, and gene flow between populations. For example, the ovoviviparous isopod *Excirolana braziliensis* has significant genetic differentiation between scattered populations over its extensive range, possibly related to extinction/recolonization events. Unfortunately, connectivity between populations of sandy-beach macrobenthos and mechanisms of larval dispersal and settlement are poorly known. Colonizing larvae may play an important role in the colonization of different beach types, especially in seeding isolated reflective beaches that lack established adult populations.

Beach Types

Between-beach variability in macrobenthic populations can be considerable, and as one would expect in a physically controlled environment, this mirrors many of the patterns displayed at community level. Whereas some species can occupy all beach types, most prefer dissipative beaches and are less common or absent toward the reflective extreme. Exceptions include highly mobile crustaceans, such as *Emerita* and *Excirolana*, which may successfully establish populations on harsh reflective beaches. Although supralittoral species may differ, the responses of intertidal sandy-beach macrofauna species to the range of beach morphodynamics can be conceptualized as three categories (Figure 8.2).

- Beach-type *specialists* that may be delicate forms and/or slow burrowers occur only on beaches displaying reasonably benign dissipative characteristics and may even extend their range through ultradissipative beaches to occur on tidal flats (e.g., most polychaetes and almost all deposit feeders).
- *Intermediate forms* that require fairly vigorous swash action are most abundant on dissipative beaches but can colonize a wide range of intermediate beaches (e.g., many mollusks, such as donacid clams).

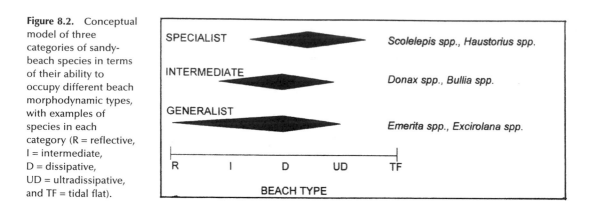

Figure 8.2. Conceptual model of three categories of sandy-beach species in terms of their ability to occupy different beach morphodynamic types, with examples of species in each category (R = reflective, I = intermediate, D = dissipative, UD = ultradissipative, and TF = tidal flat).

- Beach-type *generalists* that are highly mobile, rapid burrowers, and require vigorous swash action can establish populations on all beach types, even harsh reflective beaches (e.g., mole crabs and isopods of the genus *Excirolana*).

The hypotheses proposed to explain the exclusion of most species toward the reflective end of the beach gradient, outlined in Chapter 7, emphasize the role of increasingly harsh sediment and swash conditions. Coarse sand and harsh swash may select against small and delicate forms on reflective beaches. *Excirolana hirsuticauda*, for example, is larger on reflective beaches with coarser sand than on dissipative beaches in Chile (Jaramillo and McLachlan 1993). This increasing harshness in habitat toward reflective beaches should also affect population characteristics. Indeed, at the population level we would expect that increasing stress would cause a corresponding decline in vigor and life history traits toward reflective beaches. Individuals may divert more energy to survival and maintenance of position under harsh conditions than to growth and reproduction. This would be expressed as decreased growth rate, decreased size, and reduced reproductive output toward reflective beaches. This is the basis of the *habitat harshness hypothesis* (Defeo *et al.* 2003). There is mounting evidence in support of this. Even those generalist species that can colonize reflective beaches do, in fact, seem to respond to changes in beach type as predicted.

There are several studies that have quantified population and life history characteristics of robust generalist species over a range of beach types, showing increased abundance, biomass, fecundity, growth, and survival rates towards finer sands, flatter slopes, and dissipative states (e.g., Jaramillo and McLachlan 1993, Dugan and Hubbard 1996, Jaramillo *et al.* 2000, Brazeiro 2005, and others in Defeo and McLachlan 2005). Higher values for female somatic growth, fecundity, size at maturity, and weight at size in dissipative beaches (even in species that are successful on reflective beaches) support the *hypothesis* of a harsher habitat and less scope for growth on reflective beaches (Figure 8.3) (e.g., Dugan *et al.* 1994, Brazeiro 2005). This conforms to the *habitat harshness hypothesis*, which implies that post-settlement processes are responsible for most of the changes in population characteristics across beach types (Defeo *et al.* 2003).

Sandy-beach populations are often structured as metapopulations (i.e., they occur as benthic adult populations on separate beaches connected by pelagic larvae). In this situation, some populations may supply other populations with settling larvae. In particular, rich populations on dissipative beaches may seed sparse or ephemeral populations on

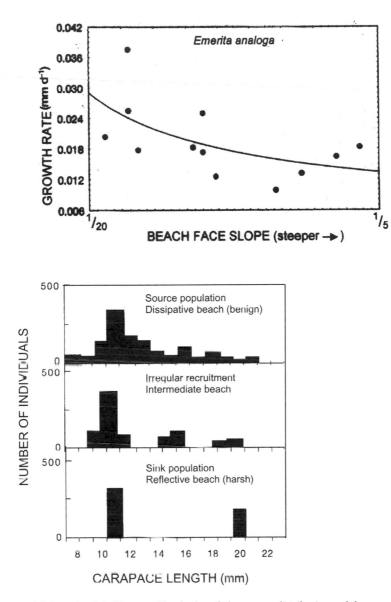

Figure 8.3. Individual growth rates of *Emerita analoga* in Chile decrease as beach slope becomes steeper and more reflective, suggesting harsher conditions toward reflective states (after Brazeiro 2005). The relationship GR = a[slope]^b is significant (p < 0.05).

Figure 8.4. The *source-sink hypothesis* is illustrated by the length frequency distributions of the mole crab *Emerita brasiliensis* from different beach types (after Caddy and Defeo 2003, Defeo and McLachlan 2005). Reproductive adults on dissipative beaches (sources, top) seed reflective beaches (sinks, center and bottom). Larval dispersal, retention in the parental spawning area, and differential post-settlement mortality rates determine the degree of connectivity between populations.

reflective beaches. This *source-sink hypothesis* (Defeo and McLachlan 2005) proposes that beaches with large and successful populations occur where conditions are optimal and water circulation is such that larvae from adult spawning are returned to the home beach. Such populations are important sources of larvae. Where environmental conditions are harsh (limiting reproduction) and/or there is no circulation to return larvae, sink populations are not able to seed themselves and are dependent on the source populations for recruitment. In sink populations, recruitment is therefore sporadic, generating irregular size structure — the cohorts present representing episodes of successful recruitment from source populations (Figure 8.4). Many sandy-beach species may exhibit such variability between well-established source populations and more impoverished sink populations, with beach length, beach type, and circulation patterns all playing a role.

It should be noted that supralittoral species may not follow these patterns. Supralittoral populations are not subject to the swash climate and may find coarser sand easier to burrow in than fine sand (Chapter 6). Furthermore, the backshore is more stable on reflective beaches than on any other beach type. It is not subject to erosion because this is an accretional beach state. In addition, being typically of low energy, reflective beaches are often associated with shelter provided by rocks, reefs, or other structures with macrophytes — which can result in large amounts of wrack on the backshore. For all of these reasons, supralittoral species may not show the same response to beach type as intertidal species. Indeed, for them the reflective beach may sometimes be more benign and the dissipative beach a harsher environment (Defeo and McLachlan 2005). Several talitrid populations support this: for example, *Pseudorchestoidea* (now *Atlantorchestoidea*) was found to have higher abundance, more gravid females, and larger individual size on reflective than dissipative beaches in Uruguay (Gomez and Defeo 1999, Defeo *et al.* 2003, Defeo and Gomez 2005), and *Orchestoidea* displayed higher abundance, more ovigerous females, and faster growth at an intermediate than a dissipative beach in Chile (Contreras *et al.* 2003).

8.3   Mesoscale Patterns

This scale concerns population patterns and processes within a beach, and we focus in this section on alongshore and across-shore patterns. Many studies examining the population biology of sandy-beach invertebrates have provided estimates of parameters such as growth, mortality, and production under a variety of conditions (see, for example, Figure 8.1). It is beyond the scope of this book to give a detailed account of these studies. Defeo and McLachlan (2005) review much of this material, and aspects of the population biology of beach clams are covered in Section 8.5. Further, Chapter 12 synthesizes work on energetics, including studies on the production or turnover of macrobenthos as well as other components of sandy-beach biota.

Alongshore

Variability within beaches is well developed in most populations in both alongshore and across-shore directions. Alongshore changes include responses to changing beach morphodynamics, such as decreases in abundance toward reflective conditions. The most typical alongshore gradients are changing exposure from one end of a beach to the other and salinity effects due to river mouths. Both can cause significant alongshore changes in populations. Although physical gradients are primarily responsible for alongshore variability, biological factors and disturbance by human activity can also play a role (Peterson *et al.* 2000, Degraer *et al.* 2003, Defeo and McLachlan 2005). A population on a beach tends to have aggregations in patches along the shore. The mode of alongshore distribution may be in the center of the beach or skewed toward one end if there are disturbances (Schoeman and Richardson 2002, Defeo and McLachlan 2005).

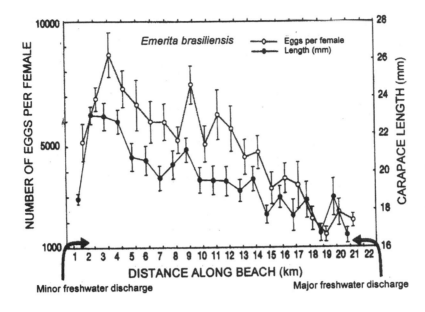

Figure 8.5. Effect of habitat unsuitability, measured as proximity to freshwater discharges, on individual size and fecundity of female mole crabs *Emerita brasiliensis* (after Lercari and Defeo 1999).

Tidal migration (displayed by many species) exposes the animals to longshore transport, which can result in significant drift and aggregations alongshore. This may vary widely between species and beaches. For example, *Emerita* has been shown to move longshore at up to 15 m·d^{-1} (Dilley and Knapp 1969), whereas tagged *Donax* showed little movement (Dugan and McLachlan 1999). Alongshore aggregation on a finer scale has been recorded in association with beach cusps and can take the form of patches that result from passive sorting by the cusp swash circulation (bivalves) or active migrations and habitat selection (hippid crabs) (McLachlan and Hesp 1984, Donn *et al.* 1986, James 1999, Gimenez and Yannicelli 2000).

A well-studied and striking example of alongshore variability concerns a 22-km dissipative beach in Uruguay with rich populations of clams (*Mesodesma* and *Donax*) and mole crabs (*Emerita*) but subject to disturbance, both as human harvesting and as freshwater discharge at each end of the beach. The freshwater discharge created unsuitable habitat through lowered salinities and increased erosion at each end of the beach, so that abundances of all species were highest toward the center of the beach. Fecundity in *Emerita* dropped off toward the ends (Figure 8.5), where mortality was highest (Defeo 1993, Lercari and Defeo 1999).

Another striking case of variability in longshore distribution concerns *Donax serra* in spiral bays in South Africa (Donn 1987, Schoeman and Richardson 2002). These clam populations tend to develop higher abundance and biomass toward the exposed eastern ends of the bays, where the morphodynamic state shifts from low intermediate to high-energy intermediate or even dissipative. There are also river mouth effects, with adult clams less abundant where river estuaries opened through the beaches. On a finer scale, this clam has a patchy distribution associated with rhythmic shorelines, tending to concentrate on flatter areas and to display semilunar migrations (Donn *et al.* 1986).

Patterns such as these have led to a *hypothesis* of habitat favorability (Defeo and McLachlan 2005), which is an extension of the *habitat harshness hypothesis*. It attempts

Figure 8.6. Illustration of the favorability aspect of the *habitat harshness hypothesis* (after Defeo and McLachlan 2005). In benign habitats (undisturbed dissipative beaches) populations are significantly influenced by density-dependent mechanisms (biological control), whereas populations in reflective or seriously disturbed beaches are entirely physically controlled. Source populations on benign dissipative beaches fluctuate less between years, whereas sink populations may disappear during years of poor recruitment.

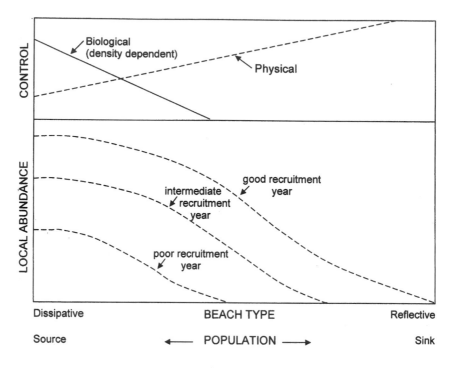

to explain some aspects of patchiness and mesoscale distribution of sandy-beach populations. In short, this proposes that in benign environments free from serious disturbances (such as pristine dissipative beaches) high macrofaunal densities can lead to biological interactions and density-dependent effects on growth and mortality, whereas in harsher or more marginal habitats densities are lower and there is less potential for biological interactions or density-dependent effects. Thus, peripheral populations and those in marginal habitats are largely physically controlled, whereas populations toward the center of a species range and in benign beach habitats will experience biological and density-dependent effects in addition to physical control (Figure 8.6).

Across Shore

Across-shore distribution at population level may include zonation in response to the intertidal aerial exposure gradient, aggregation in response to factors such as wrack input, and stratification of size classes — all made more complex and dynamic by the high degree of mobility of most species and their tidal migrations. Zonation across the intertidal gradient may be related to temperature and moisture tolerances and preferences, and to tolerance to air exposure. Although thermal tolerance and resistance in species of *Bullia* and *Donax*, for example, relate to latitude and tide level (Ansell 1983, Brown 1983a), much as has been found for rocky shores, most species have tolerance levels that exceed those necessary for survival in their habitats (Chapter 6). Ocypodid crabs are exceptions and may experience conditions close to their lethal limits. However, tolerance to air exposure is unlikely to be important in an environment in which exposure is escaped by burrowing. Vertical moisture zones in the sand must influence distribution of species on the upper shore, based at least partly on tolerances to desiccation. The ability of crustaceans, in particular, to utilize the upper intertidal and supratidal levels of sandy beaches must relate to their adaptations to avoid desiccation. These adaptations are mainly behavioral and include burrowing (to reach deep moist layers) and nocturnal activity (Chapter 6).

Maintenance of zones may be active or passive. Nel (2001) examined this in *Donax serra* by releasing live and dead adults in the mid-intertidal on an incoming tide and comparing their distribution on the next low tide. The steeper the beach face slope the greater was the tendency for dead and live clams to be washed down into the subtidal. For live clams, clam distribution was strongly correlated with the number of effluent line crossings by the swash, suggesting that on beaches with a higher proportion of swashes crossing the effluent line (i.e., tending toward reflective) more clams were displaced downshore, whereas under more dissipative conditions clams were better able to maintain position. This confirms the importance of swash climate at population level. In another example, displacement of a shallow burrowing polychate, *Armandia*, has been ascribed to passive landward transport by waves during winter (Tamaki 1987).

Many species exhibit a vertical separation of adults and recruits, possibly related to intraspecific interactions. In some cases, spat settle low on the shore and subsequently move up, but in other cases settlement is higher and followed by downshore movement. Small individuals may remain in the swash zone and be more dependent on passive sorting by the swash, whereas adults (being more mobile) may maintain position higher in the intertidal. This is not always the case, however.

An extensively studied case of size-class zonation is the clam *Donax serra* in South Africa. This species occurs on the west coast (an area marked by upwelling and temperate conditions) and along the warm temperate south and southeast coasts. On the west coast, where the species grows large (about 80 mm maximum shell length, as opposed to 70 mm in the southeast), adults occur mainly in the surf zone. Spat settle near the low-tide mark and move upshore as they grow, reaching maximum height on the shore at around 25 mm in shell length (Donn 1990a) and then moving down again later as they approach sexual maturity around 40 mm in shell length. On the southeast coast, in contrast, *Donax serra* spat seem to settle mainly in the shallow surf zone and move upshore as they grow, so that larger juveniles and adults (> 40 mm shell in length) occur on the upper midshore during spring tides but lower during neaps (Donn 1990a, Donn *et al.* 1986). These contrasting zonation patterns result mainly from different adult distributions, which in turn may result from different responses to the environment. West coast populations have flatter, more rounded valves that are thinner and lighter, resulting in lower density, which in turn implies decreased stability in the swash (Donn 1990b). Coupled with lower burrowing rates and lower temperatures (Donn and Els 1990), this means less ability to move in the swash and therefore confinement of west coast adults to the subtidal, where the ability to surf and undergo semilunar migrations is less critical. This tendency for large clams in cool waters to occur low on the shore is also exhibited in the South American clam *Mesodesma donacium*, where adults are subtidal and juveniles intertidal (Figure 8.7).

Vertical segregation of sandy-beach populations by size has been recorded in amphipods, isopods, mole crabs, and clams. This suggests both a differential capability of adults and recruits to select and maintain a suitable zone and possibly an evolutionary response to partition space and avoidance of intraspecific competition for space or food. Intertidal levels around the center of across-shore distribution probably have optimal conditions. These tend to be occupied by high densities of adults as dominant

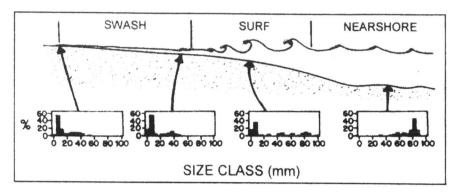

Figure 8.7. Typical across-shore zonation by size of the surf clam *Mesodesma donacium* on an exposed dissipative beach in southern Chile (after McLachlan and Jaramillo 1995).

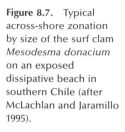

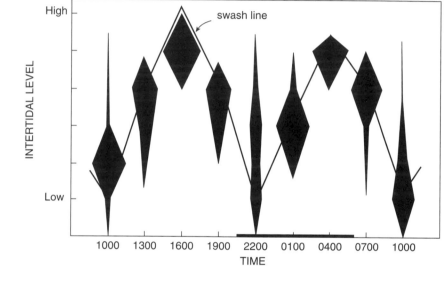

Figure 8.8. Tidal shifts in a population of the whelk *Bullia rhodostoma* on an exposed intermediate beach (after McLachlan *et al.* 1979). Solid line shows the upper position of the swash on the beach face over 24 hours.

intraspecific competitors. Smaller individuals may then be displaced up- or downshore to suboptimal microhabitats toward the extremes of the intertidal distribution range (McLachlan and Jaramillo 1995, Defeo and McLachlan 2005).

Other factors that can influence zonation are size differences between the sexes and predation. Differential zonation patterns between males and females have been recorded in *Emerita* (Veloso and Cardoso 1999). Thus, although across-shore distribution in a population is typically unimodal, a bimodal pattern may result where size segregation is pronounced. Movements of fishes and swimming crabs upshore on the rising tide and birds downshore on the falling tide can also cause shifts or modifications in zonation (Takahashi *et al.* 1999, Yu *et al.* 2003), as animals may actively select microhabitats that enable them to avoid predators (Takahashi and Kawaguchi 1998).

Temporal Changes

Whereas supralittoral species may shift their distribution in order to track the drift line and wrack deposits, many intertidal species show tidal migrations (Chapter 6) that result

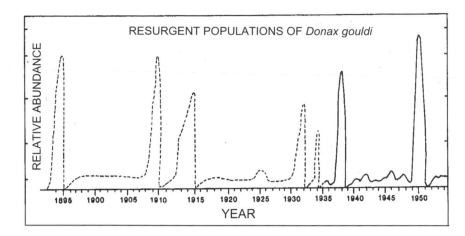

Figure 8.9. Resurgent populations of the bivalve *Donax gouldi* on a Californian sandy beach (after Coe 1955).

in shifts in distribution over the tidal cycle (Figure 8.8), as well as changes coupled to spring/neap and diel cycles and storms (Marsden 1991, McLachlan and Jaramillo 1995, Colombini and Chelazzi 2003). Orientation and activity rhythms are very important in establishing and maintaining zonation (Chapter 6).

In addition to the short-term fluctuations related to tides and storms, significant long-term temporal changes may occur in sandy-beach populations. Perhaps the most classical example of this is the resurgences in populations of *Donax gouldi* recorded on the Californian coast by Coe (1955) (Figure 8.9). Such dramatic changes (including mass mortalities) have been reported for several beach species, including *Donax serra* in South Africa (where mass mortalities are caused by red tides; de Villiers 1974), *Mesodesma donacium* in Peru following El Niño events (Arntz *et al.* 1988), and *M. mactroides* in Argentina, possibly caused by a coccidian parasite (Fiori *et al.* 2004).

Seasonal changes have been recorded for most of those species that have been studied over a period of time. On/offshore movements in winter have been reported in several species, possibly related to temperature declines and reproductive requirements, but are more likely to be an adaptation to avoid winter storms. Dugan *et al.* (2000) studied burrowing rates and behavior of three anomuran crabs and suggested that during harsh conditions intertidal species could shift to a shallow subtidal distribution, returning back into the intertidal when conditions were more benign. Species that can burrow rapidly (*Emerita*) could cope with most swash conditions, but slower burrowers (*Blepharipoda* and *Lepidopa*) experienced difficulty under harsh conditions and it was these latter species that moved into the shallow subtidal. Storms are not always associated with offshore migration, however, and some animals simply display deeper burrowing during the storm period. This is true, for example, of the mole crab *Hippa* (Shepherd *et al.* 1988).

8.4   Microscale Patterns

The microscale or quadrat scale covers interactions between individuals and responses to fine-scale environmental gradients. Perhaps the most important interactions on this scale are intraspecific interactions between adults and larvae of the same species. These

Figure 8.10. Scatter diagram of *Mesodesma mactroides* recruit density plotted against adult density in each quadrat during recruitment periods. Recruit densities peaked under low adult densities and high fishery extraction levels between 1983 and 1987 (•), but were low in 1989 to 1990 (○) as a result of fishery closure (after Caddy and Defeo 2003 and Defeo and McLachlan 2005). The broken line defines the envelope constraining stock and recruitment.

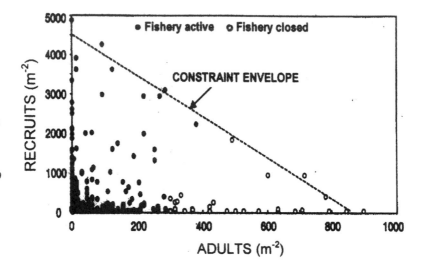

can occur as a result of intraspecific competition for space or food, or by adults filtering out larvae during feeding — in which case greatest effects will be found at highest adult densities. This has been demonstrated on a sandy beach for the clam *Mesodesma* (Defeo 1993, 1996). Maximum recruit densities in *Mesodesma* are inversely proportional to adult densities, and adult densities constrain the number of recruits (Figure 8.10) and thus probably settlement. This indicates inhibition of recruitment at high adult densities, probably due to active filtering of larvae.

Intraspecific interactions (in the form of inhibition of burrowing) have been shown experimentally in donacid clams, but this effect could only be demonstrated at extreme densities, above those recorded in the field (McLachlan 1998). Nel (2001) examined intraspecific interactions in *Donax serra* in more detail and showed significant interactions between adults and juveniles. Clams tended to burrow slowest into high background densities of their own size, and large sizes were most affected. However, experiments monitoring the numbers of juvenile clams recovered after varying times in adult patches of *Donax serra* showed no displacement of juveniles by adults. Recovery of juveniles did not change at different adult densities but rather physical factors, in particular wave energy over the study period, explained any differences in recovery (Nel 2001).

Evidence for intraspecific interference competition between adults and juveniles on sandy beaches is therefore conflicting. The third dimension (vertical distribution in the sand) gives another mechanism of space partitioning, and because recruits cannot burrow as deeply as adults this may serve to reduce intraspecific competition.

Fine-scale aggregations with positive effects have also been recorded in sandy-beach macrofauna. The tropical trochid gastropod *Umbonium vestiarium* forms conspicuous aggregations on sand flats, appearing as mounds formed by high densities of snails fixing sediment with their mucus (Huttel 1986). These aggregations may assist in feeding and may prevent snails from being washed out of their zone. Similarly, Sastre (1985) showed fine-scale aggregations in *Donax denticulatus*, which he considered mechanisms to ensure fertilization.

8.5 Invertebrate Fisheries

Sandy-beach invertebrate populations have long been harvested where the target species are sufficiently large and abundant. There are a few regions where crustaceans, typically hippid and ocypodid crabs, are collected on a small scale and eaten. However, all significant sandy-beach fisheries are based on clams. Beach clam fisheries have been reviewed by McLachlan *et al.* (1996a) and Defeo (2003). Where they are locally abundant, beach clams can attain very high biomass values and completely dominate the sandy beach macrofauna. The 15 clam species harvested (Table 8.1, Figures 8.11 and 8.12) are

Table 8.1. Summary of key features of beach clam fisheries. Under habitat I = intertidal, S = subtidal. Under fishery type A = artisanal, C = commercial, R = recreational, and * = no longer operational. Lengths are in mm, life spans in years, and landings are in tons live weight per year. (After McLachlan *et al.* 1996a.)

Species	Habitat	Length at maturity	Lifespan	Legal size	Max length	Fishery type	Maximum landings	Common name
Donax deltoides	I & S	36	3.5	None/35	60	C & R	Unknown	Pipi
Mesodesma mactroides	I	43	3.5	50	85	A, C & R	1078 (1953)	Yellow clam
Mesodesma donacium	S	50	15	50–60	100	A & C	18000 (1989)	Macha
Tivela stultorum	I & S	25	> 20	114–127	187	R & C	24000 (1945)	Pismo clam
Siliqua patula	I & S	100	5 19	85–114	170	C & R	3600 (1915)	Pacific razor clam
Donax serra	I & S	44	> 5	58	80	R & C*	20 (1968)	White sand mussel
Donax trunculus	I & S	8	1.5–> 6	30	44	R & A	Unknown	Wedge clam
Tivela mactroides	I & S	20	1.5	—	35	C & R	354 (1992)	Guacuco
Donax denticulatus	I	10	1.5	—	23	A	226 (1990)	Beach clam
Donax striatus	I	10	1.5	—	35	A	Unknown	Chipi-chipi
Donax cuneatus	I	12	< 3	—	23	A	Unknown	—
Donax faba	I	10	< 3	—	27	A	Unknown	—
Paphies ventricosa	I	40	> 10	100	160	R* & C*	185 (1940)	Toheroa
Paphies subtriangulata	I & S	Not known	> 5	None	85	R & C	131 (1993)	Tuatua
Paphies donacina	I & S	Not known	17	None	109	R & C	5 (1992)	Tuatua

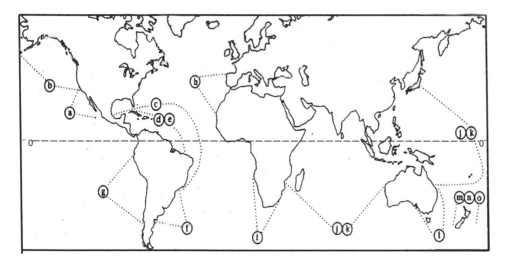

Figure 8.11. Geographical distribution of 15 species of harvested beach clams (after McLachlan *et al.* 1996a). North America: (a) *Tivela stultorum* and (b) *Siliqua patula*. Caribbean: (c) *Tivela mactroides*, (d) *Donax denticulatus*, and (e) *D. striatus*. South America: (f) *Mesodesma mactroides* and (g) *M. donacium*. Europe: (h) *Donax trunculus*. Africa: (i) *D. serra*. Asia: (j) *D. cuneatus* and (k) *D. faba*. Australia: (l) *D. deltoides*. New Zealand: (m) *Paphies ventricosa*, (n) *P. subtriangulata*, and (o) *P. donacina*.

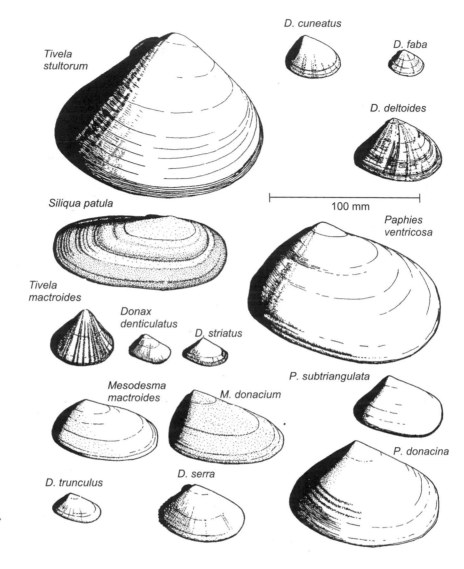

Figure 8.12. The 15 species of harvested beach clams (after McLachlan *et al.* 1996a, sketch by Dave Hubbard).

distributed around the globe but show some distinct patterns that allow them to be placed into two categories.

- Large temperate species that dominate the biomass on dissipative beaches, mostly occurring in the shallow surf zone
- Smaller warmwater species, mainly donacids, found on reflective and intermediate beaches and mostly occurring in the intertidal

The warmwater species support artisanal fisheries, whereas the larger temperate forms may support artisanal, commercial, or recreational fisheries. It is important to distinguish among these different types of fisheries: a *recreational* fishery is based on collection for bait or food without sale or dependence on the resource, an *artisanal* fishery is based on collection for subsistence or sale by individuals or groups using traditional methods, and a *commercial* fishery is collection for sale by corporate or collective organizations. These distinctions are not precise. In many cases, artisanal fisheries are also commercial (e.g., Chile, Uruguay), and recreational fisheries often also support commercial activities.

Beach clams are broadcast spawners with external fertilization, high fecundity, and planktonic larvae. In almost all cases juveniles tend to settle at a different level before migrating to the adult zone. Most stocks are structured as metapopulations with variable recruitment, as per the *source-sink hypothesis*. These populations can undergo large fluctuations due to environmental, density-dependent, and human factors such as climatic events, toxic algal blooms, viral diseases, and overexploitation.

There appears to be a good relationship between tide level occupied by adult clams and water temperature, those occurring in the warmest waters (*Donax denticulatus*, *D. striatus*, *D. faba*, and *D. cuneatus* from tropical areas and *Mesodesma mactroides*, *Donax serra*, *D. trunculus*, *D. deltoides*, and *Tivela stultorum* from warm temperate areas) being either intertidal or tidal migrants, whereas those from higher latitudes (*Siliqua patula*, *Mesodesma donacium*, *Paphies donacina*) are mixed or subtidal. Burial time increases with size in bivalves, making repositioning and burial on the beach face difficult or impossible for large species at low temperatures. The subtidal shift and absence of tidal migrations in populations in areas of lower water temperature may therefore be a consequence of large size and slower burial rates.

Bivalves from intermediate and dissipative beaches include large forms with low densities and variable shapes, in contrast to the small, dense, wedge-shaped forms adapted to cope with the turbulent swash conditions of reflective beaches (McLachlan *et al.* 1995). The species supporting substantial beach clam fisheries in temperate areas tend to be large, long-lived, and with high growth rates, in contrast to the tropical species, which are short-lived. The exceptions are *D. serra* on the south coast of South Africa, *D. deltoides* in Australia, and *D. trunculus* in Europe, which are intermediate between the bulky forms from dissipative beaches and the small dense forms from reflective beaches such as those found in the tropics. This may explain the ability of these three species to colonize a fairly wide range of intermediate beach types and to maintain intertidal positions.

Population dynamics of species supporting beach fisheries are not particularly well studied, and in many cases data on recruitment, stock size, mortality, and growth are incomplete or absent. Exceptions are *Tivela stultorum*, *Siliqua patula*, *Donax serra*, *Paphies ventricosa*, *Mesodesma mactroides*, and *M. donacium*. Recruitment is the least studied aspect of beach clam fisheries and has been shown to be stock dependent in *M. mactroides* but may be independent in *Tivela stultorum*. Most species have rapid growth to sexual maturity, and then slower indeterminate growth, and various growth models have been successfully applied. Mortality rates vary considerably between species, beaches, and years, and total mortality can be high, ranging from 0.5 to $3.1\,\mathrm{y}^{-1}$. Fishing mortality makes up a significant proportion of this, and incidental mortality (clams not taken but discarded on the surface to die) can be a major cause of mortalities. Density-dependent processes play a role through adults filtering out larvae, inhibition of growth, and increased mortality at high densities.

Most stocks are already fully exploited, overexploited, or depleted and only support small-scale artisanal fisheries — the exception being the macha fishery in Chile (Figure 8.13). Intertidal species are readily accessible and harvesting has low operating costs. Thus, they are particularly vulnerable to overexploitation and require careful management. In contrast, species occupying high-energy surf zones are relatively inaccessible and less management is required. In recreational fisheries, limits to the number of fishers are usually not possible and managers have made recourse to size and bag limits, restric-

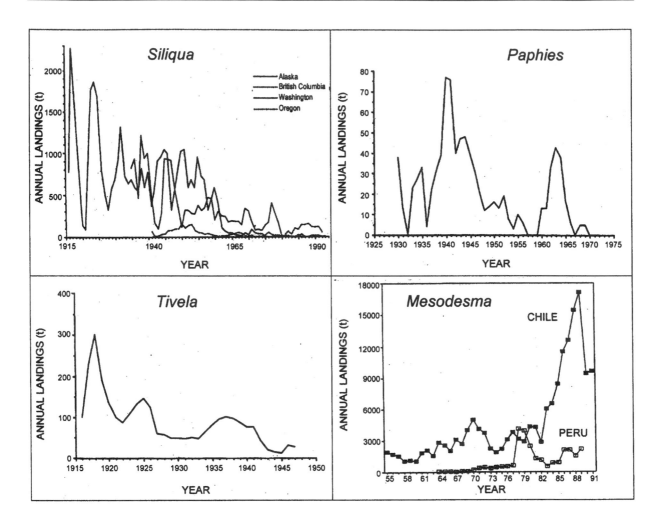

Figure 8.13. Trends in some major beach clam fishery landings: *Siliqua patula* in northwest America, *Paphies ventricosa* in New Zealand, *Tivela stultorum* in California, and *Mesodesma donacium* in South America (after McLachlan *et al.* 1996a).

tions on collecting gear, and closed seasons and areas. However, for many of these fisheries research and management are limited. In a few areas, restocking or enhancement has been tried as a management tool, but on the whole the idea of reseeding beaches has not been tested and in most cases larval biology is unknown.

Sandy beaches are scenic and ideal for the development of recreational activities, and recreational fisheries have tended to replace commercial fisheries (e.g., *Siliqua patula*, *Tivela stultorum*) as beach recreation becomes more popular and human population increases. However, recreational activities on beaches can clash with commercial fisheries, especially for intertidal species in warm climates (*Donax serra*, *D. deltoides*), where beach recreation is extremely popular. For subtidal species this is less of a problem because fishing effort is based below the intertidal beach. All beach clam fisheries include some recreational components, but artisanal fisheries predominate in South America and some tropical areas. Commercial fisheries remain important only in the case of *Mesodesma donacium* in Chile. The tendency in beach clam fisheries is away from large commercial operations and toward more recreationa/artisanal fisheries, which have fewer ecological impacts but are difficult to control.

Beach clams are used for food and bait. Best prices are obtained as food, and clams can reach U.S. $30 kg^{-1} in their shells. A problem affecting commercial fisheries is the tendency of several species to accumulate toxins produced by blooms of certain species of phytoplankton. High levels of toxins are harmful and pose health hazards, limiting the potential utilization of some stocks in cooler waters. As human impacts on coastal waters continue, blooms of toxin-producing phytoplankton could affect beach clam fisheries even more, requiring careful monitoring.

8.6 Conclusions

Sandy-beach macrobenthic populations are mobile and dynamic in space and time. They tend to exhibit similar responses to physical features and morphodynamics (as do the communities), confirming the general validity of the *autecological hypothesis* for these physically controlled environments. However, supralittoral species may show the opposite trends to intertidal species because they are not strongly influenced by the swash. Populations tend to display patchiness both alongshore and across shore. At the fringe of their distribution ranges, populations may be completely physically controlled, whereas in more optimal locations (especially under benign dissipative conditions and high densities) biological factors such as intraspecific competition can become more important. Nevertheless, there is still a considerable degree of physical control on sandy-beach populations and primary adaptations of most species are to the physical environment. In that responses at population level underlie the clear patterns at community level, it is at population level that the most useful future work is likely to be done to improve our understanding of sandy-beach ecology.

Interstitial Ecology

9

9.1   Introduction

In contrast to the wave-swept surface sand inhabited by most of the macrofauna, the lacunar interstitial system is complex and three-dimensional, often having great vertical extent in the sand body. This porous system averages about 40% of the total sediment volume. Its inhabitants include small metazoans forming the meiofauna, as well as protozoans, bacteria, diatoms, and other microorganisms. The meiofaunas are defined as those metazoan animals passing undamaged through 0.5 to 1.0-mm screens and trapped on 30- to 100-μm screens. On most beaches, the interstitial fauna is rich and diverse, even exceeding the macrofauna in biomass in some cases. On sandy beaches there may be as many as 25 interstitial species for every macrofauna species (Armonies and Reise 2000). The reason for this diversity is the greater stability and complexity of the interstitial habitat.

The interstitial environment experiences a continuum of conditions ranging between chemically and physically controlled extremes (Chapter 3). The chemical extreme is represented by sheltered beaches of fine sand, where water filtration through the sand is negligible and organic inputs may be high. Here, the consumption of oxygen by the sand fauna and flora exceeds the meagre supply, which is limited by poor wave action and low permeability. Consequently, the sediment becomes deoxygenated and steep vertical gradients in interstitial chemistry develop. Oxygenated sand occurs at the surface and reduced layers a little way below, so that the interstitial fauna concentrates at the surface and is adapted primarily to the chemical stresses of this "stagnant" environment. Low-energy dissipative beaches and sand flats exhibit these conditions.

The physical extreme occurs on coarse-grained steep beaches subject to strong wave action, where huge volumes of water are flushed through the beach face and drain rapidly back to the sea because of high sand permeability. The interstitial system here is always highly oxygenated and drains out during the low tide. Whatever the organic inputs, oxygen demand thus never exceeds supply. Here, the interstitial fauna has a deep vertical distribution — up to several meters into the sediment — and its adaptations are primarily to the physical environment (i.e., high interstitial flow rates, desiccation during low tide, and disturbance of the sediment by waves). Medium to high-energy coarse-grained reflective beaches exhibit these conditions. Most beaches are intermediate between these two extremes, with elements of both chemical and physical gradients shaping the interstitial climate.

9.2   Interstitial Climate

The major process controlling interstitial climate is the filtration of seawater through the sand. This is determined both by the wave climate and by sediment properties. As described in Chapter 3, water is introduced into the sediment either through flushing by swashes and tides in the intertidal zone or by the pumping effects of waves in the subtidal. Progressing from reflective to dissipative beaches, the filtered volumes decrease over the range 100 to less than $1 \, m^3 \cdot m^{-1} \cdot day^{-1}$, whereas the residence times of this water in the intertidal sand increase over the range 1 hour to many days. Thus, coarse-grained steep (reflective) beaches have large volumes of seawater flushing rapidly through them, whereas fine-grained flat (dissipative) beaches receive less water, which percolates more slowly.

This water input concentrates organic material in the sand. Toward the chemically controlled extreme, organic input exceeds oxygen supply and the sediment becomes deoxygenated, developing strong chemical gradients. Toward the physically controlled extreme, water and oxygen inputs exceed organic input and the interstices remain open as high-energy windows. The contrasting vertical gradients in chemistry that develop in beaches of these two types are illustrated in Figure 3.16.

The interstitial system is subject to cyclic changes related to storm/calm, tidal, diel, and seasonal cycles. In physically controlled beaches, this results in fluctuations in the water table, pore moisture content, and surface temperature, whereas toward the chemical extreme it may result in sharp changes in chemical gradients. During warm conditions, the reduced layers may move toward the surface, whereas storms or photosynthetic activity by surface diatoms can drive these layers deeper.

On exposed beaches toward the physical extreme, the great vertical extent of the system and the drainage it experiences at low tide allow subdivision of the sand body into layers or strata. In all cases, the layers range from dry surface sand at the top of the shore to permanently saturated sand lower down. The permanently saturated layers have little water circulation and tend toward stagnation, the resurgence zone has gravitational water draining through it during the ebb tide, the retention zone loses gravitational water but retains capillary moisture during low tide, and the zones of dry sand and drying sand lose even capillary moisture. The zone of retention represents optimum conditions for the interstitial fauna, there being a good balance between water, oxygen, and food input, physical stability, and lack of stagnation (see Chapter 3, Figure 3.18).

Interstitial organisms thus experience a lacunar environment of perpetual darkness. Space and oxygen are the two main limiting factors. Toward the physical extreme, strong pulsing currents, extreme changes in pore moisture and movement, and the grinding of sand grains require special adaptations such as adhesion tubes, spines, and setae in nematodes (Gheskiere *et al*. 2005). At the other extreme, in more stagnant conditions limited oxygen and toxic reduced compounds (H_2S, NH_4) need to be tolerated.

9.3   Sampling

Interstitial faunas are usually sampled by taking cores of sediment with a handheld corer. For meiofauna on ocean beaches, corers with an internal cross-sectional area of $10\,cm^2$ (internal diameter 3.58 cm) are typical, but narrower ones have also been used. Smaller corers are normally used for microfauna. Where only the surface layers are of interest, the corer may be thrust 10 to 30 cm into the sand and the core extracted. However, on exposed beaches tending toward the physical extreme deep meiofauna penetration necessitates cutting steps and taking a vertical series of cores far down into the sediment body, or using special corers capable of taking cores 50 cm to 1 m long.

Sediment extracted from the corer is treated chemically to relax (MgCl) or to fix (4% formalin or 96% alcohol) the fauna before separating it from the sand. Meiofauna is usually extracted from the sand by elutriating or decanting several times (i.e., suspending the sand in seawater and passing the supernatant through a mesh of 30 to $65\,\mu m$ to collect the meiofauna, including larger protozoans such as foraminiferans and ciliates). Bacteria and smaller protozoans tend to adhere tightly to sand grains and need to be separated by subjecting small samples to sonication or shaking before filtering on fine filter paper ($0.5\,\mu m$). Whereas meiofaunas are usually stained with a red dye, Rose Bengal, or Eosin to facilitate counting under a normal dissecting microscope, bacteria and small protozoans are typically stained with a fluorescent stain and counted under a compound fluorescence microscope. Identification of meiofauna below the level of broad taxonomic groups is complex and usually requires a specialist. Methods for the study of meiofauna are described in detail by Higgins and Thiel (1988) and Vincx (1996).

9.4   Interstitial Biota

Large numbers of microscopic organisms occupy the interstices: fungi, algae, bacteria, protozoans, and metazoans. Individual dry weights range from 10^{-10} to $10^{-12}\,g$ for bacteria, 10^{-6} to $10^{-11}\,g$ for protozoans, and 10^{-5} to $10^{-8}\,g$ for meiofauna. Analysis of the entire size spectrum of sandy-beach benthic organisms consistently shows three size peaks corresponding to bacteria, meiofauna, and macrofauna (Schwinghammer 1981, Warwick 1984), with discontinuities between grain colonizers and interstitial forms and between interstitial forms and macrofaunal burrowers. These patterns are most clear in marine beach sands (Figure 9.1), and much less clear in very fine sands and muds. Meiofaunas have a modal dry body mass of $0.6\,\mu g$ or an equivalent spherical diameter of $125\,\mu m$. Bacteria have a modal equivalent spherical diameter of $0.5\,\mu m$ and a mean dry mass around $10^{-11}\,g$. The macrofaunas are more variable but typically have equivalent spherical diameters of about 2 mm and a modal biomass around 3 to 4 mg.

Fungi occur mainly near the dune margin and do not penetrate far into the marine system. They play an important role in primary colonization and aggregation of dune

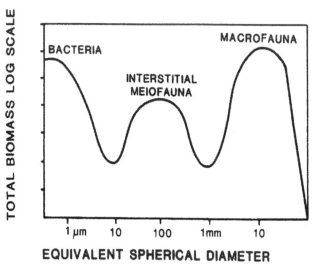

Figure 9.1. Size spectrum of benthic sandy-beach biota (after Schwinghammer 1981 and Warwick 1984).

sands. Their spores occur in sea foam and are well adapted to the beach by the attachment of fruiting bodies to sand grains. Bubbles generated in the surf scavenge the spores and hyphae of fungi and coalesce on the sea surface as foam, which is carried inland to inoculate the beach.

A diverse flora sometimes develops on the surface of sheltered sands (see Chapter 4). This may include cyanobacteria, cryptomonads, euglenoids, phytomonads, dinoflagellates, and diatoms. Living microflora in the sand can penetrate to a depth of 20 cm and extend some way into the sublittoral zone. Small diatoms attach to sand grains well below the surface, and in some cases undergo vertical migrations to and from the surface. *Hantzschia* cells move to the surface during the diurnal low tide and back into the sediments before being inundated. Similar movements have been recorded for several other diatoms. The cells of surf diatoms from exposed beaches (*Anaulus, Chaetoceros, Asterionella, Aulacodiscus*) enter the sand at night and are often abundant in intertidal and subtidal sand.

Bacteria in the sand are mostly attached to the grains. Numbers range from 10^8 to 10^{10} cells per gram dry sand and increase with finer sand and greater surface area. There are many different strains responsible for a variety of decomposition processes, and autotrophs using light or chemical energy to synthesize organic compound. The chemoautotrophs are particularly important near the boundary of the reduced layers (Fenchel and Riedl 1970). Nitrite bacteria oxidize NH_4 to NO_2, and nitrate bacteria oxidize NO_2 to NO_3, which in turn may be converted to N_2 gas by denitrifying bacteria. N_2 gas can be converted back into NO_3 by nitrogen-fixing bacteria (Appendix B). In the sulphur cycle, SO_4 is reduced to H_2S anaerobically by desulfovibrio bacteria and H_2S is oxidized to SO_4 by thiobacilli. Through these and other processes, sedimentary microorganisms play an important role in the decomposition of a wide spectrum of organic compounds and in the recycling of inorganic nutrients.

A wide range of protozoans may be present in the interstitial system, but ciliates and foraminiferans are usually the most abundant. They are often relatively large and especially important on beaches of fine sand. Amoebae and zooflagellates are less important. Protozoan numbers generally range from about 10^{-1} to 10^3 per gram dry sand, and they

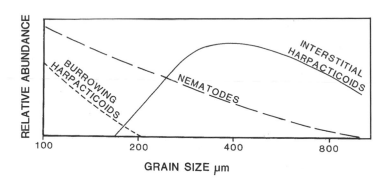

Figure 9.2. Conceptual model of response of sandy-beach meiofauna to a particle size gradient (after McLachlan *et al.* 1981a).

can occur deep into the sediments. Special adaptations may include elongate shape in larger forms (1 mm), special fields of cilia to cling to smooth surfaces, loss of light-sensory structures, and tolerance of reducing conditions (Fenchel 1978). Interstitial protozoans may feed on bacteria, diatoms, other protozoans, and even metazoans.

Meiofaunas constitute the best-studied component of the interstitial biota. These small metazoans are considered temporary meiofauna if they are larval stages of macrofaunal forms and permanent meiofauna if their entire life cycle is spent in this size category of benthos. In finer, more sheltered, sediments temporary meiofauna may be particularly abundant during certain seasons, but in more dynamic situations (where all the meiofauna is truly interstitial) temporary forms are usually unimportant. The dominant taxa of sandy-beach meiofauna are nematodes and harpacticoid copepods, with other important groups including turbellarians, oligochaetes, mystacocarids, gastrotrichs, archiannelids, ostracods, mites, and tardigrades. In capillary sediments (clean sands), these all tend to be interstitial forms, whereas in noncapillary sediments (mud and muddy sand) they may be too large for the interstices and resort to burrowing, forcing a passage through the sand by pushing the grains inside.

Body size in meiofauna tends to decrease as grain size, and consequently pore size, decreases (Swedmark 1964). However, there is a lower size limit below which the metazoan body plan cannot go. In sands coarser than 200 μm median particle diameter, virtually all the meiofaunas are interstitial forms, moving between the sand grains. Harpacticoids can remain interstitial down to a particle size of between 160 and 170 μm and nematodes (with their more slender shape) can pursue the interstitial mode in sands as fine as 100 to 125 μm. Below 100 to 125 μm mean particle diameter, no interstitial fauna remains. Purely interstitial groups (e.g., Gastrotricha) are therefore excluded from very fine sands, whereas exclusively burrowing groups (e.g., Kinorhyncha) are excluded from medium to coarse sands (Coull 1988). In sands finer than 300 μm, nematodes are usually dominant, whereas in sands coarser than 350 μm copepods are usually more abundant (these two groups often being of similar importance in sands of 300 to 350 μm). Most marine beaches have sands coarser than 200 μm, and consequently an almost exclusively interstitial meiofauna. Some of the previously cited trends are summarized in Figure 9.2.

Unlike the robust burrowing forms, interstitial meiofaunas are slender and vermiform (Figure 9.3). Being near the lower limits of body size for metazoans, they have reduced cell numbers and display simple organization. Size may be reduced to as little as 0.2 mm total length. Body walls are reinforced by cuticles, spines, or scales, and protection

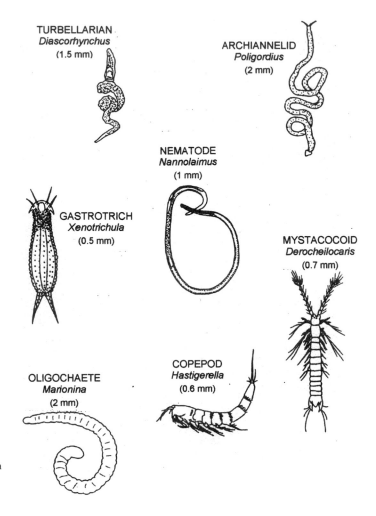

Figure 9.3. Some characteristic meiofauna genera found in sandy beaches.

against abrasion is also afforded by the ability to contract. Adhesive organs allow many forms to anchor to sand grains in the face of turbulence. Locomotion can be by ciliary gliding, writhing caterpillar-like (in the nematodes), or crawling (in the crustaceans). Because of low cell numbers, gamete production is low and consequently fertilization is often internal, brood protection common, and pelagic larvae absent. A wide variety of feeding types occur among the meiofaunas, including omnivores, predators, diatom and bacterial scrapers, detritus feeders, and suspension feeders.

Meiofauna numbers typically average $10^6 \, m^{-2}$ in marine sediments, and sandy beaches are no exception. They may display numbers as low as $0.05 \times 10^6 \, m^{-2}$ or as high as $3 \times 10^6 \, m^{-2}$, lowest values generally being obtained from beaches near the ends of the physical-chemical gradient and highest values from intermediate beaches that have reasonable organic inputs.

9.5 Distribution of Interstitial Fauna

There is little information on large-scale patterns in sandy-beach meiofauna distribution, such as latitudinal effects or overall responses to the range of beach types (but see Section 9.7). However, within beaches horizontal and vertical distribution shows clear patterns.

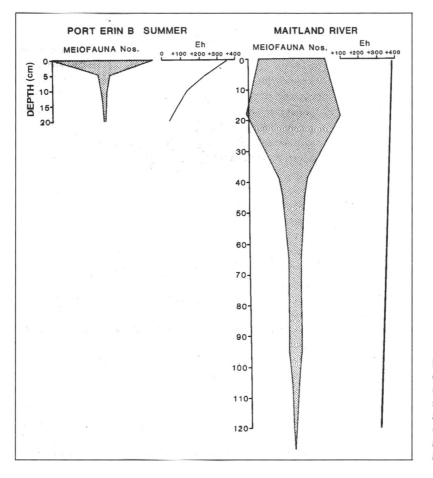

Figure 9.4. Vertical distribution of meiofauna in a sheltered (Port Erin) and an exposed beach (Maitland River) (after McLachlan 1978 and McLachlan *et al.* 1979b).

The vertical distribution of meiofauna is determined by the degree of drainage and oxygenation of the sediment. In low-energy environments and muddy areas the interstitial fauna is concentrated in a thin layer controlled by the depth of the RPD zone. This is also the case for beaches near the chemical extreme (i.e., flat fine-grained beaches). Oxygen is the major controller of the vertical chemical gradients in such systems. Different meiofaunal groups differ in their tolerances of reduced oxygen tensions and consequently in their vertical extent across such gradients. Harpacticoid copepods are the most sensitive taxon to decreased oxygen tensions, whereas nematodes are more tolerant (Giere 1993). Some forms (e.g., gnathostomulids, gastrotrichs, and nematodes) can penetrate below the RPD.

Toward the physical extreme, where the interstices are well oxygenated, the interstitial fauna may extend as deep as oxygenated conditions persist — sometimes several meters. Figure 9.4 contrasts the vertical distribution of meiofauna in these two different beach types. In sublittoral sediments, including surf zones, vertical distribution is more compressed, although meiofauna may still be abundant 10 to 30 cm into the sediment.

Horizontal distribution may take the form of zones (sheltered beaches) or layers (exposed shores) (Figure 9.5). The layers of exposed beaches correspond closely to the layers or strata shown in Figure 3.18, whereas the zones across sheltered shore apply essentially to changes in the fauna across the surface layers. In the exposed case (Figure

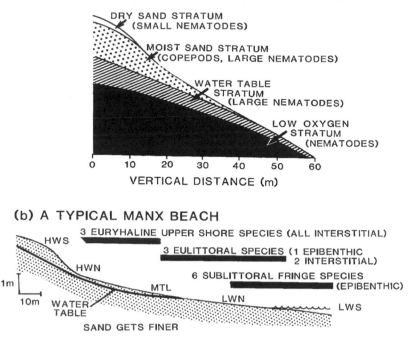

(a) A TYPICAL EAST COAST CAPE BEACH

DRY SAND STRATUM
(SMALL NEMATODES)

MOIST SAND STRATUM
(COPEPODS, LARGE NEMATODES)

WATER TABLE
STRATUM
(LARGE NEMATODES)

LOW OXYGEN
STRATUM
(NEMATODES)

0   10   20   30   40   50   60
VERTICAL DISTANCE (m)

(b) A TYPICAL MANX BEACH

3 EURYHALINE UPPER SHORE SPECIES (ALL INTERSTITIAL)

HWS

3 EULITTORAL SPECIES (1 EPIBENTHIC
2 INTERSTITIAL)

HWN

6 SUBLITTORAL FRINGE SPECIES
(EPIBENTHIC)

1m

MTL

LWN

10m   WATER
TABLE   LWS

SAND GETS FINER

Figure 9.5. Comparison of vertical and horizontal distribution of meiofauna in (a) a well-drained exposed intermediate beach and (b) a sheltered macrotidal beach where distribution tends to be three-dimensional and two-dimensional, respectively (after McLachlan 1980a and Moore 1979). In both cases, copepods are concentrated above the low-tide water table. On the sheltered (Manx) beach, the interstitial fauna is mainly concentrated high on the shore, where some drainage occurs. Nevertheless, meiofaunal zonation on this sheltered beach is horizontal rather than vertical.

9.5a), faunal distribution, abundance, and composition within each layer or stratum are not uniform and some horizontal changes occur in each layer. The richest fauna is found in the moist sand, or retention zone, in the mid to upper part of the beach, where optimal physical conditions occur and microbial activity is also greatest.

The meiofauna of mud, including copepods and nematode worms, has been shown to occur in the water column, due both to passive erosion processes (Savidge and Taghon 1988) and to active behavioral emergence (Ullberg and Olafsson 2003). This has also been demonstrated for sandy-beach meiofauna and is almost certainly of great importance in dispersal, particularly of those species lacking planktonic larvae. Fegley (1988), working on a sheltered sand flat in North Carolina, has shown that the estimated density of meiofauna settling from the water column greatly exceeded the observed density of meiofauna recolonizing a defaunated area. This suggests that other factors, such as meiofaunal behavior or microbial dynamics, are more important in determining recolonization than the magnitude of meiofaunal drift. Passive dispersal may, in fact, be of greater importance on surf-swept beaches and active dispersal prevalent in less turbulent, sheltered areas (de Patra and Levin 1989).

9.6 Temporal Changes

Movements of interstitial fauna through the sand have been recorded on tidal, diel, and seasonal time scales. Tidal migrations of meiofauna are mostly coupled to changes in the

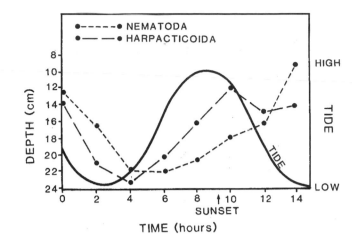

Figure 9.6. Tidal migrations of interstitial nematodes and harpacticoid copepods in the upper 24 cm of sand on an exposed sandy beach (after McLachlan *et al.* 1977). Broken lines show movements of the centers of gravity of the populations in response to the drying and wetting of the sand.

moisture content of the sediment, the animals often moving upward and upshore as the tide rises and the sand fills with water (Figure 9.6). This seems primarily related to avoidance of desiccation in the surface layers during the low tide. Similar vertical movements have been noted in response to rain and wave disturbance. These tidal movements are most marked where sands are well drained and the fauna has a deep vertical distribution. In waterlogged fine-sand beaches little vertical movement of the meiofauna occurs in time scales as short as those of tidal cycles.

Longer-term movements may relate to diel temperature changes, storm-calm cycles, or other factors. Because benthic metabolism is increased by higher temperatures, warmer conditions increase oxygen consumption in the sediment and raise the redox layer toward the surface (Giere 1993). This may cause the fauna to migrate upward so as to remain in the oxygenated layers. Alternatively, where there is high primary production by epipsammic diatoms and other microflora the generation of oxygen during daylight may drive the RPD downward. Epipsammic diatoms also undergo vertical migrations in the upper sediment.

Similar vertical movements of the RPD have been related to seasonal changes in temperature and wave energy. Lower temperatures and more vigorous wave energy during winter force the RPD down, and the fauna moves with it (Figure 9.7). In temperate areas experiencing such effects, the meiofauna occurs in lower abundance and moves deeper into the sediment. In warmer areas, seasonal changes are less clear and in sublittoral areas they may be more complex. Nevertheless, most meiofaunal communities exhibit some seasonality, with greatest abundance in the warmer months and little evidence for periodicity of more than a year (Coull 1988).

9.7   Meiofaunal Communities

The meiofaunal communities of sandy beaches are diverse in taxonomic composition and have complex three-dimensional patterns. Where sufficient wrack is cast ashore to create a distinct drift line, rich communities may develop high on the shore. Large oligochaetes and nematodes dominate such areas. Elsewhere, where the sand at upper tide levels dries out during ebb tide, these layers are occupied by diverse groups, although small nematodes capable of moving in water films around the sand grains predominate. Oligochaetes, being possibly of terrestrial origin, can tolerate the desiccation experienced here.

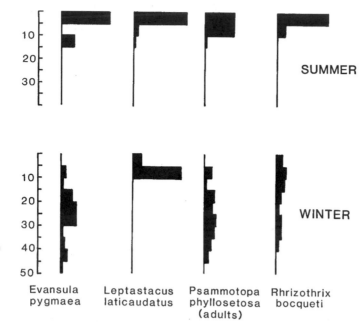

Figure 9.7.
Comparison of vertical distributions of four species of copepods in a sheltered sandy beach in summer and winter (after Harris 1972), showing the downward movement in winter. Depth in cm.

Moving to the midshore, or zone of retention (where much water movement occurs in the sand but some capillary moisture is retained during low tide) a very rich fauna develops. Nematodes and harpacticoids are usually dominant, but many other groups may be represented. Turbellarians and harpacticoids need thicker water films in which to move, and these they find in this zone.

Further downshore or deeper into the sediment, where the sand is saturated but some circulation of interstitial water still occurs, the meiofauna is less abundant, although nematode worms become more important (Gheskiere *et al.* 2002). This includes the surface layers out into the surf zone. Below this, in saturated sand where water circulation is more restricted, oxygen tensions drop. On very exposed beaches of coarse sand, conditions may never become anoxic, but under less dynamic conditions this layer may include the RPD and black layers. On very sheltered fine-grained beaches, only this layer may be represented. Here, nematodes are usually dominant.

There have been few studies concerned with the species composition of sandy-beach meiofaunal communities. Species diversity is much higher than that of the macrofauna, with many beaches supporting more than 100 meiofauna species (Figure 9.8). Complex vertical and horizontal gradients in physical and chemical features create many spatial niches and allow the development of greater diversity than on muddy shores. Diversity is usually greatest on the mid- to upper shore and decreases downshore, the opposite trend from that of the macrofauna. Indeed, meiofaunas are species rich in surf-zone sediments where macrofaunas are scarce or absent. Gheskiere *et al.* (2004) found nematode densities to increase downshore across a dissipative beach, with highest diversity in the midshore, where there was an optimum balance between submergence, oxygen supply, and sediment stability. Nematodes are usually represented by the greatest number of species, whereas harpacticoid copepods seldom have more than 5 to 10 species on any one beach. On the other hand, individual copepod species may be very abundant.

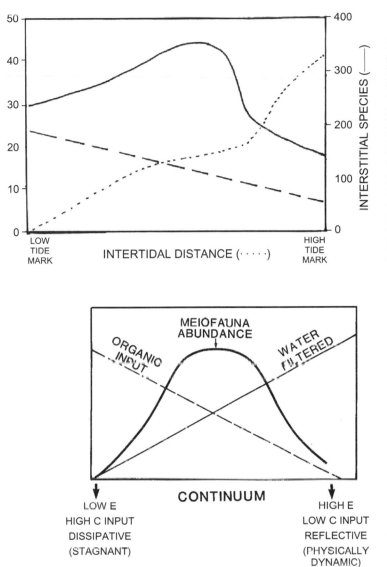

Figure 9.8. Distribution of species richness of interstitial fauna (solid line; 89% meiofauna, 11% protozoans) and macrofauna (broken line) across a North Sea beach (after Armonies and Reise 2000). Dotted line shows the beach profile. Macrofauna species richness peaks on the lower shore, whereas interstitial fauna is richest near the break in slope. Note the differences in scale.

Figure 9.9. Conceptual model of meiofauna community response to gradients of wave energy and organic input.

The rich fauna above the water table or RPD, depending on the energy level of the beach, can detect and select certain points in environmental gradients of oxygen, moisture, grain size, reduced chemicals, and water flow. The more vertical this habitat, the greater is the space available for partitioning the habitable parts of these gradients. In temperate shallow-water communities, there may be clear temporal gradients in the form of seasonal changes in temperature, wave energy, water percolation and oxygenation, benthic primary production, and related factors. In such cases, there are commonly successional changes in dominance with the seasons (Fenchel 1978).

The richest meiofaunal communities, in terms of diversity and abundance, seem to develop where there is an optimal balance between organic input and oxygenation (Figure 9.9). Excess organic input on sheltered or dissipative beaches can result in an impoverished fauna due to steep chemical gradients and compressed vertical extent of the oxygenated zone. Conversely, very reflective beaches with dynamic interstitial water circulation tend to have fewer large-bodied forms. It is therefore expected that diversity of interstitial meiofauna should generally be greatest on intermediate beaches of medium sands (250 to 500 μm) (see Rodriguez 2004, Gheskiere *et al.* 2005a), but a comprehensive study of meiofaunal community response to a full range of beach types has yet to be undertaken. Preliminary studies suggest that meiofaunal response is opposite to that of the macrofauna (i.e., increasingly rich communities toward reflective beaches) (Rodriguez *et al.* 2003).

Latitudinal variations in sandy-beach interstitial meiofauna diversity and distribution have received little attention. There is no reason to expect that biodiversity will vary from the usual trend of increase toward the tropics, and limited data for nematodes support this (Nicholas and Trueman 2005). However, Kotwicki *et al.* (2005) could not demonstrate this for higher taxa of true meiofauna, and only found such a pattern when small and juvenile macrofauna were included. They found turbellarians more important at higher latitudes and nematodes at lower latitudes. This is an area that warrants further study, especially for the interstitial nematodes and harpacticoids, but needs to include consideration of beach type.

The discharge of groundwater through beaches, as unconfined aquifers, provides a path for interstitial beach fauna to evolve landward to terrestrial and freshwater habitats. Toward the backshore of most sandy beaches a rapid change in interstitial salinity occurs, as percolating seawater gives way to inflowing fresh groundwater. Here, a narrow brackish transition zone may occur (see Chapter 13). Brackish water meiobenthic communities typically have lower densities and fewer species than either pure marine or pure freshwater communities (Remane 1933, Gerlach 1954).

9.8 Trophic Relationships

In sandy beaches the interstitial system represents a complete food web, with virtually no exchanges with the macrofauna — except in very sheltered situations of fine sand, where the presence of large burrowing forms on or close to the surface represents a food source available to macrofauna. The interstitial system of exposed beaches includes a wide variety of organisms, from primary producers to top meiofauna predators and decomposers. Feeding niches may be highly specialized. Individual meiofaunal species may be adapted to feed on diatoms, cyanophytes, flagellates, heterotrophic bacteria, detritus, other meiofauna — or they may even be filter-feeders. Many species select prey by size, and may partition food resources by displacing food size niches by one standard deviation.

Most turbellarians are predatory. Oligochaetes feed on detritus and bacteria, nematodes feed on a variety of foods (including bacteria, diatoms, and detritus), and copepods eat bacteria. Feeding categories for nematodes include selective and nonselective deposit feeders, epigrowth feeders, and omnivore/predators, a conclusion based on buccal morphology. Nonselective deposit feeders are often the dominant group, with epigrowth feeders more important at higher tide levels.

Food inputs to the interstitial system consist of (1) primary production by interstitial microflora (photoautotrophs), (2) dissolved and particulate organics flushed or pumped into the sand by wave action, and (3) significant chemoautotrophic synthesis in sheltered beaches with pronounced reduced layers. There may be a tendency for the interstitial system of micro- and meiobenthos to depend mainly on detritus in the upper intertidal, more autotrophic activity in the subtidal, and with the lower beach and swash zone a transitional area (Vassallo and Fabiano 2005). Primary users of material from these sources are the bacteria, both in the interstitial water and attached to sand grains. Heterotrophic bacteria utilize dissolved and particulate organic materials flushed into the sediments. They mostly coat the sand grains, but also occur in the interstitial water. Autotrophs, cyanobacteria, diatoms, and flagellates (as well as these heterotrophic bacteria) are taken by ciliates and meiofauna — which may ingest them whole, filter them from the interstitial water, or scrape them from sand grains. These animals are in turn taken by meiofauna and ciliate predators at the top of the interstitial food chain.

There are no links with the macroscopic food chain, such as occur in finer sediments from low-energy environments. The interstitial fauna of beach sand is too dispersed vertically (and not concentrated at the surface) to form a readily available food resource. However, deposit feeders or sand swallowers (such as *Arenicola* and *Callianassa*) that may occur on sheltered beaches will ingest some meiofauna and other interstitial forms (thus having some impact on the interstitial system).

The interstitial system should, therefore, be seen as a carbon sink — mineralizing the organic inputs it receives from the sea. This material is utilized by the interstitial food chain, whereas nutrients are recycled. The larger the organic input and the greater the volume of water filtered by a beach the more nutrients it will recycle. This is covered in some detail in Chapter 12.

9.9   Biological Interactions

Interactions involving competition, predation, and disturbance and their effects on macrofaunal communities have been widely studied (see Chapter 7), but much less attention has been paid to the meiofauna. As with trophic interactions, biological interactions between interstitial and macrofaunal biota are more prevalent toward the sheltered end of the scale and apparently absent from coarse-grained exposed beaches.

The most extensive work on such interactions has been that of Reise (1985) on the sand flats of the Wadden Sea, and experimental studies in the Baltic by Aarnio (2001). Important interactions Reise (1985) demonstrated include the effects of macrofaunal and fish predation on the meiofauna, and the impact of macrofaunal burrows and irrigation on sediment properties and on the interstitial fauna. Although absent from exposed beaches, these effects were marked in the low-energy sand flats of the Wadden Sea.

Numerous small fish and invertebrates become active predators on meiofauna at high tide: gobiid fish, juvenile flounder, juvenile shore crabs, and brown shrimp (Reise 1985, Aarnio 2001). These forms primarily sort the surface sand and remove juvenile macrofauna and meiofauna, their predatory effects seldom extending more than 1 cm into the sediment. In the case of predation by juvenile crabs, Reise (1985) showed significant reductions in the numbers of nematodes, platyhelminthes, and harpacticoids due to predation. Excluding all predators from the sediment, he found large increases in nematode numbers, although they were less affected than juvenile macrofauna. He considered their small size and deeper vertical distribution to be protection mechanisms shielding the meiofauna from predation by macrofauna.

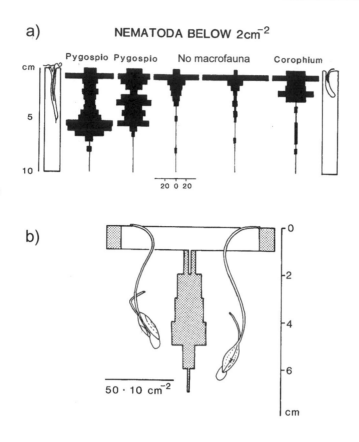

Figure 9.10. Effects of macrofauna on the vertical distribution of meiofauna (after Reise 1985). (a) Distribution of nematodes (individuals below 2 cm² in intervals of 0.5 cm) in the presence of the polychaete *Pygospio elegans* and the amphipod *Corophium arenarium*, and in the absence of macrofauna. (b) Distribution of platyhelminth abundance in normal sediment (white) and within experimental aggregates of *Macoma balthica* (shaded).

Reise (1985) also demonstrated several effects of macrofaunal burrows and actions on the sediments, which promote meiofauna. Many of these involve macrofauna irrigating and oxygenating otherwise reduced sediments and thus rendering the interstitial climate more conducive to the development of meiofauna. Figure 9.10a illustrates the effects of a polychaete and an amphipod on vertical nematode distribution in a sand flat, and Figure 9.10b shows similar effects due to the bivalve *Macoma*. Increased numbers of meiofauna around macrofaunal burrows raises the question of how the sedimentary environment is improved. Irrigation and oxygenation of the sediment are the obvious and primary benefit. Further, excretory or secretory products of the macrofauna may enhance bacterial growth, which in turn attracts meiofauna. This activity might also result in the release of extra nutrients and promote better diatom growth on the sediment, thereby also providing more available food for the meiofauna.

Reise (1985) demonstrated significant effects of burrows of the lugworm *Arenicola* on the meiofauna. Nematodes avoided unstable funnels and fecal mounds, although the presence of burrows generally increased meiofaunal densities. At the same time, it should be remembered that deposit feeders such as *Arenicola*, *Callianassa*, and others ingest sediment containing meiofauna and thus feed directly on the latter. Meiofaunal patchiness has been widely documented, with localized food supply and macrofaunal activity two of the contributing factors (Coull 1988). Biogenic structures, predation, and disturbance (such as the feeding mounds of rays) may all exert local effects on meiofaunal abundance and distribution.

9.10   Meiofauna and Pollution

Meiofaunal species are known to be sensitive indicators of environmental perturbations and pollution (Coull and Palmer 1984, Schratzberger *et al.* 2000). The nematodes are an obvious taxon to use as ecological indicators for benthic environments because they reach high densities (even when macrofauna is sparse); are easily sampled; have ubiquitous distribution, high diversity, and short generation times; are restricted to sediments; and include a range of species from very tolerant to very sensitive (Heip *et al.* 1985, Sanduli and de Nicola 1991, Gheskiere *et al.* 2005b). A disadvantage of using nematodes is the specialized nature of species identifications. However, Vanaverbeke *et al.* (2003) demonstrated that nematode biomass spectra could be used for pollution studies without *a priori* species identification. The disadvantages of using meiofauna in environmental studies have been reviewed by Schratzberger *et al.* (2000).

Because of the difficulties and tediousness of species identifications for many meiofaunal groups, pollution studies have tended to look for broader taxonomic categories. The two usually dominant groups, nematodes and harpacticoid copepods, have been the focus of most of these studies. Sandy beaches can tolerate quite high organic loads and are able to maintain their function without becoming seriously deoxygenated. However, as loading increases and oxygen tensions reduce (creating stronger chemical gradients) changes in the proportions of these two groups may occur. Nematodes as a whole, not considering individual species, are less sensitive to reduced oxygen and may even increase under organic loading. Harpacticoid copepods, on the other hand, are very sensitive and rapidly decrease in abundance when the interstitial chemistry changes in response to such disturbances. Thus, using meiofauna (and specifically the ratio of nematodes to copepods) has been suggested as a tool in pollution monitoring (Rafaelli and Mason 1981).

The nematode/copepod ratio as a pollution tool has been criticized and has not been really successfully applied in pollution studies. Nevertheless, it may have some potential if developed further and used with caution (Shiells and Anderson 1985). Nematodes are more tolerant of enrichment and other forms of pollution that reduce oxygen tensions, and harpacticoid copepods are very sensitive. But both display considerable variability in distribution and density within any beach, so that their ratio will vary across a beach and with depth into the sand. However, if the entire intertidal zone is sampled down to the depth in the sand where meiofauna numbers drop off the nematode/copepod ratio for an entire beach could be useful because it tends to be fairly constant, depending on sand particle size (Figure 9.2) and beach morphodynamics. An example of distribution of these groups across an intertidal beach is shown in Figure 9.11. This warrants further study on exposed sandy beaches subject to many forms of pollution (see Chapter 14).

9.11   Conclusions

Open sandy beaches generally have fine to medium sands in the range 200 to 500 μm, providing ideal pore space for the development of rich interstitial faunas dominated by bacteria, protozoans, and meiofauna. The overall climate of this three-dimensional habitat is a function of grain size, wave energy, and organic input. In beaches of fine sand, low wave energy, and high organic input, reduced conditions occur: oxygen is limited, steep vertical chemical gradients occur, and the fauna is concentrated in the surface oxygenated

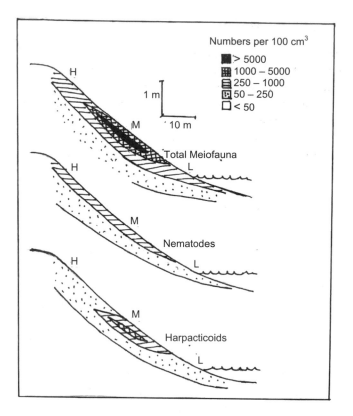

Figure 9.11.
Distribution of total
meiofauna, nematodes,
and harpacticoids
across an intermediate
microtidal beach in
South Africa (after
McLachlan *et al.* 1981a).
H, M, and L refer to
high-, mid-, and low-
tide levels, respectively.

layers. In beaches of coarse sand, high wave energy, and low organic input, the sediment body is vigorously flushed with water, is well drained and oxygenated, and with little vertical chemical gradation has organisms extending deep into the sediment. Between these extremes, intermediate beaches represent more favorable conditions and usually have the richest interstitial faunas.

The prime function of the interstitial system is the processing of organic materials flushed into the sand. This material, both dissolved and particulate, is mineralized by a food chain having heterotrophic bacteria at its base and predatory meiofauna at the apex. In the process, nutrients are recycled to the sea. Sandy beaches thus function as great digestive and incubating systems, purifying coastal waters (Pearse *et al.* 1942, see Chapter 12). On most beaches, therefore, the interstitial system functions as a biological filter that mineralizes organic materials and has little (if any) direct trophic interactions with other food chains.

Surf-zone Fauna

10

10.1   Introduction

Sandy beaches are essentially marine systems and virtually the entire resident fauna is of marine origin. Furthermore, the major exchanges of organic materials and nutrients are with the sea, the pathway for such exchanges being the surf zone. In addition to its importance in shaping the beach and in transporting materials, the water envelope of the surf zone supports a varied fauna of zooplankton and fishes. Many of the species found are endemic or restricted to surf zones and may be considered residents, whereas others are visitors merely passing through this turbulent zone. Because of the difficulties experienced in working in high-energy surf zones, the fauna of these areas has received less attention than zooplankton and fish faunas in other coastal ecosystems. Nevertheless, sufficient information is available for some generalizations to be made.

10.2   Zooplankton

Composition

The surf-zone zooplankton consists almost entirely of crustaceans, as opposed to gelatinous forms that may be sporadically important. The following categories may be recognized among the surf-zone zooplankton.

Table 10.1.   Zooplankton in the Eastern Cape, South Africa (based on a variety of surf-zone studies).

Category	Subcategory		No. of species	% numbers	% biomass
Residents:	(a) Planktonic		2	18	80
	(b) Benthoplanktonic		6	1	5
Nonresidents:	(a) Holoplankton	(i) Micro	100	60	4
		(ii) Meso & macro	80	20	9
	(b) Meroplankton		20	2	2

- Resident species that may be truly planktonic or benthoplanktonic forms
- Nonresidents that may be holoplankton (micro-, meso-, and macro-zooplankton) or meroplankton (larval stages)

True planktonic forms resident in surf zones may include mysid shrimps, small prawns, and other larval stages of some sandy-beach animals. Residence in this context implies regular rather than permanent occurrence. Benthoplanktonic forms include aquatic isopods, amphipods, and mysids that occur in the sand but migrate into the plankton, particularly during nocturnal high tides. Holoplanktonic animals, on the other hand, spend their entire lives in the plankton. The nonresident holoplankton of surf zones is divided into two categories: the micro-zooplankton such as tintinnid ciliates and copepod nauplii (which pass through a 200-μm mesh) and the meso- and macro-zooplankton, including copepods, cladocerans, chaetognaths, and even jellyfishes that may be transported into the surf zone by winds and surface currents. Finally, the meroplankton (or temporary nonresident zooplankton) comprises the larval stages of non-planktonic animals from other areas (for example, the larvae of crabs, bivalves, polychaetes, and fishes from nearby estuaries, rocky shores, or offshore ecosystems).

The resident zooplankton tends to be made up of relatively large forms that dominate the zooplankton biomass, although they generally represent few species. Mysids are frequently the most abundant. Nonresidents are mainly small in size, with copepods the most characteristic forms. They contribute most of the species in surf zooplankton samples and may also be numerically dominant. The nonresidents are highly variable in occurrence, their sporadic appearance in the surf zone being due to advection by wind-, wave-, and tide-induced currents. Although they may at times be of considerable importance, especially as a food source, they are not discussed further here.

The average zooplankton composition in well-studied sandy-beach surf zones in the Eastern Cape, South Africa, is outlined in Table 10.1. There is, regrettably, insufficient information from the surf zones of other parts of the world to attempt a full comparison. However, studies that have been undertaken elsewhere suggest that the values given in Table 10.1 may not be atypical for sandy-beach surf zones in general.

Sampling

Surf-zone zooplankton is difficult to sample, particularly in exposed situations. Furthermore, because large forms are a feature of surf-zone zooplankton large nets must be employed. Otherwise, the community (and particularly mobile species such as swimming prawns and mysids) will be undersampled. Ideally, four different strategies are required to sample the zooplankton effectively: (1) filtration of buckets of water to collect the microplankton, (2) the use of standard plankton nets for mesoplankton, (3) the rapid towing of large coarse-mesh nets to capture forms such as prawns and mysids, and (4) pulling sleds through the surf zone to collect benthoplanktonic forms on or close to the sand surface. Diel variability dictates that sampling should be undertaken both during the day and at night.

Adaptations

Surf zones are extremely dynamic turbulent habitats and it is hardly surprising that a common feature of the zones' planktonic animals is their relatively large size. Moreover, the largest species tend to be characteristic of the most turbulent areas. Clutter (1967) examined four inshore mysid species at La Jolla, California, and found that the largest of them dominated inside the surf zone. Similarly, in the Eastern Cape the surf zone is dominated by a penaeid prawn that reaches 5 to 10 cm in length, whereas in the swash zone a benthoplanktonic mysid some 2 cm long dominates. Outside the breakers, the zooplankton is dominated by a mysid only 1 to 2 cm in length. Associated with size is mobility, and all surf-zone residents are of necessity strong swimmers.

Another feature of resident surf-zone zooplankton is that all of its members either display brood protection or leave the surf zone to spawn in quieter waters. All of the benthoplanktonic forms (mysids, isopods, and amphipods) have brood pouches, whereas the true planktonic forms display one strategy or the other — a necessity in such a dynamic environment. In the prawn *Macropetasma*, the adults move offshore to spawn — juveniles returning to the surf zone, which they exploit as a nursery area (Cockcroft and McLachlan 1986).

Further adaptations common to many surf-zone zooplankton residents are opportunism and omnivory, in which they resemble the residents of the intertidal zone (see Chapter 6). Most zooplankton species will, in fact, switch to whatever food becomes available, be it detritus, phytoplankton, or microzooplankton. A final feature of the resident species is their tendency to form large swarms, which results in extreme patchiness of distribution. The value of this swarming behavior is uncertain and may relate to feeding, avoidance of predators, or facilitation of reproductive contact.

Migrations

Migrations are common among the surf-zone zooplankton. They may be associated with spawning, with the tides, or with the day/night regime. The prawn *Macropetasma* undergoes spawning migrations, as noted previously — adults moving offshore into water of 9 to 18 m depth to spawn. When the juveniles reach a length of about 33 mm, they return to the surf zone, which they utilize as a nursery area (Cockcroft and McLachlan 1986). On the other hand, the benthoplanktonic amphipod *Amphiporeia*, which has been studied on an exposed sandy beach in Maine, exhibits a tidal cycle of entry into the plankton. It undergoes a daily tidal migration over the shore, swimming activity increasing before high tide and peaking during ebb (Hager and Croker 1980) (Figure 10.1).

Many species show increased activity at night (de Lancey 1987), benthoplanktonic forms most commonly entering the plankton then. A good example of such a diel migration pattern is that displayed by the planktonic mysid *Mesopodopsis*. This species swarms

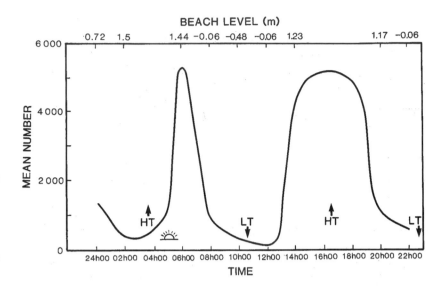

Figure 10.1. Mean numbers of the amphipod *Amphiporeia virginiana* in the plankton over a sandy beach during 24 hours, showing entry into the plankton at high tides (after Hager and Croker 1980).

1 to 2 km offshore near the bottom, during the day. In the afternoon, the animals move inshore to concentrate just behind the breakers near the sea bed, after dark. Here, the mysids feed on phytoplankton and detritus being outwelled by rip currents. They then disperse after daybreak and move back offshore (Figure 10.2). Diel vertical migrations are typical of most zooplanktonic species, so that vast differences are evident between day and night samples from surf zone — the latter being much richer.

Migrations related to storm/calm cycles are also probably common, and many species may abandon the surf zone during very rough conditions. However, the difficulty of sampling under such conditions has prevented accurate study of this phenomenon.

Distribution

Resident surf-zone zooplankton frequently exhibit clear distribution patterns or zonation, both on/offshore and alongshore. Furthermore, this may be at either an inter- or intraspecific level. Clutter (1967) demonstrated an offshore zonation of four mysids, a large species dominating the surf zone, *Metamysidopsis* dominating the rip-head zone (the area just outside the breakers), and two other species further offshore. The body-size gradient (which was apparent) was clearly related to the gradient of turbulence. Wooldridge (1983) recorded no less than 14 species of mysids along transects out to a depth of 20 m off sandy beaches in the Eastern Cape. Only two of these were abundant, however — *Gastrosaccus* dominating the swash and surf zones and *Mesopodopsis* the rip-head zone and the area seaward of it, only entering the surf zone during calm conditions.

Intraspecific zonation at right-angles to the coast has also been reported in benthoplanktonic forms. In *Gastrosaccus*, for example, brooding females remain close inshore (against the beach), whereas nonbrooding females, males, and juveniles are more abundant in the turbulent water of the surf-zone proper. As the larvae develop within the brood pouch, the females bearing them move still closer inshore until females with young ready to emerge occur near the upper edge of the swash zone (Wooldridge 1983).

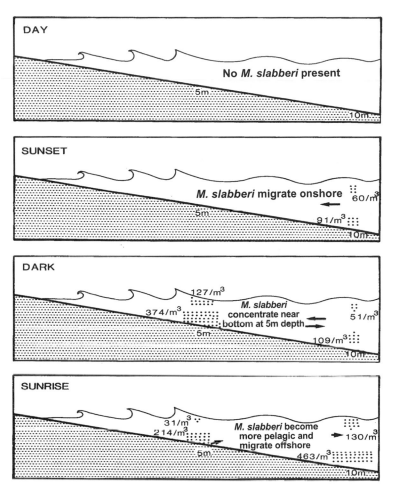

Figure 10.2. Diel migratory cycle of the mysid *Mesopodopsis slabberi* (after Webb 1987).

Longshore distribution variations are associated with rhythmic shorelines and surf-zone patterns. McLachlan and Hesp (1984) showed that the surf-zone zooplankton (benthopelagic amphipods and mysids) of a reflective Australian sandy beach concentrated at times off the cusp horns, although this trend was less pronounced among the benthos. Wooldridge (1989) has similarly demonstrated a patchy distribution of the mysid *Gastrosaccus* alongshore, related to the spacing of rip currents and other rhythmic shoreline features.

Biomass and Abundance

A limited number of studies provide estimates of biomass and abundance for the surf-zone zooplankton. A feature that emerges from these few quantitative studies is the very high biomass typical of the zone. Indeed, both biomass and abundance appear to be far greater than in deeper water beyond the surf zone: this despite the much greater turbulence the surf zone presents. This biomass differs greatly between day and night, however, and is patchy due to swarming. In the Eastern Cape, recorded dry biomass figures have varied over the range 0.01 to 150 $g \cdot m^{-3}$, with average values around 1 to 2 $g \cdot m^{-3}$. The highest values recorded were due to swarms of mysids or prawns. Mysids generally make up 20 to 80% of the biomass, prawns 0 to 80%, and small crustaceans belonging to other

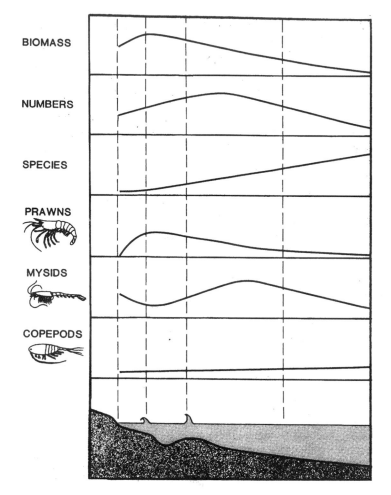

Figure 10.3. Profiles of zooplankton biomass, abundance, diversity, and taxonomic composition across an Eastern Cape surf zone, averaged over day and night periods.

taxa about 10%. Numbers range from 20 to 40 m^{-3} for surf-zone prawns, and average about 2,000 m^{-3} for all zooplankton in the vicinity of surf phytoplankton blooms. Mysid density displays an extreme range, being as low as 10 m^{-3} when the animals are scattered but reaching up to 15,000 m^{-3} when in dense swarms. Biomass tends to be highest inside the breakers, slightly lower in the rip head zone, and much lower further offshore. A typical on/offshore profile of zooplanktonic features (Figure 10.3) shows some generalized trends. This is, however, an average picture and takes into account neither the extreme patchiness commonly encountered nor diel variation.

Food and Feeding Relationships

Water circulation in the surf zone concentrates particulate food, such as detritus and phytoplankton. Clutter (1967) suggested that the surf zone was characterized by much detrital food in suspension and that this decreased offshore, thus explaining the greater zooplankton biomass inshore. Certainly, many zooplanktonic species are attracted to phytoplankton blooms in the surf, to feed on the diatoms (Webb *et al.* 1987). Although they may be attracted primarily to the phytoplankton (and also feed on detritus), the larger forms of zooplankton are usually omnivorous and switch readily to an animal diet, whereas smaller forms (such as many copepods) are purely herbivorous. The zooplankton is, in turn, a major component of the food of surf-zone fishes and thus occupies a key position in the center of the surf-zone food chain.

10.3   Fishes

Sampling

Surf-zone fish communities have been sampled by seine netting in widely different areas and types of beach, but particularly on sheltered beaches in the Northern Hemisphere. Where whole communities have been sampled, the number of species recorded generally ranges from 20 to 150, with a few species dominant and marked seasonal fluctuations often apparent. To adequately sample surf-zone fishes, it is essential that studies cover extended periods and use different types of sampling gear, at least different mesh sizes of seine net (Table 10.2). Studies on flatfishes have (in addition to seine nets) used various push nets and dredges to properly sample these bottom dwellers. This is important because seine nets tend to pass over fishes flush with, or half buried in, the sediment (especially small flatfishes).

Larvae, Juveniles, and Nursery Areas

The most distinctive features of surf-zone ichthyofaunal communities are their great variability in both space and time, their opportunistic feeding behavior, and the high proportion of larvae and juveniles. Further, the community composition may be influenced by other nearby habitats. These fishes may be grouped in different ways: (1) adults, juveniles, and larvae on the basis of age, (2) residents, seasonal migrants, and strays on the

Table 10.2.   Examples of sampling strategies from studies that used seine nets for surf-zone fishes.

Study	Study period	Beach type	Net measurement
McDermott 1983	24 months 293 hauls	Medium energy, intermediate	15 m × 1.5 m × 6.5 mm mesh
Lasiak 1984	26 months 130 hauls	Medium energy, intermediate	a. 60 m × 2 m × 40 mm mesh b. 30 m × 2 m × 17 mm mesh
Robertson & Lenanton 1984	30 months 66 hauls	Low energy, reflective	15 m × 1.6 m × 10 mm mesh
Ross et al. 1987	13 months 120 hauls	Low to medium, reflective/ intermediate	50 m × 1.8 m × 3.2 mm mesh
Romer 1986	9 months 190 hauls	High energy, intermediate	a. 90 m × 1.5 m × 50 mm mesh b. 60 m × 2 m × 23 mm mesh
Suda et al. 2002	5 years 1,007 hauls (fine mesh) 365 hauls (coarse mesh)	Low energy	a. 5 m × 1 m × 1 mm mesh b. 26 m × 2 m × 4 mm mesh
Strydom 2003	10 months 192 hauls	Medium energy, intermediate	4.5 m × 1.5 m × 0.5 mm mesh for larval fishes
Suda et al. 2005	36 months 75 hauls	Low to medium energy, reflective	26 m × 4 m × 8 mm mesh

basis of temporal association, (3) benthic feeders, planktivores, piscivores, and omnivores on the basis of diet, or (4) pelagic and demersal species on the basis of habitat or behavior.

It is becoming increasingly clear that estuaries, seagrass meadows, and other sheltered habitats are not the only or always the most important coastal nursery areas for fishes. Surf zones are also used extensively as fish nursery areas. Some species even spawn in the intertidal zone; for example, the Californian grunion *Leuresthes* and the Atlantic silverside *Menidia* (Middaugh *et al.* 1983). The grunion are rhythmic spawners that deposit their eggs in the upper intertidal at night just after the highest high tides, so that subsequent high tides result in sand being deposited over the eggs. The embryos are washed out of the sand by the next series of spring highs, two weeks later. The silverside deposits its eggs in protected situations such as crab burrows, detrital mats, seagrass plants, or scarps during daylight spring high tides.

Most coastal species spawn outside surf circulation cells. In several species (e.g., *Trachinotus*, *Rhabdosargus*, and *Pomadasys*), juveniles and larvae inhabit the surf zone, whereas adults occur almost exclusively in deeper water. Suda *et al.* (2002) grouped surf-zone fishes into five categories based on developmental stages present in the surf zone.

- Type I is represented by postlarvae (postflexion or late larvae and early juveniles). These may be anadromous fish that enter rivers as juveniles.
- Type II is represented by juveniles (and larvae), and the surf zone serves as a transient habitat (e.g., sparids).
- Type III is sporadically recorded as adults.
- Type IV is resident in the surf as adults (e.g., *Sillago*).
- Type V spawns in the swash zone of the beach (e.g., grunion).

Several species may spawn in or near surf zones, and a variety of fish larvae have been recorded in these areas. Ruple (1984) collected larvae from Gulf of Mexico surf zones, finding more in the outer surf zone than the inner zone, and significantly greater numbers at night than during the day. Species frequenting the surf zone as juveniles were not abundant as larvae, however. The occurrence of a variety of small juvenile stages of four species implied that these used the surf zone as a nursery area. In most cases, spawning probably occurred offshore, with the larvae later moving into the surf zone.

Surf zones are utilized as nursery areas by milkfish in Asia, and by *Chanos* (Senta and Hirai 1981, Morioka *et al.* 1993) and a variety of species (including anchovies) in Brazil (Pessanha and Araujo 2003, Silva *et al.* 2004). Temperate South African surf zones are important as nursery and transient areas for fish larvae (Strydom 2003). Vast accumulations of detached macrophytes are utilized as nursery areas for juvenile fishes in Western Australian surf zones (Lenanton *et al.* 1982). The early developmental stages of four species utilized these accumulations both for shelter from predators and to provide food (largely in the form of rich amphipod populations). However, where the occurrence of macrophytes in the surf in South Africa was more sporadic and less dense they did not attract significant numbers of juvenile fishes (van der Merwe and McLachlan 1987), and similarly juveniles dominated catches in False Bay, South Africa, but were not clearly associated with detached macrophytes (Clark *et al.* 1996).

In South Africa, Watt-Pringle and Strydom (2003) recorded 37 species of larval fish in the inner surf zone, many in depressions where water was less turbulent. Seventy-nine percent of these marine species were estuarine dependent (and as late larvae they are capable swimmers that actively select these habitats in the surf zone). Also in South Africa, Lasiak (1981) recorded 30 species of juvenile fishes using Eastern Cape surf zones as nursery areas (half of these also being recorded as adults). Most were in the size range 2 to 10 cm, and virtually all of these juveniles were zooplankton feeders. In other coastal areas, however, benthic microflora, meiofauna, detritus, benthos, and zooplankton can all be important food sources for larval and juvenile fishes. Late larvae and juvenile fishes nearly always occur in shoals, which are quite sporadic in occurrence in the surf zone. Rich zooplankton food, and possibly shelter from predation, may be the most important features of sandy-beach nursery areas. Lasiak (1983) also showed that most of the juvenile fishes she studied remained in the surf zone for only short periods and were not permanent residents.

In temperate waters (such as those around the British Isles, Japan, Europe, and the Oregon coast of the USA), vast numbers of juvenile flatfish utilize sheltered sandy beaches and their sublittoral zones as nursery areas. This phenomenon has received considerable attention in Japan (Yamamoto *et al.* 2004) and in British and Dutch waters, where plaice (*Pleuronectes platessa*), dabs (*Limandia limandia*), and turbot (*Scophthalmus maximus*) are abundant and commercially important. Spawning and larval development in these species occur in deeper water. Pelagic eggs are produced, followed by larval stages lasting one to two months. The larvae may undergo passive but selective horizontal transport by swimming up from the sea bed during flood tides but remaining on the bottom during ebb (Rijnsdorp *et al.* 1985). In summer, after metamorphosis, vast numbers of juveniles move into shallow sandy bays — preferring water about 3 m deep (Figure 10.4).

Plaice are by this stage 1 to 2 cm in length, averaging 14 mm at metamorphosis (Figure 10.5). They start arriving in the surf zone in April or May, peak in June (when settlement ends), and decline rapidly in abundance thereafter. Population densities can be extremely

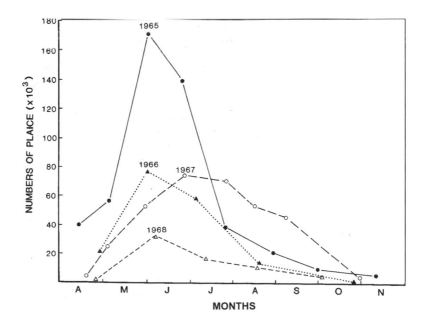

Figure 10.4. Estimates of plaice populations (1965 to 1968) at Firemore Bay, Scotland, showing arrival of juveniles in April and May (after Steele *et al.* 1970).

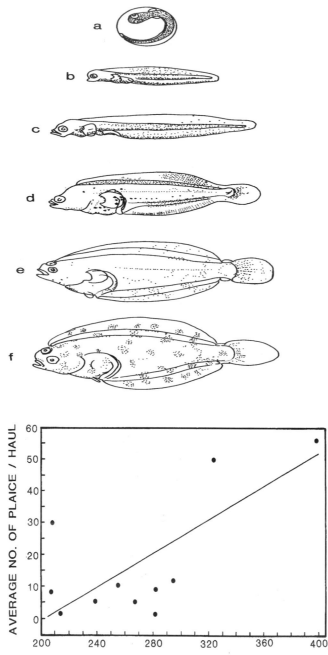

Figure 10.5.
Development of
plaice from egg to
metamorphosing
post-larva.

Figure 10.6.
Relationship between
average number of O-
group plaice per haul
and the median grain
size of sediments (after
Poxton and Nasir 1985).

high in some areas (2 million in sandy bays around the Firth of Forth in midsummer 1979/80 and 10 to 20 million in the Clyde Sea area (Poxton and Nasir 1985). They appear to select coarser sediments (Figure 10.6), being more abundant on medium-grained than on fine sands (where they feed on demersal plankton and benthos, including mysids, polychaetes, and the siphons of *Tellina*). Mortality rates are high, in the region of 40% per month, due mainly to predation. Numbers thus decline rapidly, the survivors moving into deeper water at the onset of winter (Figure 10.4). This movement may occur once the juveniles are large enough to be less susceptible to predators. The movement of the

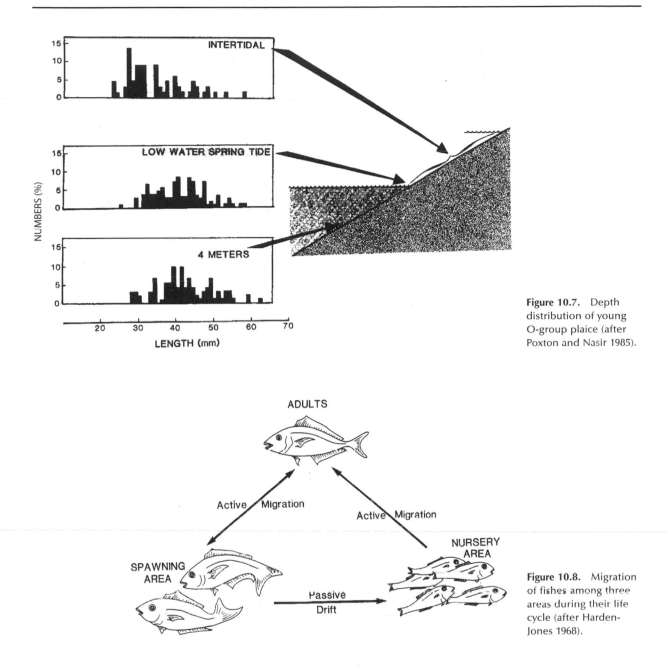

Figure 10.7. Depth distribution of young O-group plaice (after Poxton and Nasir 1985).

Figure 10.8. Migration of fishes among three areas during their life cycle (after Harden-Jones 1968).

larger fish into deeper water during September/October results in a decrease in the mean size of the remaining fish. There are seasonal changes in diet and growth rates. Most active feeding occurs during the day, especially on high tides, when the intertidal residents become accessible as prey. Larger fish tend to occur in deeper water and smaller fish in very shallow water (Figure 10.7).

It is generally considered that recruitment patterns must influence community structure to the extent that in the long term species recruiting successfully into an area will come to dominate it. However, surf-zone fish assemblages are so dynamic, and so few species are truly resident, that this may not be the case here. Harden-Jones (1968) envisaged fish migrating among three areas during their life cycles (Figure 10.8). Larval fishes move from spawning to nursery areas by passive drift, and later by active swimming. Many larval fishes recorded in surf zones may be estuarine-dependent species (Strydom and d'Hotman 2005), moving to or from estuaries associated with sandy beaches. Such estuary-dependent marine larval fishes may increase in abundance in response to cues from the outflow plumes when temporarily closed estuaries discharge strongly after rains (Strydom 2003). In the nursery area, they enjoy shelter and food, the rich zooplankton of the surf zone being a key attractive feature — with the turbidity, shallowness, and turbulence of the zone providing protection from predators. As these predators hunt by sight, capture success is related to the transparency of the water. Predators may also have difficulty maneuvering in the shallows.

Surf-zone Fish Communities

Seining studies on surf-zone fish communities have been undertaken in several countries, but chiefly in the United States, Japan, Australia, and South Africa. In these studies, the main emphasis has been on trophic relationships, although valuable information has also been obtained on community structure and variability in time and space. Various workers have employed different net and mesh sizes (see Table 10.2), and different studies are not always comparable in terms of beach exposure. Nevertheless, some generalizations do emerge.

First, it is clear that fish assemblages associated with sandy beaches are by no means impoverished but can include a wide diversity of species. A few species are usually numerically dominant, however, and occur in shoals. The number of species reported off a single beach ranges up to 160, of which few are usually resident. Diversity generally decreases with exposure — higher-energy surf zones harboring fewer species, which display greater dominance (Clark *et al.* 1996). Biomass structure tends to be more evenly distributed, however, as solitary species are often represented by relatively large individuals. If all studies had utilized the same seining techniques, capable of sampling all sizes of fish, it is likely that the range of number of species recorded would be consistently above 50. Thus, surf zones support highly diverse ichthyofaunas, despite the fact that generally only about 10% of the species may be resident. In most cases, only between 1 and 5 dominant species are true residents, defined as species occurring in the surf zone all year round and in sufficient numbers to be recorded on most sampling occasions (Suda *et al.*'s Type IV).

Typical residents of sandy-beach surf zones include members of the genera *Trachinotus*, *Umbrina*, *Menticirrhus*, *Anchoa*, *Mugil*, *Liza*, *Diplodus*, *Lithognathus*, *Engraulis*, *Trachurus*, *Pomatomus*, *Pelsartia*, *Sillago*, *Cnidoglanis*, *Crapatalus*, *Amphistichus*, various flatfishes, and elasmobranchs such as *Rhinoabatus* and *Myliobatis*. The flatfishes and batoid elasmobranchs (skates and rays) make up the demersal component, which is probably far more important than generally supposed because they tend to be missed by seine-netting techniques. Indeed, it is the dorso-ventrally compressed shape of these fishes that allows them to enter very shallow water to feed on intertidal prey. They can only be sampled effectively by nets that consistently scrape the bottom and do not lift. Nonresident fish species include strays and seasonal migrants — species that move into the surf zone in relatively large numbers but only for a few months of the year. Many larvae and juveniles fall into the latter category. Occasionally, quite unexpected fishes may be found in some numbers in the surf zone. The holocephalan *Callorhynchus capen-*

sis is common just offshore in False Bay, South Africa, despite the fact that holocepha-lans are characteristically found in much deeper water. This fish includes large numbers of the gastropod *Bullia* in its diet. Surf-zone fishes may exhibit some offshore zonation related to the exposure gradient or to the zonation of benthic or zooplankton prey, but this has been little studied.

Most studies have expressed abundance as estimates per haul and not per unit of surf zone, so that generalizations about absolute numbers or densities cannot at present be achieved. However, many studies report highly variable (but often huge) numbers in seine catches, sometimes on the order of 10^3 to 10^4 per haul. This is particularly true when larvae and juveniles are abundant. Lasiak (1984) and Ross *et al.* (1987) have attempted estimates of actual abundance and biomass based on netting efficiency and the area sampled. More work needs to be done along these lines in other areas before the fish community structure of surf zones can be fully appreciated.

Temporal Variability

Several authors have reported significant diel variability in surf-zone fish abundance. There is a pronounced diel variation and spatiotemporal interaction between juveniles and fish predators. Much of this variability lacks any clear pattern. It is not associated with the tides or with photoperiod. In fact, short-term variation may even exceed long-term seasonal changes. In several cases, fish abundance has been shown to increase on high tides, especially in the evening and after dark (Lasiak 1984, Ross *et al.* 1987, Layman 2000, Pessanha and Araujo 2003) (Figure 10.9). This is mainly due to larger fishes moving into the shallows at night. An opposite trend has been recorded from other beaches, however, including Japanese beaches and tropical beaches in Brazil used by anchovies (Senta and Kinoshita 1985, Silva *et al.* 2004).

Much of the short-term variability not linked to tides or photoperiod appears to result from two factors: (1) movement of fishes, including large-scale migrations, not related to changes within the surf zone, and (2) physical changes in the surf zone, especially

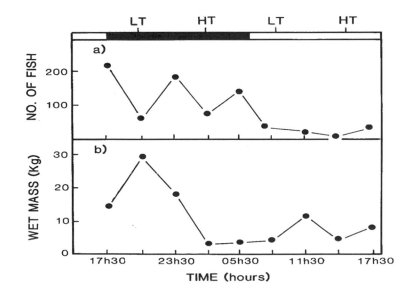

Figure 10.9. Fluctuations in total number and mass of fish caught at Kings Beach, in the Eastern Cape, by Lasiak (1984), showing elevated abundance at night and especially just after dark. LT = low tide, HT = high tide.

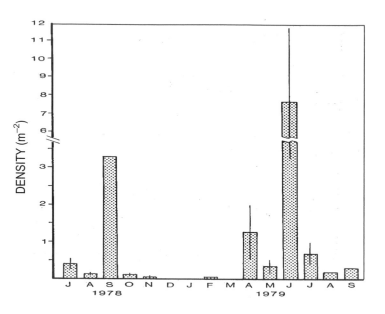

Figure 10.10. Seasonal changes in the density of fishes in the surf zone at Horn Island, Gulf of Mexico (after Ross *et al.* 1987).

changes in wave action and turbulence associated with winds, and temperature. The studies of Lasiak (1984) demonstrated the importance of wind, which causes changes to the dimensions and dynamics of the surf zone and hence affects the community structure of surf-zone fishes. In fact, these changes affect the surf-zone ichthyofauna more than seasonal changes. Nevertheless, some fishes can be locally resident in surf zones, and Ross and Lancaster (2002) recorded juvenile pompano (*Trachinotus*) and kingfish (*Menticirrhus*) remaining for several weeks at sites where they were tagged in a North Carolina surf zone.

Seasonal changes are also important and have been demonstrated in several cases, particularly in the Northern Hemisphere (Modde and Ross 1981, Senta and Kinoshita 1985, Ross *et al.* 1987, Layman 2000) (Figure 10.10). Greatest abundance and diversity generally occur in the warmest months, and some beaches are almost devoid of fishes in winter, although this is not true at all latitudes (Pessanha and Araujo 2003). Much of the seasonal fluctuation is more evident for individual species than for the fish community as a whole, and the seasonal appearance and disappearance of migrants can be very marked (Figure 10.11).

A feature of note is the utilization of macrophyte accumulations for feeding and shelter in certain situations. In some west Australian surf zones characterized by large macrophyte accumulations, Robertson and Lenanton (1984) demonstrated a significant correlation between the quantity of macrophytes and the numbers of fish caught — the ichthyofauna being 2 to 10 times more abundant in surf zones with floating weeds. Many fishes left the wrack at night and moved to open sand, presumably because the need for refuge from predators was then reduced. This again emphasizes the importance of predators in sandy-beach surf zones.

Trophic Relationships

There are five ichthyofaunal trophic groups: (1) benthic feeders, (2) zooplankton feeders (including most larvae and juveniles), (3) herbivores, (4) piscivores, and (5) omnivores.

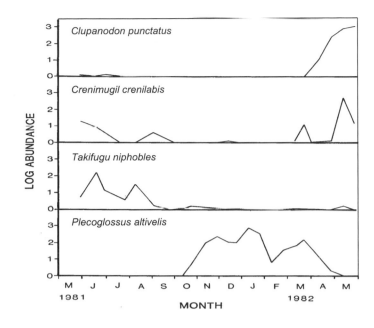

Figure 10.11. Seasonal occurrence of four dominant species of fish in Japanese surf zones (after Senta and Kinoshita 1985).

The main features of the feeding relationships of surf-zone fishes are (1) the high level of opportunism displayed, (2) the dominance of species usually classified as benthic and zooplankton feeders, and (3) the importance of zooplankton in the diets. Benthic feeders are usually the most important component in terms of both number of resident species and biomass, being uniquely adapted to feed on the resident benthos. This group is often dominated by batoids and flatfishes, which tend to be underestimated due to poor sampling methods when only seine nets are used. Their flattened shape allows movement into very shallow water, so that even benthic invertebrates undergoing tidal migrations may be preyed upon. At the same time, these demersal forms can rise readily into the water column to take larger zooplankton when the latter is abundant.

Zooplankton feeders often dominate numerically because of the preponderance of juveniles. Most trophic groups will, in fact, switch to a diet of zooplankton when mysids and other large forms are swarming in the surf. Plankton feeders also include mullet, which exhibit considerable dietary plasticity, taking not only zooplankton but surf diatoms when the latter are concentrated at the surface of the water (Romer and McLachlan 1986). In South Africa, *Liza richardsonii* shoals in the shallows and feeds directly on surf phytoplankton accumulations. Herbivores may occur off beaches as strays or when the beaches are associated with macrophyte accumulations or rocky outcrops. However, on most open beaches devoid of a source of macrophytes herbivores can safely be ignored as a normal component of the ichthyofauna. The macrophyte-dominated surf zones studied by Robertson and Lenanton (1984) differ from those previously studied in the relative unimportance of planktivores and the dominance of benthic feeders utilizing food sources associated with the macrophytes.

Some surf-zone fishes are so opportunistic as to be labeled omnivores, and will readily take benthos, zooplankton, or even macrophytes. However, they do not usually form an important component of the icthyofauna. Piscivores may also not occur in large numbers, nor display much species diversity or account for a major portion of the biomass. They are nevertheless important in view of their impact on other fishes, both preying on them

Table 10.3.   Summary of some studies on surf-zone ichthyofaunal diets.

Study	Main feeding groups	Chief prey items
Robertson & Lenanton 1984	Benthic feeders	Weed-associated amphipods, polychaetes
Lasiak 1983	Benthic feeders, planktivores	Mysids, prawns, benthos
McDermott 1983	Benthic feeders	*Scolelepis*, zooplankton, benthos
Ross *et al.* 1987	Benthic feeders	Benthos
McFarland 1963	Planktivores	Zooplankton
Romer 1986	Planktivores	Zooplankton
Du Preez *et al.* 1990	Benthic feeders, zooplankton feeders	Zooplankton
Inoue *et al.* 2005	Zooplankton feeders, benthic, and epiphytic feeders	Calanoid copepods, mysids

and affecting their spatial and temporal distribution patterns. It may therefore be stated that the major feeding groups of surf-zone fishes, in order of importance, are benthic feeders, planktivores, and piscivores. All of these may, however, be highly opportunistic. The benthic feeders are dominated by demersal forms, whereas planktivores include many larvae and juveniles utilizing the surf zone as a nursery area. Planktivores tend to increase in importance in more exposed surf zones, possibly as a result of the increased availability of suspended food. The most important food items are usually benthic species and crustacean zooplankton (particularly mysids, copepods, and prawns). There are about equal numbers of studies showing benthos or zooplankton constituting the key items (Table 10.3), and again this may vary in relation to exposure.

Du Preez *et al.* (1990) reviewed the bioenergetics, trophic relations, and roles of fishes in surf zones — especially the temperate Eastern Cape region of South Africa, where more than 60 fish species occur and play an important role in energy flow. They showed that fishes were the main predators on macrofauna, important transformers of carbon and nitrogen within the beach and surf zone ecosystem, and key agents for the transfer of materials across the nearshore boundary of the surf zone.

10.4   Other Groups

In addition to zooplankton and fishes, several other groups utilize surf zones and intertidal beaches for feeding or reproduction. Terrestrial invaders, especially shorebirds, are dealt with in Chapter 11 (together with marine turtles). Cetaceans and dugongs are transient, although the latter may be important where there are extensive seagrass meadows in the shallow subtidal. Other animals that enter the intertidal beach through the surf zone include the xiphosurid *Limulus*. Female horseshoe crabs invade some beaches in North America in vast numbers in order to lay eggs near the high-water line. When the nest is inundated on very high tides, rupturing of the eggs is facilitated by osmotic shock and disturbance caused by the swash, thereby ensuring that the hatching trilobite larvae enter the water column during periods of high-water levels (Ehlinger and Tankersley 2003).

10.5   Conclusions

Being an open system, the surf zone has a rapid flux of zooplankton through it, many species being deep-water forms swept inshore. There is, however, a unique component typical of, and resident in, surf zones. This comprises crustacean zooplankton of relatively large body size and high mobility, and includes both planktonic and benthoplanktonic forms. These residents can attain very high abundance and biomass values and be a major food of fishes. The zooplankton community displays a high level of dominance by a few species. There is a strong tendency for species to swarm, creating marked patchiness, which is further accentuated by day/night differences. They characteristically exhibit on/offshore zonation patterns as well as diel and other rhythms. Surf zooplankton deserves more research effort.

Surf-zone ichthyofaunas are highly variable, but are usually dominated by larval and juvenile fishes, indicating that these areas are important nursery grounds. Spawning and early larval development usually occur elsewhere, recruitment to the surf zone being apparent at the post-metamorphosis, settlement, and juvenile stages. The nursery function of the surf zone is enhanced by abundant food availability and possibly by shelter from predators. Surf-zone fish communities are generally diverse but contain few resident species — usually 2 to 5 in a total of typically 50 to 100 species. However, these few residents may be unique to surf zones. There is considerable variation in distribution in both space and time, and extreme patchiness. Seasonality is often only weakly developed but may be marked in some individual species, as has been clearly shown from studies in the Northern Hemisphere. Tidal cycles, photoperiod, wind, and seasonal temperature changes are important abiotic factors affecting community composition, but wave action may override all of these. Because the surf-zone fish assemblage is so dynamic, biological interactions are largely limited to the effects of predation. Most resident species display a high degree of dietary opportunism. Benthic feeders (especially batoids and flatfishes) and zooplanktivores dominate, although piscivores are also important. Because of their motility and migratory behavior, fish play a significant role in the export of energy out of the beach/surf-zone system.

Other marine invaders, such as xiphosurids (which pass through the surf zone), may be locally important in some regions but are of little significance for sandy-beach ecology globally. The role of surf-zone fauna in beach ecosystems is examined in Chapter 12.

Turtles and Terrestrial Vertebrates

11

Chapter Outline

11.1 Introduction

Several groups of vertebrates make use of sandy beaches for foraging, nesting, and breeding. Turtles, as marine vertebrates, nest on the backshore of sandy beaches. Terrestrial reptiles and even amphibians may occasionally stray onto the backshore when foraging. Birds are the most important vertebrates commonly encountered on sandy beaches, both in terms of abundance and diversity and their role in beach ecosystems. Some birds (such as cormorants, pelicans, and penguins) use beaches only for roosting and feeding at sea. Other birds (including shorebirds, waders, and raptors) forage in the intertidal and supralittoral, and several nest on the backshore. Two groups of mammals utilize sandy beaches. Seals and other pinnipeds, as marine mammals, haul onto sandy beaches in several areas for molting, nesting, breeding, and raising pups. Although there are not many such seal rookeries, their local impacts on the beaches can be considerable. The other group of mammals utilizing beaches are terrestrial forms that descend onto the beach to forage. This category includes otters that cross the beach to reach the sea and several other species that forage on stranded carrion along the drift line or take intertidal invertebrates. Typical examples include jackals, coyotes, wild cats, mongooses, raccoons, baboons, monkeys, and even lions. Domestic stock, such as sheep and cattle, may also visit beaches to feed on stranded kelp. Dune systems are often rich in small mammals, rats, mice, gerbils, and moles, which may occasionally forage on the backshore. In terms of their regular use of beaches and the impacts they have, turtles and birds are most important and are covered in more detail.

11.2 Turtles

Marine turtles have been around for 100 million years and are much larger than their land relatives. All nest on sandy beaches. There are seven species of marine turtles in the world (Figure 11.1), divided into two families, Cheloniidae and Dermochelyidae: Kemps

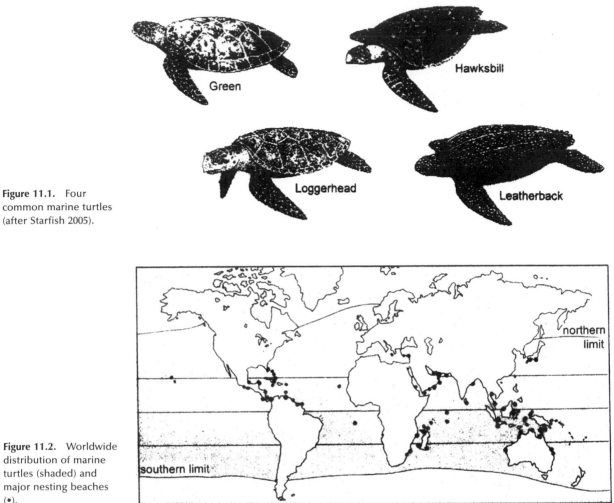

Figure 11.1. Four common marine turtles (after Starfish 2005).

Figure 11.2. Worldwide distribution of marine turtles (shaded) and major nesting beaches (•).

(Atlantic) Ridley turtle (*Lepidochelys kempii*), olive (Pacific) Ridley turtle (*Lepidochelys olivacea*), flatback turtle (*Natator depressus*), green turtle (*Chelonia mydas*), hawksbill turtle (*Eretmochelys imbricata*), leatherback turtle (*Dermochelys coriacea*), and loggerhead turtle (*Caretta caretta*).

All seven species occur in both hemispheres except the flatback, which is only found around Australia. Distribution of the major nesting beaches is shown in Figure 11.2. Turtles are found all around the world in tropical and warm temperate regions. Most migrate over long distances to reach their favored breeding sites from feeding areas. Turtles are mainly carnivorous and feed on jellyfish, tunicates (ascidians, sea squirts), sponges, soft corals, crabs, squids, and fishes — but greens and Kemps Ridley also feed on seagrasses and algae.

It is for nesting that turtles depend on sandy beaches. Breeding occurs in cycles that vary from 1 to 5 years. Turtles live and mate at sea, but during the reproductive season to lay their eggs the females return to the beach on which they themselves emerged as hatchlings. As hatchlings, they become imprinted to the earth's magnetic field — and possibly to the odor of the waters adjacent to the nesting beach — which allows them to successfully complete their migration and return to breeding areas. A single female usually lays several batches of eggs at two- to three-weeks intervals. When ready to lay eggs, the female turtle crawls out of the sea to above the high-water mark, usually about one hour before to about two hours after the nocturnal high tide. Turtles are built for swimming, and thus appear very slow and clumsy when emerging from the sea. In preparation for nesting, the female turtle scrapes away loose sand with all four flippers to form a wide shallow body pit. In this shallow depression she then excavates a vertical pear-shaped egg chamber with the hind flippers. Often, the sand is unsuitable for nesting (especially if it is too dry), and the turtle moves on to another site. For most turtle species, digging the nest takes about 45 minutes. About 50 to 150 spherical eggs are deposited. After laying, the turtle fills the egg chamber with sand using the hind flippers, and then fills the body pit using all four flippers before finally crawling back to sea, entering the surf a few hours after emerging. While on the beach, fluid hangs from the turtle's eyes. This is a concentrated salt solution that helps to remove excess salt ingested by the turtle (from drinking seawater) and to wash the eyes free of sand.

Incubation time is usually about two months, but this and the sex of the hatchlings depend on the temperature of the sand. Dark warm sand produces mostly females and the eggs hatch in seven to eight weeks. Eggs laid in cool white sand take longer to hatch and mostly result in males. The hatchlings take a few days to dig their way through the sand to the surface. When leaving the nest, usually at night, hatchlings head for the low-elevation horizon of the ocean. Hatchlings can easily be disoriented and attracted to bright lights such as street and house lights. This contributes to many hatchling deaths. Most hatchlings reach the sea, although crabs and sea birds attack them on the beach. During their first few hours in the water, these young turtles face heavy predation by sharks and other fishes. Young marine turtles drift and feed in the open ocean. When they are about dinner-plate size, turtles collect near inshore feeding grounds. Marine turtles grow slowly and take between 30 and 50 years to reach sexual maturity. They live for years in one place before they are ready to make the long breeding migration of up to 3,000 kilometers from the feeding grounds to nesting beaches. After reaching sexual maturity, marine turtles breed for several decades, although there may be intervals between breeding of one to five years.

Sea turtles nest on a variety of beach types as long as the beaches are accessible from the sea, the backshore is high enough not to be flooded by sea- or freshwater, temperatures are suitable for egg development, and the sand is coarse enough to facilitate gas diffusion but fine enough to be reasonably stable for excavation of the egg cavity. Horrocks and Scott (1991) found that hawksbill turtles in the Caribbean preferred steeper beaches with lower wave energy, which reduced travel distance for females and hatchlings. These turtles tended to nest in dune vegetation where the sand was less compact and hatchlings enjoyed higher escape success. Mortimer (1995) surveyed the beaches used by green turtles on Ascencion Island in the Atlantic Ocean, finding that longer beaches without obstacles had more clutches laid than other beaches. The presence of rock and artificial lighting seemed to be the greatest hindrances to nesting. Kikukawa *et al.* (1999) examined beach selection by nesting loggerhead turtles in Japan and found that the most important factors enhancing beach selection were increasing sand softness, increasing distance from human settlement, presence of a lagoon in front of the beach, shorter beach length, and higher beach profile. Karavas *et al.* (2005) found that loggerheads nesting in Greece preferred sand that was well sorted and not too fine (thus too compact), and well aerated but able to hold moisture. Further, nesting stopped short of the vegetation line.

The previous suggests that turtles generally select for undisturbed beaches, that are reasonably steep with sand not too fine or compact. Biotic factors may also be of some importance (for example, the presence of dune vegetation and the absence of heavy predation on eggs and hatchlings by mammals and birds). Human disturbance by vehicles or trampling can also limit nesting. In areas where turtles nest in high densities, they can have considerable impacts on the beach, enhancing swash excavation and turnover of sand and causing significant organic input. This must have effects on supralittoral macrofauna and meiofauna. However, despite the great attention paid to turtle conservation and protecting their nesting areas, knowledge of their selection of nesting sites or impact on the beach remains limited.

All marine turtle species are experiencing serious threats to their survival. The main threats are pollution and changes to important turtle habitats, including coral reefs, seagrass beds, mangrove forests, and especially threats to nesting beaches caused by human disturbance in forms such as development, recreational activities, and even beach nourishment (Rumbold *et al.* 2001). Other threats include accidental drowning in fishing gear, overharvesting of turtles and eggs, and predation of eggs and hatchlings by foxes, feral pigs, dogs, reptiles, and even mole crickets (Maros *et al.* 2003). There are only a few large nesting populations of the green, hawksbill, and loggerhead turtles left in the world.

11.3 Birds

Birds to be found feeding intertidally on sandy beaches include Laridae (gulls and terns), Scolopacidae (sandpipers, sanderling, snipes), Chionididae (sheathbills), Haematopodidae (oystercatchers), and Charadriidae (plovers) — all members of the order Charadriiformes. Less commonly encountered in this habitat are members of the order Ciconiiformes, including Ardeidae (herons), Scopidae (Hamerkop), and Threskiornithidae (ibises and spoonbills). The Phalacrocoridae (cormorants) feed offshore by diving, as do the terns, and these may have considerable impact on surf-zone fishes where the birds are abundant. Other orders may be represented in specific areas or under certain conditions. For example, in Chile the falconiform chimango (*Milvago chimango*) feeds intertidally on sand mussels. In South Africa, swallows (*Hirundo rutisca*) feed on sand hoppers (Amphipoda; Talitridae). Swallows, flycatchers, pipits, and other forms may also be encountered at times, especially when large numbers of insects are blown onto the beach.

Birds occurring on sandy beaches may be foraging, roosting, or nesting on the backshore. Foraging birds can be divided into three general groups: (1) probers that use tactile cues to locate prey below the surface, including deep probers with long bills such as oystercatchers and whimbrels, and shallow probers with short bills such as sanderling; (2) surface feeders and peckers that forage using visual cues, such as many small plovers and even wagtails and passerines; and (3) scavengers that feed on stranded carrion, typically gulls and raptors. Birds that commonly roost on beaches but do not feed in the intertidal zone inlcude terns, cormorants, pelicans, and flamingos, which feed offshore or in other coastal habitats. Birds that nest on the backshore include species that also

forage on the beach, such as plovers and oystercatchers and others, such as terns. Typical beach birds are illustrated in Figure 11.3.

The presence and abundance of birds on sandy beaches shows considerable fluctuation due to movements of migrants and variations in numbers of resident species. Peak abundance of shorebirds on sandy beaches can exceed 1,000 individuals per kilometer during migration periods and annual abundance can exceed 100 per kilometer. However, in many areas numbers may be declining due to loss of coastal habitats and disturbance due to human activity. Many shorebirds are totally or partially migratory, visiting Southern Hemisphere beaches for part of the year (for example, in the austral summer), so that avian densities are in general greater there in summer than in winter. In the Northern Hemisphere, peak abundance may be in fall and winter during the nonbreeding season (Figure 11.4). Some species may peak in abundance on beaches they use as stopover points during migrations in spring and fall, whereas on their breeding beaches they will peak in summer. Abundance is strongly related to prey availability and is generally greatest on beaches with highest macrofauna biomass. Dugan *et al.* (2003) showed that two species of plovers on Californian beaches were most abundant when wrack and their food, wrack-associated macrofauna, were most abundant. Long-term effects can include decreases in shorebird abundance in years following El Niño Southern Oscillation events (Hubbard and Dugan 2003).

Overwintering or nonbreeding populations commonly have a far greater impact per bird than do breeding populations. Birds are less active as predators or scavengers at night than during the day, but it should be noted that although nocturnal foraging (if it occurs) is generally of shorter duration than diurnal feeding it may in some species be of greater intensity. Birds constitute by far the most mobile component of the sandy-beach ecosystem, moving easily and rapidly not only from one part of the beach to another but from beach to beach in search of food. Opportunistic foraging appears to be the rule, highly specialized diets the exception, and mixed-species nonbreeding assemblages tend to select the most abundant prey species. This is even more apparent on sandy beaches than on rocky shores, because of the greater importance of tactile and chemical cues in the former environment.

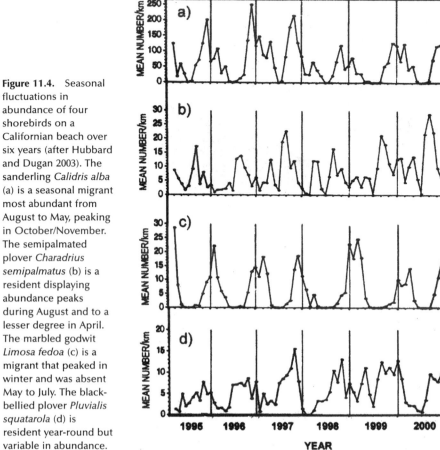

Figure 11.4. Seasonal fluctuations in abundance of four shorebirds on a Californian beach over six years (after Hubbard and Dugan 2003). The sanderling *Calidris alba* (a) is a seasonal migrant most abundant from August to May, peaking in October/November. The semipalmated plover *Charadrius semipalmatus* (b) is a resident displaying abundance peaks during August and to a lesser degree in April. The marbled godwit *Limosa fedoa* (c) is a migrant that peaked in winter and was absent May to July. The black-bellied plover *Pluvialis squatarola* (d) is resident year-round but variable in abundance.

Although the diets of shorebirds may be catholic, different sizes of prey are commonly taken by different species, which may also display some spatial separation up the shore. For example, sanderling (*Calidris alba*) typically feed as close to the water as possible, running down the slope between waves to probe the sand, whereas plovers forage higher up the shore. Temporal separation of feeding activities also occurs, some species feeding on a falling tide or at a specific time of day. Not only do shorebirds tend to select prey of a certain size range but of a particular sex of invertebrate prey species, where pronounced sexual dimorphism is apparent. They may even select a particular color morph. This is, however, more common on rocky shores than on sandy beaches, where chromatic polymorphism is less common and foraging does not rely primarily on vision.

Some limitation on possible prey is imposed by beak morphology and the depth to which the beak can conveniently penetrate the sand. Thus, the burrowing bivalve *Donax serra*, on South African beaches, is safe from most birds — which cannot reach it in the sand and in any case cannot open it to feed on the flesh within. The chief exception is the African black oystercatcher (*Haematopus moquini*, Figure 11.3), a deep prober with a robust bill — although other birds may detach and eat protruding *Donax* siphons. In some cases, however, behavioral adaptations have developed that effectively overcome such morphological limitations. Thus, the kelp gull *Larus hartlaubi* has learned to "paddle" the water-saturated sand of the foreshore with its feet, bringing individuals of *Donax* to the surface. It then commonly flies to some height with its prey, eventually dropping the clam onto a hard surface such as a rocky outcrop or a roadway to break the shell and facilitate extraction of the flesh. Dropping clams to break them has also been recorded in several other sandy-beach birds.

Predation by birds on the intertidal invertebrates of soft substrata in general has been estimated to range from 6 to 44% of annual production, and Myers *et al.* (1982) — working on a sandy beach in California — estimated a 41% reduction in available prey energy due to avian predation. Hockey *et al.* (1983) calculated that on southern African sandy beaches birds consume from 2 to 65% of invertebrate standing crop and from 10 to 49% of production annually, although the latter figure may well be an overestimate.

It has been suggested that such impact is on the whole negligible in the tropics (Schneider 1985). This apparent decrease in avian predation pressure with decreasing latitude is a trend opposite that generally attributed to predatory fishes and invertebrates. The greatest predatory impacts by birds are apparent in migratory nonbreeding avian populations, these tending to concentrate at discrete localities. The extraordinarily high densities of birds that may then be encountered in some areas lead to predation pressures that almost certainly could not be sustained by the prey were it not for the temporal refuge afforded by the seasonal migration of the birds. Even birds maximizing their overwintering fitness must of necessity predate their prey at a level no higher than the maximum sustainable rate. Feedback mechanisms are clearly important in protecting the prey from overexploitation. Shorebirds respond to a decrease in prey availability, which they themselves may have brought about, by spatial redistribution of foraging, by diet switching, or by increased tolerance of less favorable prey size.

The extent to which prey depletion stimulates avian migration is not understood, although it is known that migratory departure may coincide with low densities of prey. It has also been proposed that the timing of migration in Charadrii is maintained by selective forces originating in the seasonality of prey production. Relationships between migration patterns and prey densities are likely to be complex, and more work is required before any firm conclusions as to cause-and-effect relationships can be attained. However this may turn out, it should be noted that in not a single instance has poor feeding conditions been associated with prey depletion by the birds themselves. Neither has any density-dependent mortality of shorebirds ever been directly attributed to depressed prey densities, despite several observations that shorebird numbers may be correlated with such densities.

The apparent complexity of prey-predator interactions is increased by consideration of their relative mobilities. Where the prey are immobile, the behavior of the predators should predominate. Where the prey are relatively mobile and a spatial refuge exists, the prey responses should dominate. Where both are mobile and there is no spatial refuge, no clear pattern may emerge. In view of the great mobility of invertebrates on high-energy beaches, as compared with sheltered shores, this may have an effect on prey-predator relationships.

The predator-prey arms race is likely to result in the evolution of anti-predator behavior being followed by a compensatory change in the behavior of the predator, and vice versa. However, cause-and-effect relationships in this regard are difficult or impossible to demonstrate. Thus, several bivalve mollusks bury themselves most deeply in winter, when the risk of predation by birds is highest. Although it is true that deep burrowing affords protection from such predation, it may be the result of other factors, predator avoidance being fortuitous. Roberts *et al.* (1989) demonstrated a tidal rhythm of vertical migration in the clam *Mercenaria* through sand (and other habitats), the animals being buried most deeply at the time of low water, and have ascribed this behavior to avoidance of avian predators. The mass nocturnal emergences of Crustacea such as the amphipod *Talitrus* and the isopod *Tylos* (which remain hidden during the day) have also been interpreted as prey avoidance, because the kelp gulls, turnstones, and other birds that prey on them tend not to feed at night. However, it could equally well be argued that the predominantly nocturnal emergence of semiterrestrial Crustacea is linked to the avoidance of high temperatures during the day and the threat of desiccation, the talitrid amphipods in particular showing little resistance to water loss.

It is often forgotten that although birds certainly exploit intertidal invertebrates as food they also return to the beach quite considerable quantities of organic material in the form of feces, feathers, and carcasses. Hockey *et al.* (1983) calculated that the South African sandy-beach avifauna consumes about 20% of the total energy consumed on these beaches. Not all of this is returned to the intertidal beach, of course, and much of it must enrich the dune system. It may also be noted that although birds such as cormorants do not normally feed intertidally they do roost on beaches and consequently contribute nutrients to the system.

11.4   Conclusions

Turtles play minor roles in sandy-beach ecology globally but can be locally important through the disturbance their nesting activity causes. Further, because they are threatened in many areas they confer on the beaches where they nest a high-conservation status. Birds, on the other hand, are an important component of the sandy-beach macrofauna, playing a significant role as predators in beach food chains, especially where macrofauna is concentrated at the top of the shore around wrack deposits. Like turtles, they raise the conservation status of beaches where they nest along the dune/beach interface.

Energetics and Nutrient Cycling

12

Chapter Outline

12.1   Introduction

This chapter examines the food sources and the main trophic pathways for sandy beaches, the patterns and processes of energy flow and nutrient cycling in beach/surf-zone systems, and the exchanges between beaches and adjacent ecosystems. The main biotic components have been covered in the preceding chapters, and this section brings together their trophic interactions by examining beach and surf-zone ecosystems as a whole.

In sandy beaches, the macrofauna and the interstitial fauna comprise distinct communities, with few or no trophic links. In terms of trophic relationships, the macrofauna form part of a larger food web that includes zooplankton, fishes, and birds, whereas the interstitial fauna in principle constitute a discrete food web within the sand. In fine sands on sheltered beaches, some trophic links between the two systems may, however, occur: finer sediments have smaller pore spaces and lower oxygen tensions, resulting in large burrowing meiofaunal forms concentrated at the surface. Under such circumstances, and where there are significant numbers of macrofauna deposit feeders, the meiofauna is available as food to the macrofauna, juvenile fishes, and even zooplankton. Such conditions are not typical of exposed beaches, where the sand is well drained and the meiofauna are all truly interstitial, penetrating deep into the sediment and thereby being unavailable to macrofauna. Although the macroscopic and interstitial food chains are therefore virtually distinct on exposed sandy beaches, both faunas may utilize common food sources in the form of dissolved and particulate organic matter in the surf water.

The surf zones of sandy beaches affect both beach energetics and nutrient cycling. Beaches are marine systems, despite having some terrestrial affinities, and their food input is derived mainly from the sea. Consequently, size of the beach, proximity of food sources, and surf-zone characteristics are very important in determining the food supply. Two main types of beach may be recognized, based on the presence or absence of a surf zone: (1) reflective, sheltered beaches without surf zones (which comprise sand bodies only, have faunas that are entirely benthic, and are totally dependent on food imports, mainly from the sea) and (2) high-energy or exposed beaches of the intermediate or dissipative types, where the ecosystem comprises the beach and a surf zone to the outer limits of surf circulation cells (that is, a sand body and a water envelope), and whose faunas typically include benthos, plankton, and nekton. Of course, there are intermediate types.

On exposed beaches with surf zones, a third biotic component may be distinguished in addition to the macroscopic fauna and the meiofauna: water-column microbes. These include bacteria and their predators, flagellates and ciliates. This microbial loop is the least studied of all beach/surf-zone food chains and warrants more attention.

12.2   Food Sources

Beaches lack attached macrophytes below the drift line, except in rare cases of sheltered beaches with seagrass meadows extending onto the lower shore. Food chains mainly begin and end in the sea, the land playing a relatively minor role. Food inputs to beaches may be divided into the categories listed in Table 12.1. The resident primary producers on beaches are epipsammic diatoms. On sheltered beaches and flats of fine sand these may contribute to a measurable, but never high, primary productivity. Values range from 0 to $50\,gC·m^{-2}·y^{-1}$, the highest values being recorded in the most sheltered situations. In beaches experiencing wave action, values are less than $10\,gC·m^{-2}·y^{-1}$, and practically zero in exposed situations. This limited food source is available to the benthic macrofauna and meiofauna.

Surf-zone primary production rates are less well known but are certainly more variable. Where surf diatom accumulations occur, rates may be exceptionally high — on the order of 500 to $2,000\,gC·m^{-3}·y^{-1}$ and peaking to 5 to $10\,gC·m^{-3}·h^{-1}$ where consistent accumulations occur (see Chapter 4). In the absence of such diatom patches, primary production rates in surf waters are much lower and probably range from 20 to $200\,gC·m^{-3}·y^{-1}$ in different systems. The highest rates occur in dissipative surf zones with consistent phytoplankton patches, and the lowest rates in lower-energy intermediate surf zones. In the

Table 12.1.   Food sources for sandy beaches.

Food type	Source	Remarks
Benthic microflora	Beach sand	On sheltered beaches
Surf diatoms, flagellates	Surf water	In well-developed surf zones
Stranded macrophytes	The sea	Near kelp beds, seagrass meadows, estuaries, etc.
Carrion	The sea	Especially near seabird and seal colonies
Particulates	The sea	—
Dissolved organics	The sea	—
Insects	The land	Particularly during strong, offshore winds
Organic detritus	The land	From dune vegetation

Eastern Cape, South Africa, surf diatoms produced $120,000 \, gC \cdot m^{-1} \cdot y^{-1}$ within the surf zone (250 m), whereas mixed phytoplankton (mainly autotrophic flagellates) produced $110,000 \, gC \cdot m^{-1} \cdot y^{-1}$ in the rip head zone (250 m) (Campbell and Bate 1987). When surf phytoplankton is concentrated into foam, it represents an important food source for benthic and planktonic filter-feeders, and even fishes (Romer and McLachlan 1986). It is also a more consistent food resource than some other marine inputs, such as carrion, which are highly erratic in supply. Nevertheless, primary production can be widely variable on a daily basis.

Particulate organic material, or detritus, is usually at least as common as phytoplankton and represents another relatively constant food source of marine origin. It may be derived from the breakdown of plants and animals, feces, messy grazing, or aggregations. There are no published values that allow generalizations to be made, but in the Eastern Cape values of 1 to $5 \, gC \cdot m^{-3}$ have been recorded (McLachlan and Bate 1984) and this range probably covers most surf-zone waters.

Several studies have examined macrophytes brought to beaches by the sea. These may be salt-marsh grasses from estuaries, seagrasses from sheltered subtidal sands, or kelps and other algae from rocky shores and subtidal reefs. Most beaches receive a small amount of such inputs, but in some situations (for example, sandy coves adjacent to rocky shores or beaches fringing seagrass meadows) the input may be substantial, especially after winter storms. In such cases, wrack cover may be extensive and input values of between 20,000 and $300,000 \, gC \cdot m^{-1} \cdot y^{-1}$ have been recorded (Griffiths et al. 1983, McLachlan and McGwynne 1986, Dugan et al. 2003) — this material totally dominating the sandy-beach food chains (for review, see Colombini and Chelazzi 2003). The macrophytes are fed upon directly by scavengers associated with the drift line, such as talitrid amphipods, isopods, and insects. However, much decomposition is accomplished by bacteria, and breakdown can be completed in days or weeks — with seagrasses taking longer than kelps and other algae (Koop and Lucas 1983, Griffiths et al. 1983, Jedrzejczak 2003). Following breakdown, large quantities of dissolved and particulate material may leach into the sand and become available to the interstitial fauna. These detached macrophytes may not all end up on the beach, however, and significant quantities may wash around in the surf zone — where they provide food as well as shelter from predators (Robertson and Lenanton 1984).

Carrion represents a highly erratic food supply, usually mainly of marine origin and common to all beaches. Jellyfish, siphonophores, bivalve mussels, tunicates, fishes, turtles, seabirds, seals, cetaceans, and other animals are all represented at various times. When other food (for example, surf phytoplankton or stranded macrophytes) is of any significance, carrion usually represents a minor food source. However, in the absence of other major inputs and on beaches adjacent to seal or seabird colonies carrion may be of major importance. Carrion input has been estimated in the Eastern Cape at about $120 \, gC \cdot m^{-1} \cdot y^{-1}$, and similar rates have been recorded in a few other studies (Colombini and Chelazzi 2003). On reflective beaches without surf zones, where life for filter-feeders is harsh because of the steep beach face and the brief swash periods, scavengers and predators dominate the fauna. These are often supralittoral forms, and stranded carrion may be their main source of food.

Dissolved organic materials (materials passing through a 0.5-μm filter) occur in high concentration in all seawaters and have various origins: phytoplankton exudates, faunal excretion, mucus breakdown, leachates from damaged plants and animals, and material generated through messy feeding by members of the macrofauna. This material may be concentrated by the waves into a rich, yellow storm foam, which can accumulate in the surf or on the beach. To what extent this foam is available to macroscopic life is uncertain, although it appears to be utilized by *Donax serra*. It is certainly of great importance to water-column microbes and to the interstitial fauna. When deposited on the sand surface, it may also be taken by members of the macrofauna, such as sand lickers or deposit feeders. Typical concentrations of dissolved organic matter in seawater are 0.1 to $5\,gC \cdot m^{-3}$ (or $mg \cdot \ell^{-1}$).

Finally although not usually of great quantitative significance, two organic inputs from land, insects and plant litter or detritus, are often found on the beach and in surf waters. They are potentially of importance where winds blow offshore, particularly during certain seasons (for example, during insect hatches in spring). Their value as a food for birds, fishes, and invertebrates has often been recognized but never successfully quantified for a sandy beach. Terrestrial plant litter mainly contributes to the drift line, where (if refractory) it may remain for a long time.

A wide variety of food materials, mainly of marine origin, thus impinge on sandy beaches. Some organic materials may be produced within the beach/surf-zone system, whereas others are imported. Depending on the relative quantities of these different resources available on the beach, a number of different trophic pathways may be present.

12.3   Macroscopic Food Chains

The living components of macroscopic food chains comprise benthos, zooplankton, fishes, and birds (Figure 12.1). Among the benthos, filter-feeders, herbivore scavengers, and carnivorous scavengers/predators are the main groups. Deposit feeders also occur on sheltered shores. Where the sediment is sufficiently stable and organic detritus abundant in the sediment, deposit feeders such as *Arenicola*, *Callianassa*, and the polychaete worm *Scolelepis* may occur in large populations. Here, they can turn over the sediment and feed on epipsammic microflora and even meiofauna. On exposed open beaches, however, this component is either absent or insignificant. The organic content of beach sand, including microflora and meiofauna, ranges from 0 to $5\,mg \cdot g^{-1}$.

Where beaches provide good feeding conditions in the swash, filter-feeders are usually dominant and often very abundant. Feeding on surf phytoplankton and/or particulate detritus, they may represent the main pathway of energy flow through the benthos, accounting for up to 90%. The flatter a beach and the more the surf zone leans toward the dissipative state the greater the tendency to develop surf phytoplankton accumulations. Scavengers and predators seldom attain great abundance. The notable exceptions are beaches with large macrophyte inputs, where scavenging herbivores may dominate the fauna. Here, benthic biomass is concentrated around the drift line and decreases downshore — a trend opposite that found on the majority of beaches. This phenomenon may be accentuated by the fact that where large macrophytes pass through the system and wash around on the beach, filter-feeding bivalves find it impossible to extrude their siphons above the sand surface, whereas the behavior of carnivorous scavengers such as *Bullia* is completely disrupted.

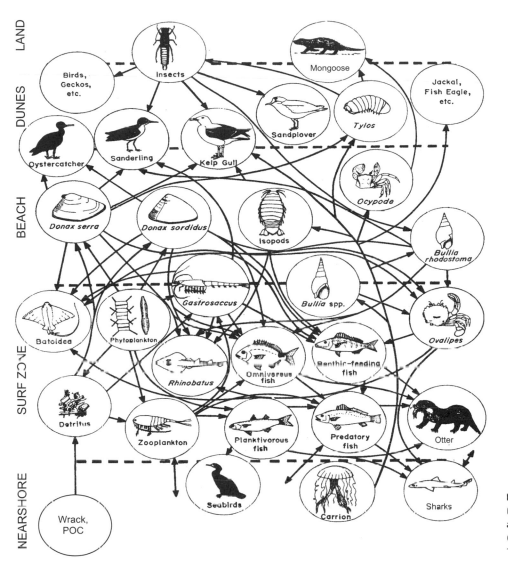

Figure 12.1. Food web for the macrofauna on an Eastern Cape beach (after McLachlan *et al.* 1981b).

LAND

DUNES

BEACH

SURF ZONE

NEARSHORE

Birds, Geckos, etc.

Insects

Mongoose

Jackal, Fish Eagle, etc.

Sandplover

Tylos

Oystercatcher

Sanderling

Kelp Gull

Ocypode

Donax serra

Donax sordidus

Isopods

Bullia rhodostoma

Gastrosaccus

Bullia spp.

Batoidea

Phytoplankton

Ovalipes

Rhinobatus

Omnivorous fish

Benthic-feeding fish

Detritus

Zooplankton

Planktivorous fish

Predatory fish

Otter

Wrack, POC

Seabirds

Carrion

Sharks

Feeding on primary food sources and being prey in turn for top predators, the benthic macrofauna occupies a key position in the center of sandy-beach food chains. Where surf zones are present, they share this position with the zooplankton. The zooplankton of surf zones is dominated by relatively large forms (for example, mysid shrimps, prawns, and benthoplanktonic forms, see Chapter 10). Little is known of these surf-zone animals, and only in the Eastern Cape of South Africa have they received detailed attention. Feeding groups include filter-feeders, predators, and omnivores. Swarms of mysids and prawns may filter phytoplankton from surf diatom accumulations or take detritus. They can switch to capturing smaller zooplankton at night, when the phytoplankton disappears from the water column. The aggregations of these zooplankters into swarms make them the chief prey items of fishes (Lasiak and McLachlan 1987). Because of their relatively small size and high level of activity, zooplankters are generally characterized by higher turnover rates than the benthos. Consequently, they may be of great importance in energy flow through macroscopic food chains.

This raises the question of different turnover rates of faunal components on various beach types. Scottish workers have highlighted great differences in the turnover rates of benthos between temperate and tropical beaches (Ansell *et al.* 1972, 1978). They showed that turnover, or production/biomass values, of the benthos on tropical Indian beaches were about 10 times greater than on temperate Scottish beaches. For example, temperate bivalves (*Tellina* and *Donax*) have P/B ratios of less than $1\,y^{-1}$, whereas tropical *Donax* have ratios of about $10\,y^{-1}$. On intermediate beaches, in warm temperate situations, values of 1 to $2\,y^{-1}$ are typical (McLachlan *et al.* 1981b). Generally, smaller forms such as amphipods and isopods have higher turnovers, and the zooplankton still higher ratios — on the order of 5 to $10\,y^{-1}$, even in temperate regions. The importance of populations in energy flow is therefore dependent not only on their abundance or biomass but on their turnover rate. Thus, populations of small tropical invertebrates may consume greater quantities of food than larger populations of temperate species of larger body size.

Sandy-beach macroscopic food chains end in fishes, carnivorous invertebrates, and birds. Most birds can only gain access to intertidal and supralittoral prey and are therefore of significance only where the benthos of these zones is abundant. On beaches receiving high inputs of macrophytes, vast supralittoral populations of talitrid amphipods, oniscid isopods, and insects may dominate the benthos and be preyed on almost exclusively by birds. Birds may also be significant predators on intertidal populations such as clams (see Chapter 11), although marine predators such as fish, crabs, and gastropods are usually more important. A simple gradient may thus be envisaged in which terrestrial predators (mainly birds, but also invertebrates such as arachnids, ghost crabs, and insects) are dominant at the top of the shore and marine predators at the bottom of the slope. Which group of predators is more important thus depends on the location of the center of gravity of the benthic biomass. As this usually lies in the lower intertidal (or even in the sublittoral) on most beaches, marine predators generally have the greater impact.

Invertebrate marine predators include gastropods and crabs, the former being more important in sheltered situations. Slow-moving predatory gastropods are limited in their prey and generally attack clams such as *Donax* and *Mesodesma*. Typical genera include *Natica*, *Olivancillaria*, and *Oliva*. Crabs, both supralittoral ocypodids and sublittoral portunids, operate over a wider range of beach types and may be important even in exposed situations. Ocypodids are both scavengers and predators, and at times capture intertidal benthos. On coarse-grained reflective beaches with depauperate faunas they may be the only abundant members of the macrobenthos. Portunids are voracious predators, moving into the intertidal at high tide to prey on bivalves and gastropods, as well as other invertebrates. The common portunid genera are *Ovalipes*, *Arenaeus*, *Matuta*, and *Callinectes*.

Fishes are the only higher predators of zooplankton, and on most beaches they are the main predators of intertidal and subtidal benthos as well (for review of surf-zone fish bioenergetics see Du Preez *et al.* 1990). Juvenile fishes are an important component of the ichthyofauna, preying on zooplankton and the smaller benthos. Juvenile flatfishes crop the siphons of bivalves and the palps of polychaetes, and take smaller prey whole. Benthic feeders dominate fish biomass, and their most successful members are often the flatfishes and the batoid rays. These fishes move into very shallow water and feed on benthos, but may at times also take zooplankton from the water column when it is abundant. Typical benthic feeders include *Rhinobatos*, *Myliobatos*, *Umbrina*, *Lithognathus*,

Menidia, *Menticirrhus*, *Dasyatis*, and *Pomadasys*. Planktivores include both phyto-plankton and zooplankton feeders. Mullet of the genus *Liza* graze surf diatom accumu-lations directly (Romer and McLachlan 1986), whereas most juvenile fish and adults of *Monodactylus* and *Pomadasys* feed on zooplankton. However, most grazing on zoo-plankton is by opportunists that normally feed on the benthos. Zooplankton is thus a less regular component of the diet of surf-zone fishes, although when abundant it may supercede all other foods. This food tends to be taken intermittently because of its patchy occurrence and swarming nature. When taken, however, it may be consumed in huge quantities.

Predatory fishes, cetaceans, seals, and diving birds (such as cormorants) constitute the top of the beach/surf-zone food chain. Of these components only predatory or piscivo-rous fishes are generally important. Typical predators include *Pomatomus*, *Trachinotus*, and the carangids — the two former mainly in temperate waters and the latter largely tropical. These are highly motile species that traverse the surf zone and are not consid-ered true residents. Even these fishes take benthic and zooplankton prey, however, high-lighting once again the opportunistic feeding tendencies of fishes associated with the surf zone. With these introductory comments on general feeding relationships and compo-nents of sandy-shore food chains, some specific examples of such food webs can be considered (Figure 12.2).

Examples of Macroscopic Food Chains

(1) *India*. Tropical Indian beaches, studied by Steele (1976) and Ansell *et al.* (1978), are low-energy reflective to intermediate beaches during calm weather but erode severely during monsoon storms. The surf zone is mostly negligible and there is no pronounced drift line. These workers divided the benthos into particulate feeders and carnivores, the former consuming on average about $1,000 \, \text{gC} \cdot \text{m}^{-1} \cdot \text{y}^{-1}$ and the latter $70 \, \text{gC} \cdot \text{m}^{-1} \cdot \text{y}^{-1}$, more than 90% thus being taken by the filter-feeders. Crabs were the chief predators in both the supralittoral and sublittoral, followed by fishes, the impact of birds being negligible. Man was also a major predator. Despite a low biomass (5 to $10 \, \text{gC} \cdot \text{m}^{-1}$), these beaches nevertheless process large amounts of organic material derived from the sea because of the high turnover rates of the invertebrate fauna, the overall P/B being $20 \, \text{y}^{-1}$. These food chains end again in the sea, the production being cropped and exported by marine predators.

(2) *Scotland*. In low-energy dissipative Scottish beaches studied by the same workers, biomass was higher (at about $240 \, \text{gC} \cdot \text{m}^{-1}$), but turnover was slower, the average P/B being about $2 \, \text{y}^{-1}$. Total consumption was thus about $500 \, \text{gC} \cdot \text{m}^{-1} \cdot \text{y}^{-1}$, approximately half that in India. Most of this was again due to particulate feeders, especially filter-feeding bivalves, although deposit feeders and carnivores were also present. Another contrast was pre-sented by the predators. Fishes, particularly juvenile flatfishes, were more important on Scottish than on Indian beaches. These demersal fishes prey on *Tellina* siphons, poly-chaete palps, and whole small crustaceans.

(3) *Western Australia*. The coast around Perth, in Western Australia, is characterized by microtidal low-energy reflective beaches without surf zones, many of them receiving high macrophyte inputs. Their intertidal faunas are impoverished, consisting of small populations of donacid bivalves and hippid crabs — with sparse populations of

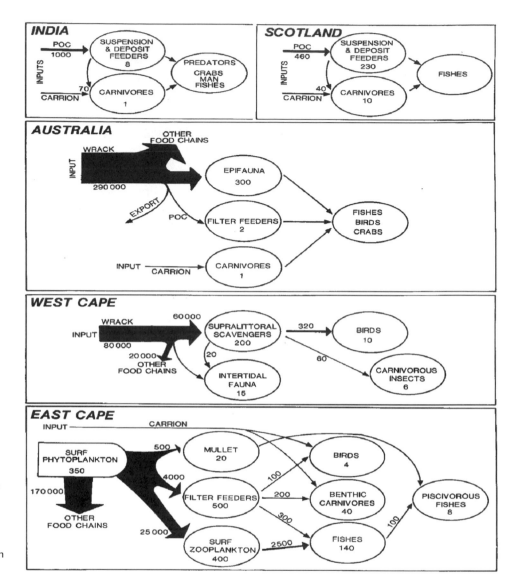

Figure 12.2.
Macroscopic food chains and carbon flow through some sandy-beach systems. All values are given in gC·m⁻¹·y⁻¹. For a full list of references see Brown and McLachlan (1990).

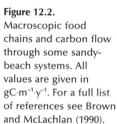

ocypodids occupying the drift line. Total biomass is only about 1 gC·m⁻¹. The energetics of these populations have not been quantified, although they are known to consume small quantities of particulate material and stranded organisms cast ashore. The beaches are thus dependent on food inputs from the sea. However, stranded wrack (input of which totals 290,000 gC·m⁻¹·y⁻¹) accumulates on these beaches, with an average standing biomass of 10,000 to 50,000 gC·m⁻¹ (Robertson and Hansen 1981, McLachlan 1985). This wrack (kept moist by the swash) supports large populations of epifauna, notably the amphipod *Allorchestes compressa* — which comprises more than 90% of this fauna and averages about two individuals per gram of wrack, reaching densities of 100 per gram of carbon. The wrack is broken down by physical processes, microbial degradation, and amphipod grazing of the fragments (Robertson and Lucas 1983). Not only stranded wrack but large wrack deposits in the shallows are inhabited by these amphipods. As there is no true surf-zone fauna, and because intertidal populations are sparse, predators are not numerous on these Western Australian beaches. They include wading birds, portunid and ocypodid crabs, and fishes. The great abundance of amphipods associated with wrack, not only on the intertidal beach but in the shallows, constitutes a valuable food supply for fishes — which are, in fact, the main predators. Moderate populations of small fishes inhabit the swash and shallows, and most of these are benthic feeders, with *Allorchestes* as the chief dietary item. These Australian beaches thus serve as interfaces, processing organic materials (predominantly macrophytes) derived from the sea. Most of the production returns to the sea via marine predators (Robertson and Lenanton 1984).

(4) *Western Cape, South Africa*. In contrast to the Western Australian beaches, where much of the macrophyte input remains in the shallows, on the sandy beach near Cape Town studied by Griffiths *et al*. (1983) the macrophytes were all cast up on the drift line. This shore differs from the Australian sites in having a mesotidal regime, in displaying strong wave action, and in its close proximity to rocky shores and reefs — thereby ensuring high inputs of the kelps *Eklonia* and *Laminaria*. Inputs average about 80,000 gC·m^{-1}·y^{-1}. The intertidal fauna is sparse, although an abundance of insects, talitrid amphipods, and oniscid isopods inhabit the drift line — with an average biomass of 200 gC·m^{-1}. This supralittoral fauna grazes the kelp directly, whereas physical processes and microbial degradation aid in the breakdown of the material. The supralittoral macrofaunas are estimated to consume up to 70% of the kelp, the remainder being processed by interstitial fauna and microorganisms. Much of this consumption is returned to the detritus pool as feces. The production of the drift line fauna (400 gC·m^{-1}·y^{-1}) is consumed by birds (80%), by carnivorous coleopterans (15%), and by carnivorous intertidal isopods (5%). These beaches thus also process large quantities of macrophytes imported from the sea. Although much of the carbon and nutrients contained in the kelp may ultimately be recycled to the sea, the food chains end mainly in terrestrial predators and much secondary production is exported to the land. In terms of gross carbon inputs, however, this amounts to only 0.5% of the original input going to terrestrial predators. The remaining 99.5% is either respired or returned to the sea.

(5) *California*. In Southern California, Hayes (1974) estimated kelp inputs at 30,000 gC·m^{-1}·y^{-1}, of which only 4 to 5% was consumed by the supralittoral isopod *Tylos*. Dugan *et al*. (2003), working in the same area, also showed the great importance of wrack and the key role of birds as predators on wrack-associated macrofauna.

(6) *Eastern Cape, South Africa*. Exposed sandy shores in the Eastern Cape are of the high-energy intermediate type, tending toward the dissipative extreme and becoming fully dissipative during rough conditions. Characterized by accumulations of surf diatoms, they support rich faunas both on the beach and in the surf zone. They are, however, not as rich as some Brazilian or Washington State beaches, which are dissipative and have continuous heavy diatom patches. The Eastern Cape beaches probably fall near the center of the range of beach types supporting surf diatoms, and may be considered almost at the midpoint between ultraproductive diatom-dominated dissipative beaches and less productive intermediate beaches with sparse diatom populations.

The ecology and energetics of these beaches have been reviewed by McLachlan and Bate (1984), McLachlan and Romer (1990), and Heymans and McLachlan (1996). Food inputs include stranded carrion and surf diatom production, as well as a variety of minor sources. The diatoms — although not as rich as on sandy shores in Washington, Brazil, and New Zealand — are consistent and extremely productive. Campbell and Bate (1987) estimated diatom production at 120,000 gC·m^{-1}·y^{-1} within a surf zone 250 m wide and

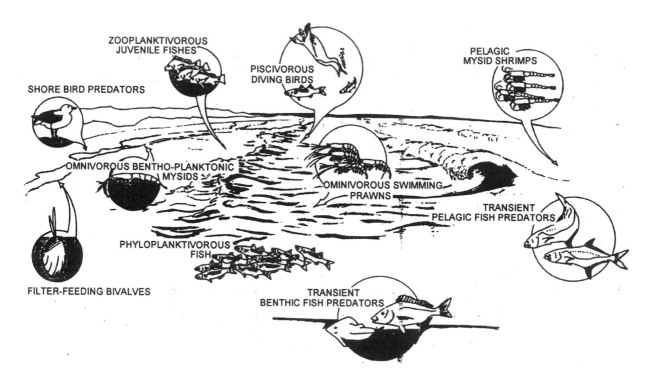

Figure 12.3. Major macrofaunal components of the surf phytoplankton-associated grazing food chain on Sundays River Beach in the Eastern Cape, South Africa (after Romer 1986).

0 to 3 m deep. This production is equivalent to $500\,gC \cdot m^{-2} \cdot y^{-1}$. Production was, however, patchy and highly variable and included exudates and the production of copious mucus. Primary production outside the breakers, in the rip-head zone, was due both to diatoms and flagellates and totaled $110,000\,gC \cdot m^{-1} \cdot y^{-1}$ in a zone 250 m wide and 3 to 10 m deep. Carrion input was estimated to be in the range 100 to $200\,gC \cdot m^{-1} \cdot y^{-1}$.

The center of this food chain is occupied by benthos, zooplankton, and fishes (Figure 12.1). The benthos is dominated by the filter-feeder *Donax*, constituting over 90% of the biomass, with scavengers and predators making up the difference. Supralittoral populations are sparse due to the absence of wrack deposits. Total benthic macrofaunal biomass is in the region of $540\,gC \cdot m^{-1}$, of which some 90% is intertidal and the rest occurs in and beyond the surf zone out to a depth of 10 m, 500 m from shore. Zooplankton biomass is dominated by omnivorous prawns (45%) in the surf zone and mysids (32%) in the rip-head zone, both of these taxa grazing surf diatoms. Zooplankton biomass totals some $400\,gC \cdot m^{-1}$. One fish, the mullet *Liza richardsoni*, also grazes diatom accumulations and its biomass averages $20\,gC \cdot m^{-1}$. It has been estimated that about 25% of the surf-zone primary production (or 15% of surf- and rip-head zone total primary production) is consumed by these groups (Figure 12.3). Zooplankton are the most important consumers (80%), followed by benthic filter-feeders (16%) and mullet (4%). Fishes (with a biomass of about $130\,gC \cdot m^{-1}$), birds ($4\,gC \cdot m^{-1}$), and crabs ($6\,gC \cdot m^{-1}$) are the predators, cropping respectively about 89%, 5%, and 6% of the production of zooplankton and benthos. These fishes, in turn, are taken by predatory fishes (with a biomass of some $10\,gC \cdot m^{-1}$).

This system is thus largely self-contained if the beach, surf, and rip-head zones are viewed as a single unit. Only a small proportion of primary production are routed through the macroscopic food chain, where zooplankton and filter-feeding benthos are the chief components. Some primary production may be exported directly, and secondary production is exported via birds and top fish predators moving out of the system. By far the bulk of energy flow, however, occurs within the system.

(7) *Brazil.* The Brazilian beaches described by Gianuca (1983) are similar to the system outlined previously, although they have been less intensively studied. They exhibit high diatom production, huge populations of intertidal filter-feeders, and probably large zooplankton populations as well. Fishes, birds, and crabs are again the chief predators, but neither predation nor biomass have been quantified.

12.4 Interstitial Food Chains

The interstitial food chain consists of bacteria, protozoans, and meiofauna within the sand, and except for very sheltered fine-sand beaches has no significant trophic links with the macrofauna. It is fueled by dissolved and particulate organic materials flushed into the sand by wave and tide action. To a large extent, therefore, the importance of the interstitial system in processing organic materials is determined by four factors: (1) the volume of water filtered by the sand body, (2) the residence time of this water in the sand, (3) its organic content, and (4) its temperature, which controls the metabolic rate of the interstitial biota. The interstitial system may thus be seen as a huge natural filter, cleansing and purifying the surf waters (Pearse *et al.* 1942). It mineralizes the organic materials it receives and returns the nutrients to the sea.

Water filtration by beach and surf-zone sand is described in Chapter 3. Essentially, the volume of water filtered by a beach increases from dissipative to reflective beach states from fine to coarse sands with decreasing tidal range and with increasing wave energy. The residence times of this water in the sand show the reverse trends. Coarse-grained exposed reflective beaches rapidly filter the most water, with rates generally exceeding $100 \, \mathrm{m^3 \cdot m^{-1} \cdot d^{-1}}$. Fine-grained sheltered dissipative beaches slowly filter the least, often at rates slower than $1 \, \mathrm{m^3 \cdot m^{-1} \cdot d^{-1}}$.

The volumes of water pumped through the bed of the surf zone are greatest in the presence of coarse sand, heavy wave action, and shallow water. They range from 0.01 to $0.5 \, \mathrm{m^3 \cdot m^{-2} \cdot d^{-1}}$ for most surf zones. As coastal waters contain 0.1 to $10 \, \mathrm{gC \cdot m^{-3}}$ in particulate organic form (POC) and similar levels of dissolved organic carbon (DOC), these processes introduce vast quantities of organic matter into the sand. This material is entirely or partially mineralized by the interstitial food chain during its passage through the sand. The longer the residence time the greater is the proportion of organics likely to be consumed and mineralized.

In addition to organic inputs, on sheltered sandy beaches the interstitial system may include its own primary producers in the form of benthic diatoms and autotrophic flagellates. Even surf diatoms may enter the sediment at night, although it is uncertain to what extent they remain in the sand and whether they are consumed by the meiofauna to any significant degree.

In its simplest form, the interstitial food web consists of bacteria utilizing POC and DOC, protozoans preying on bacteria (as well as consuming POC and DOC), meiofauna feeding on all of these, and carnivorous meiofauna taking other meiofaunal forms. Where epipsammic diatoms occur, they are also consumed by members of the meiofauna.

Bacteria take up DOC, not only from the interstitial water but that adsorbed onto sand grains. The rate at which they utilize DOC and POC is limited by oxygen and nutrient availability. These sources of nutrition are utilized to a lesser extent by higher trophic levels, which are consequently dependent on initial processing by the bacteria.

Interstitial protozoans utilize both primary foods and bacteria, and in some cases may even prey on small members of the meiofauna. Generally, however, they occupy an intermediate trophic position between bacteria and meiofauna and are usually less quantitatively important than either.

The meiofauna may broadly be divided into herbivores/detritivores, bacterivores, and predators. Most taxa have representatives in all three categories, only the Hydrozoa and the Turbellaria being exclusively predatory. Nematodes and harpacticoid copepods, the dominant groups in beach sand, are more versatile — the nematodes displaying all three feeding types and copepods lacking only predatory forms. In addition to preying on bacteria, the meiofauna may stimulate bacterial activity by the mechanical breakdown of detritus (making it more favorable for bacterial colonization), by excreting assimilable nutrients, by secreting mucus (which promotes bacterial growth), and by stirring the sand (bioturbation) — thus aiding the movement and diffusion of oxygen and nutrients.

Because of the small scale of the interstitial environment and the minute size of most of its members, studies of interstitial energetics have tended to adopt the black-box approach, measuring overall interstitial activity rather than attempting to quantify individual feeding relationships. This has been achieved by measuring benthic metabolism or oxygen demand of the sand *in situ*, in cores, or in sand columns in the laboratory and partitioning this demand between the various components of the interstitial biota. A variety of sand-column systems have demonstrated that the interstitial biota subsist entirely on DOC inputs or on DOC and POC combined. Such studies indicate that bacteria generally account for 90 to 95% of total interstitial oxygen demand and meiofauna some 5%. A high proportion of the total carbon input to the interstitial system may be consumed in this way.

Detailed studies of interstitial energetics have been undertaken by only three groups of workers: on Scottish and Indian beaches (Munro *et al.* 1978); on the macrophyte-loaded beaches of the Western Cape, South Africa (Koop and Griffiths 1982, Koop *et al.* 1982, Griffiths *et al.* 1983); and on exposed diatom-fueled beaches in the Eastern Cape (Dye 1981, McLachlan *et al.* 1981b,c, Malan and McLachlan 1991). The findings are as follows:

(1) *Scotland and India.* The comparative study of Scottish and Indian beaches revealed a number of important differences. The temperate beach had abundant epipsammic diatoms, which were absent from the tropical beach. DOC comprised 80% of the input in India but only 39% in Scotland. Community respiration on the tropical beach ($164\,gC{\cdot}m^{-2}{\cdot}y^{-1}$) was about four times that on the temperate beach ($42\,gC{\cdot}m^{-2}{\cdot}y^{-1}$) per surface area. However, respiration was similar if summed for the entire beach width, as the temperate beach was 200 m wide, compared with only 40 m in the case of the tropical Indian beach. On this basis, community respiration on the Indian beach accounted for $6{,}500\,gC{\cdot}m^{-1}{\cdot}y^{-1}$, compared with some $8{,}000\,gC{\cdot}m^{-1}{\cdot}y^{-1}$ on the Scottish beach.

Other differences included microbial production (estimated at 72 and $15\,gC{\cdot}m^{-2}{\cdot}y^{-1}$ on the tropical and temperate beaches, respectively) and meiofaunal biomass, which was an order of magnitude higher in Scotland ($0.2\,\text{gC}\cdot\text{m}^{-2}$) than in India ($0.02\,\text{gC}\cdot\text{m}^{-2}$). It was postulated that the more vigorous wave action on the Indian beach boosted interstitial metabolism and stripped bacteria from the sand.

It may be concluded that the tropical beach — experiencing higher temperatures, high rates of water flushing, and higher concentrations of organic matter in surf waters (0.9 to 3.7 as opposed to 0.5 to $1.0\,\text{mgC}\cdot\ell^{-1}$) — enjoyed much greater interstitial activity and that this was mainly microbial. The proportion of microbial production stripped by the waves was not estimated in this, or any other study, but the relatively low meiofaunal biomass on the Indian beach suggests that it may be considerable and thus worthy of quantitative investigation.

(2) *Western Cape*. In the Western Cape, rich interstitial populations develop in association with high kelp inputs to the beach. On a beach 60 m wide, meiofaunal biomass was about $200\,\text{gC}\cdot\text{m}^{-1}$ (or 3 to $4\,\text{gC}\cdot\text{m}^{-2}$) and bacterial biomass close to $300\,\text{gC}\cdot\text{m}^{-1}$ ($5\,\text{gC}\cdot\text{m}^{-2}$). Bacteria were concentrated at lower tidal levels, subsisting most probably on kelp leachates and breakdown products draining through the sand. The fact that meiofaunal oligochaetes and nematodes concentrated at the top of the shore suggests that their grazing pressure depressed bacterial numbers there. Bacteria not only assisted kelp breakdown on the drift line but rapidly assimilated breakdown products (up to $5\,\text{gC}\cdot\ell^{-1}$) draining into the sand. Bacteria accounted for more than 90% of the carbon input into the sand, converting it to bacterial biomass with about 30% efficiency. Total carbon input to the interstitial system from leachates and macrofaunal feces was estimated as $70,000\,\text{gC}\cdot\text{m}^{-1}\cdot\text{y}^{-1}$, or 85% of the total input into the beach. Based on this input and bacterial biomass, it was calculated that the bacteria might have a turnover (P/B) of $70\,\text{y}^{-1}$, whereas meiofaunal food requirements might be met by 30 bacterial turnovers a year. Whatever the bacterial turnover rate, it is clear that this beach processes a vast amount of kelp material via its interstitial system. This material enters the sand as the products of leaching and decomposition, as well as herbivore feces, and is mineralized mainly by bacteria.

(3) *Eastern Cape*. On the high-energy beaches in the Eastern Cape, where there is a well-developed surf zone with diatom accumulations, interstitial energetics have been examined both in the intertidal and subtidal. The biomass figures for bacteria, protozoans, and meiofauna in the 60-m-wide intertidal beach average approximately 10, 3, and $15\,\text{gC}\cdot\text{m}^{-1}$, respectively. Water filtration flushes some $10\,\text{m}^3\cdot\text{m}^{-1}\cdot\text{d}^{-1}$ through the system, introducing large quantities of POC and DOC. Surf diatoms also migrate into the sand at night and may exude significant amounts of mucus there. Based on oxygen consumption, the system consumes on average $3,200\,\text{gC}\cdot\text{m}^{-1}\cdot\text{y}^{-1}$, partitioned among bacteria (60%), protozoans (20%), and meiofauna (20%). The abundance and activity of the meiofauna suggests that it may not only graze bacteria but consume the primary food inputs. The interstitial fauna is abundant in the upper 1 to 2 m of the intertidal sediment but becomes concentrated into the top 20 to 50 cm in the subtidal zone.

Across the surf and rip-head zone, some 500 m wide, bacterial and meiofaunal biomass figures are 95 and $150\,\text{gC}\cdot\text{m}^{-1}$, respectively, whereas the volume of water pumped through the sea bed by wave action is about $80\,\text{m}^3\cdot\text{m}^{-1}\cdot\text{d}^{-1}$. Benthic metabolic studies indicate a total consumption of 30,000 to $40,000\,\text{gC}\cdot\text{m}^{-1}\cdot\text{y}^{-1}$. Thus, the interstitial system of the surf zone is considerably more important than that of the beach, and the two together mineralize huge quantities of organic materials, equivalent to 20% of total primary production by surf phytoplankton in this ecosystem.

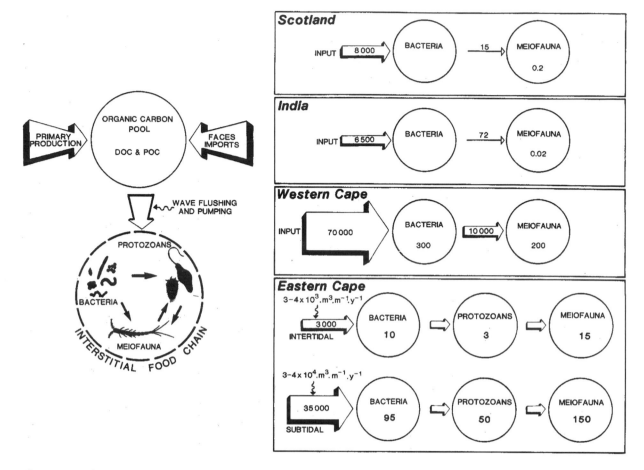

Figure 12.4. The interstitial food chain with examples from four areas. All values are given in gC·m⁻¹·y⁻¹.

Some of these interstitial processes are summarized in Figure 12.4. From this and what has already been said, it is apparent that on all beaches (other than a few dissipative systems with exceptionally rich macrofaunas) the interstitial food web consumes more carbon than does the macroscopic food web. Approximate proportions (interstitial: macrofaunal) are 77:23% on the Scottish beach, 63:38% on the Indian beach, 97:3% on the South African Western Cape beach (although this is probably not typical of Western Cape beaches in general), and 63:37% on Eastern Cape beaches, excluding the surf zone.

12.5   The Microbial Loop in Surf Waters

The microbial loop comprises bacteria, flagellates, ciliates, and other micro-zooplankton (less than $200\,\mu$m in diameter) in the water column. Its significance has only recently been appreciated, and it has been investigated in a sandy-beach/surf-zone context solely in the Eastern Cape (McLachlan and Romer 1990, McGwynne 1991). This component is considered to subsist largely on phytoplankton exudates and other forms of DOC. These materials are utilized by water-column bacteria, which generally average $10^6\,$mℓ^{-1} in coastal waters. The abundance of these microbes and their rapid utilization of DOC leads to the conclusion that they consume a major part of primary production in marine ecosystems. Because of the limited information available from sandy shores,

Table 12.2. Mean abundance and general features of components of the microbial loop in Eastern Cape surf-zone waters.

	Bacteria	Flagellates	Proto- and micro-zooplankton
Numbers	$7 \times 10^5\,\text{m}\ell^{-1}$	$1 \times 10^3\,\text{m}\ell^{-1}$	$1 \times 10^1\,\text{m}\ell^{-1}$
Size	0.2 to 2 μm	3 to 5 μm	50 to 150 μm
Composition	70% cocci, small rods	70% autotrophs	50% tintinnids
	30% large rods	30% heterotrophs	35% nauplii
Biomass	43 mgC·m^{-3}	10 mgC·m^{-3}	1 mgC·m^{-3}

the scenario is described here only for Eastern Cape surf zones and some tentative generalizations are made.

The average composition and abundance of components of the microbial loop in surf and rip-head waters in the Eastern Cape are summarized in Table 12.2. Bacterial numbers actually proved to be somewhat below average for coastal waters, but the individuals displayed a large mean size, resulting from the preponderance of large rod-shaped forms. Bacterial biomass was 43 mgC·m^{-3}, or 100 gC·m^{-1} over a distance of 500 m. A small proportion were associated with particulate matter. Included in the calculation are autotrophic cyanobacteria, which make up about 5% of the total bacterial numbers.

Flagellate numbers were average for coastal waters and represented both autotrophic and heterotrophic forms, mainly in the 3- to 5-μm size range. The heterotrophic flagellates feed on bacteria, which they filter from the water at rates some five times their own body mass per day. The top of the microbial-loop food chain in surf water consists of more than one trophic level and includes loricated ciliates (tintinnids) and the larval stages of mesozooplankton in the size range 20 to 200 μm. These forms feed on both bacteria and flagellates and may also take up POC and DOC directly from the water.

As in the case of the interstitial system, the microbial loop is visualized as being discrete from other food chains. There are, however, two possible trophic connections that could enable energy flow from this to other food webs: flagellates are sufficiently large to be consumed by filter feeders, such as *Donax*, and the filter-feeding zooplankton may also take some of the micro-zooplankton. It is uncertain how much carbon enters the macroscopic food chain via these links, but it is thought to be small. Whereas flagellates are sufficiently abundant to form a significant component of the food of bivalve filter-feeders, which can extract particles down to 3 μm, the densities of micro-zooplankton are too low to represent an important food source for the larger zooplankton. As at least four steps are required in a food chain progressing from primary producers via bacteria, flagellates and micro-zooplankton to larger organisms, very little energy probably remains to pass to the last step. We may therefore continue to view the microbial loop as a distinct trophic system, with only limited exchanges with macroscopic food chains.

The surf-zone microbial loop is illustrated in Figure 12.5, and trophic relations within this in Figure 12.6. Estimates based on biomass, turnover ratios, and dialysis-bag experiments—measuring bacterial production and flagellate grazing rates—indicate that this food web in the Eastern Cape could consume more than 100,000 gC·m^{-1}·y^{-1}, or about 50% of total surf-zone primary production. It appears that mucus, which is produced by surf-zone diatoms as part of their daily vertical migration activity, may represent up to 40% of total primary production and could be a major input of DOC to water column bacteria once it is sloughed from the diatom cells. Although quantitative data are avail-

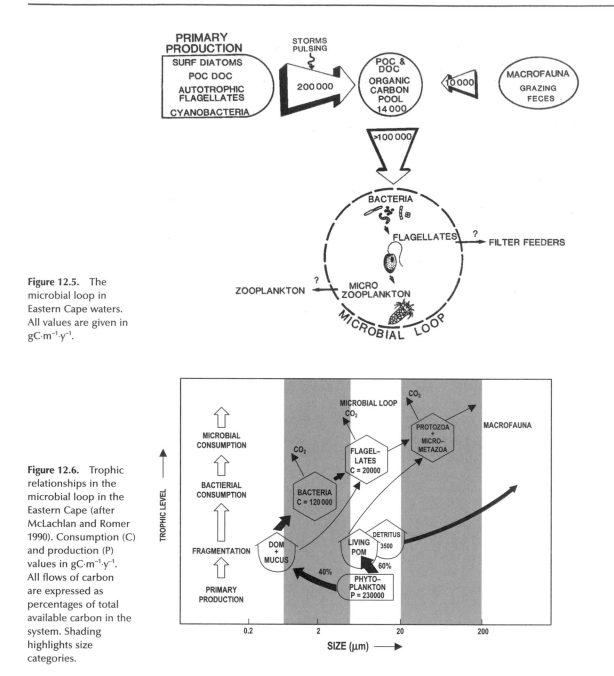

Figure 12.5. The microbial loop in Eastern Cape waters. All values are given in gC·m⁻¹·y⁻¹.

Figure 12.6. Trophic relationships in the microbial loop in the Eastern Cape (after McLachlan and Romer 1990). Consumption (C) and production (P) values in gC·m⁻¹·y⁻¹. All flows of carbon are expressed as percentages of total available carbon in the system. Shading highlights size categories.

able only from Eastern Cape beaches, it may be concluded that despite their small size and scientific neglect the organisms comprising the microbial loop represent an extremely important trophic system in surf waters in general.

12.6 Energy Flow in Beach and Surf-zone Ecosystems

The beach and surf-zone biota consist not only of meiofauna and macrofauna but all food chains so far discussed: diatoms, water-column microbes, interstitial organisms, and macrofauna. The presence of these various components and their relative importance in the system differ with beach type and particularly with the degree of coupling between the beach and the surf zone. In this respect, two distinct types of sandy-shore ecosystem may be recognized, representing the two ends of a continuum: (1) beaches with little or no surf zone, which are dependent on food imports from the sea and (2) beaches with extensive surf zones sustaining sufficient primary production to be self-supporting. This fundamental difference between open and closed beach systems revolves primarily around incident wave energy and surf circulation patterns.

Reflective beaches and beaches with low-tide rocky platforms have no true surf zones. The ecosystem consists only of an intertidal sand body, its associated fauna being fueled by marine inputs. On the other hand, beaches toward the dissipative extreme have well-developed surf zones with cellular circulation patterns (see Chapter 2). In such surf zones, diatoms provide a rich source of primary production, the circulation cells tending to retain this and other organic materials — so that the ecosystem is not dependent on marine inputs from beyond the surf. Naturally, intermediate states occur between these two extremes, which may be termed the interface and the self-sustaining beach ecosystems, respectively.

Figure 12.7 summarizes the components and key processes of these two types of ecosystem. Characteristic features of the interface type are as follows.

- Dependence on marine inputs
- The presence of only two food chains, interstitial and macrofaunal
- The great importance of the interstitial system
- The importance of wave energy in driving the interstitial system via water filtration
- Large macrophyte inputs in some cases
- Low macrofaunal biomass in most cases
- Export of macrofaunal production to marine and terrestrial predators

These features contrast with those of self-sustaining beach/surf-zone ecosystems, which are as follows.

- A wave-driven surf circulation that retains materials
- The presence of no less than four biotic systems (phytoplankton, microbial loop, interstitial system, and macrofauna)
- High primary production
- The great importance of the microbial loop
- The significance of wave energy in controlling primary production, the interstitial system, the export of surplus production, and so on
- Generally high biomass

Intermediate beaches with moderately developed surf zones and negligible primary production are transitional and include elements of both of these types. On intermediate beaches, one expects limited primary production, the presence of zooplankton, and some microbial activity in the surf water. Beaches toward the lower end of the energy scale (interface types) have their energetics dominated by their interstitial systems, and macrofauna is only abundant where there are large macrophyte inputs. Examples of such beaches have been summarized in this chapter.

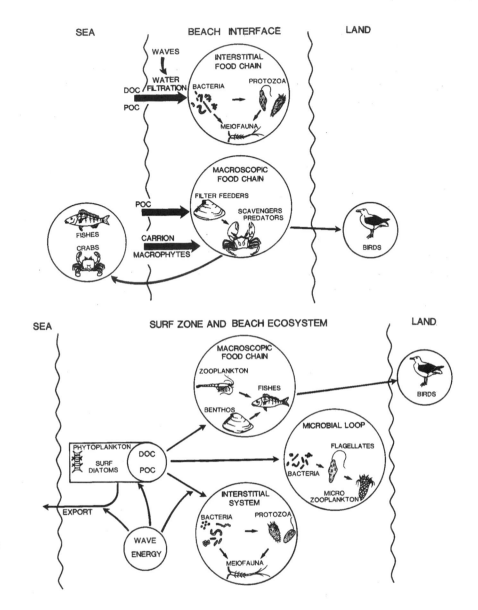

Figure 12.7. The two main types of beach ecosystem, based on carbon flux.

12.7   Case Study: Sandy Beaches of the Eastern Cape

The only sandy-beach/surf-zone ecosystem for which all components have been studied is in the Eastern Cape, South Africa. Aspects of this system have already been outlined and it is now described as a whole, as an example of the structure and dynamics of such ecosystems. The account is based on reviews by McLachlan *et al.* (1981b), McLachlan and Bate (1984), McLachlan and Romer (1990), and Heymans and McLachlan (1996).

The Sundays River Beach is a high-energy intermediate to dissipative beach with a well-developed surf zone averaging 250 m in width (see Figure 13.17). The rip-head zone, which lies beyond this to the outer limit of the surf circulation cells, extends a further 250 m. Water turnover in the surf zone takes only a few hours, but turnover of water between the combined surf and rip-head zones and the nearshore zone takes days — so that the outer limit of the surf cells forms a boundary, often clearly demarcated. The ecosystem thus comprises the sand body of the beach and the water envelope of the surf and rip-head zones, its boundaries being the dune/beach interface and the outer limit of surf circulation cells.

The chief primary producer in the system is the surf diatom *Anaulus*, but flagellates are also important in this regard. *Anaulus* collects in the foam on the water surface during the day, being concentrated into patches by surf circulation. Shoreward advection of surface water by wave bores, coupled with rip currents discharging through the surf every 200 to 300 m, results in the diatoms accumulating in areas of reduced water movement close to the beach and adjacent to the rip currents. In the afternoon, the cells sink out of the foam and develop thick mucus coats, by which they attach to sand grains. This enables them to concentrate on the sediment surface, or even to enter the sand at night. During the early morning, the mucus coat is lost, and if there is sufficient wave energy the cells are stirred into the water column, where they attach to bubbles and thereby enter the surface foam once more. This concentrates them toward the beach and prevents their loss from the surf zone. Wave energy thus controls the quantity of diatoms in the water column and thereby the primary production of the system. Maximum productivity occurs during and after high-energy events.

Total production in the surf zone is 120,000 gC·m^{-1}·y^{-1}, 90% of this due to *Anaulus*. Of this production, 40% is in the form of mucus and 5% as exudates, the remaining 55% being accounted for by cell division. In the outer surf zone or rip-head zone, primary production is estimated at up to 110,000 gC·m^{-1}·y^{-1}, due to a great number of different species, which include an abundance of flagellates. If mucus is sloughed off in such a way as to produce both particulate and dissolved organic carbon, a conservative estimate of available primary production in this ecosystem as a whole is 50,000 gC·m^{-1}·y^{-1} as DOC and 150,000 gC·m^{-1}·y^{-1} as POC, including both dead particulates and living *Anaulus* and other phytoplankters.

The interstitial system, driven by wave flushing and pumping of water through the sand, filters 3 to 4×10^4 m^3·m^{-1}·y^{-1} over a 500-m width of intertidal (10%) and subtidal (90%) zones. Studies of oxygen consumption indicate that the total carbon metabolized in this process is about 40,000 gC·m^{-1}·y^{-1}, the subtidal consuming 90%. As the meiofauna is entirely interstitial and relatively deeply dispersed through the well-oxygenated sand, there is no significant interaction between this component and the macrofauna. It is not known to what extent *Anaulus* and/or its mucus coat are directly available to the interstitial fauna during the diatom's nocturnal sedimentary phase.

The macroscopic food chain, outlined previously, consists of mullet, zooplankton, and filter-feeding bivalves consuming phytoplankton directly, whereas scavengers take stranded carrion. These animals are eaten by swimming portunid crabs, fishes, and birds, with piscivorous fishes forming the top of the marine food chain. Estimates indicate an input of about 30,000 gC·m^{-1}·y^{-1} into this food chain, most originating from phytoplankton. With an average assimilation efficiency of about 70%, some 10,000 gC·m^{-1}·y^{-1} is probably recycled as feces into the water column, where it is directly available to the microbial loop and other food chains.

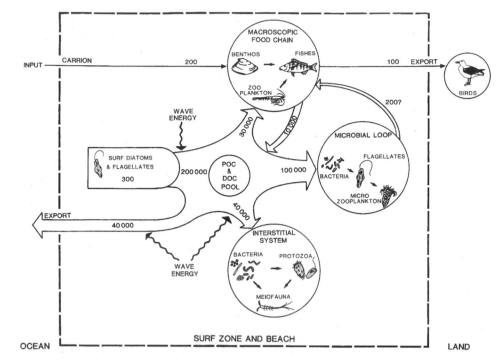

Figure 12.8. Simplified model of carbon flow through the food chains of the Sundays River Beach ecosystem. All values are given in gC·m⁻¹·y⁻¹.

The microbial loop is well developed. Although bacterial numbers are not high, large individual size results in high biomass figures. The bacteria are grazed by flagellates, which in turn are consumed by ciliates and other micro-zooplankton. However, the latter is so sparse in these waters that for all practical purposes the microbial loop can be considered in terms of only bacteria and flagellates. Total bacterial utilization of POC and DOC is estimated at 100,000 gC·m⁻¹·y⁻¹. Some flagellate production may be utilized by macrofaunal filter-feeders.

Thus, in this system (Figure 12.8) primary production exceeds faunal requirements and the surplus is probably exported as living and dead POC across the boundary of the ecosystem. The microbial loop is the most significant of the food chains, consuming about 50% of carbon production, followed by the interstitial system (20%) and the macroscopic food chain (15%). All three food chains are richly represented in this ecosystem. Despite its hostile characteristics and dynamic nature, this sandy-beach and surf-zone ecosystem supports abundant life, albeit of relatively low diversity.

Heymans and McLachlan (1996) updated this model and undertook a network analysis of carbon flow that should be referred to for the most detailed and final synthesis of this ecosystem study. Although the detail provided in their network analysis is beyond the scope of this book, it does give some additional insights to the working of this well-studied system. Table 12.3 lists the effective trophic level of each biotic component in this system, based on network analysis (where primary producers are trophic level 1, herbivores level 2, predators on herbivores level 3, and so on). This shows bacteria, mullet that graze phytoplankton, and benthic scavengers near the bottom of the food chain (the latter because they feed on carrion, a primary input). Microplankton and predatory

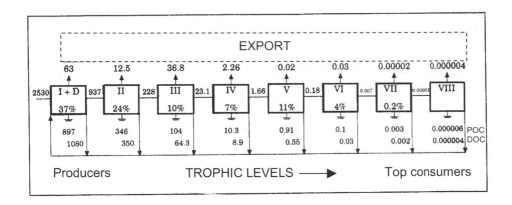

Figure 12.9. Trophic aggregation of the food web into a Lindeman Spine, a linear series of trophic steps (after Heymanns and McLachlan 1996). The autotrophs are merged with the detrital pool (I + D), the trophic levels are designated by Roman numerals, flows given in mgC·m^{-2}·day^{-1}, percentages in the boxes represent the daily trophic efficiencies, POC = particulate organic carbon, and DOC = dissolved organic carbon.

Table 12.3. Effective trophic level of each living compartment in the beach ecosystem (after Heymanns & McLachlan 1996). DOC, dissolved organic carbon; POC, particulate organic carbon.

Compartment	Effective trophic level
1. Phytoplankton	1.00
2. Free bacteria	2.00
3. Flagellates	3.00
4. Micro-zooplankton	3.72
5. Zooplankton	2.04
6. Benthic filter feeders	2.11
7. Benthic scavengers and predators	1.22
8. Mullet	2.00
9. Predatory fish	3.82
10. Omnivorous fish	3.02
11. Birds	2.88
12. Sediment bacteria	2.00
13. Protozoa	2.61
14. Meiofauna	2.90
15. DOC	1.00
16. Suspended POC	1.00

fish are at the top of their respective food chains and score near the maximum value of 4.

Meiofauna enjoy a lower effective trophic level, despite being at the top of the interstitial food chain. This is because the meiofauna include several feeding categories. Collapsing the feeding network into a finite number of linear steps produces a Lindeman spine (Figure 12.9), an abstract food chain that merges the detrital pool with the primary producers to represent the first trophic level and results in eight compartments, which is a high number for an ecosystem. The diagram indicates the amount of energy passing from each level to the next. For example, of the 228 mgC·m^{-2}·day^{-1} consumed at the third level only 23.1 mg is transmitted to level 4, giving a trophic efficiency of 10%. Trophic efficiency at the first level is 37%, and this reduces to 0.2% at level 7. The system has a high detritivory:herbivory ratio of 13:1, meaning that 92% of the carbon entering the second trophic level is recycled as detritus rather than phytoplankton being grazed directly. It shows that detritivory is more important than herbivory in this system, with most primary production being broken down to detritus. The average length of trophic pathways in this system is 2.3 steps, and the overall P/B ratio is 0.4. Further analysis of other indices and properties of this system suggests that it is well organized and with significant internal stability and resistance, a feature expected from a physically controlled environment. Overall the system was concluded to be unstressed, mature, physically controlled, and stable.

The beach described previously is of the high intermediate type, becoming dissipative during storms (see Chapter 2). It is predicted that on beaches that are more dissipative (e.g., southern Brazil and Washington State) diatom productivity and biomass will be greater and the relative importance of the microbial loop will increase. On more reflective beaches, primary productivity will decrease rapidly to zero. Biomass will also decrease, and the interstitial system will dominate. In very reflective situations only the interstitial system should be of any quantitative importance (see Figure 12.7).

12.8   Nutrient Cycling

Nitrogen and phosphorus are the most important nutrient elements in the majority of marine systems, with nitrogen more often limiting to plant growth than phosphorus. Nitrogen budgets have been investigated for only a few sandy-beach systems. Nevertheless, this work reveals many parallels between nitrogen cycling and energy flow, as described in the previous sections.

In beach/surf-zone ecosystems, nitrogen requirements may be considered to be those of the surf-zone phytoplankton and benthic diatoms. Only in surf zones displaying high primary productivity are these requirements significant. Most diatoms take up NH_4-N preferentially, but the presence of large amounts of the enzyme nitrate reductase in *Anaulus* indicates that NO_3-N is also utilized. Bacteria can also absorb inorganic nitrogen and may compete with plants in this respect. However, their needs are not considered to be part of the normal requirements of the primary producers of the system.

There are various forms of nitrogen input to beaches and surf zones. Groundwater seepage represents one of the most important and ubiquitous nitrogen sources. Johannes (1980) reviewed this literature. Groundwater draining from the land discharges into confined or unconfined aquifers, which usually drain near the low-tide mark. Discharge flows are naturally highly variable, but typical discharge rates are thought to range from 1 to $10\,m^3 \cdot m^{-1} \cdot day^{-1}$. Although flow rates are frequently low, the discharge is significant in terms of nitrogen input because of generally high nitrogen concentrations, often in the region of 1 to $5\,mgN \cdot \ell^{-1}$ and mainly in the form of NO_3-N.

Rain introduces a small amount of nitrogen, usually below $1\,mg \cdot \ell^{-1}$, also chiefly as NO_3-N. Although this concentration is not high, the input may be significant in regions of high rainfall. Other inputs take the form of inorganic nitrogen, as well as living and dead organic nitrogen, from other systems. These include inorganic nitrogen from upwelling or from estuaries, particulates, and zooplankton and from fishes entering the system from beyond the surf.

Sandy beaches have long been considered active in nutrient recycling through their role in mineralizing organic materials emanating from the sea and have been labeled "great digestive and incubating systems" (Pearse *et al.* 1942). All animals recycle nutrients in the form of urine and feces. Zooplankton, benthos, fishes, and even birds recycle about 30% of the nitrogen they absorb from their food. This is typically made up of 50%

NH_4-N, 5 to 10% urea–N, and 35% particulate–N. In urine and gill excretions, NH_4-N is the chief form, making up 60 to 90% of the total — followed by urea (5 to 15%) and amino acids (5 to 15%). Fecal nitrogen, which may exceed excreted nitrogen, is predominantly particulate organic nitrogen. The NH_4-N is utilized by phytoplankton, as are amino acids, whereas bacteria utilize both these and fecal material. As nitrification does not occur in the photic zone, little of this NH_4-N is converted to NO_3-N.

The interstitial system is considered more important than the macrofauna in nutrient recycling. As water is filtered through the sand, the interstitial biota consumes the dissolved and particulate organic matter and recycles the nutrients. This has been investigated in experimental sand-column systems (Munro et al. 1978, McLachlan et al. 1981c), as well as by modeling (McLachlan 1982). Nitrification is the dominant process in well-oxygenated sands, most organic nitrogen being mineralized to nitrate. Increasing the organic load and/or decreasing water filtration through the sand decreases both the redox potential and oxygen tensions, so that reduction to NH_4-N occurs. Nitrate is absent in reduced sediments. On most open oceanic beaches, however, conditions are such that nitrate is the main form of nitrogen produced. The percentage of organic nitrogen mineralized in passing through laboratory sand columns 10 to 50 cm deep ranges from 35 to 100%, depending on the nitrogen source, the grade of sand, the depth of the column, and other factors. It has also been demonstrated that the interstitial biota in such columns can survive on dissolved organic materials only, confirming the idea that the sequence DOC-to-bacteria-to-meiofauna is a major trophic pathway and mineralization mechanism.

Short-period studies, both in the laboratory and on the beach itself, have led to suggestions that sandy beaches may accumulate nutrients rather than recycling them (Koop and Lucas 1983). It is true that net retention may occur in the short term, or even the longer term in exceptional cases where beaches are accreting. However, longer-term experiments indicate that accumulation is not typical and that beaches characteristically recycle nitrogen to the sea (McLachlan and McGwynne 1986).

In the microbial loop, bacteria may either excrete nitrogen or take it up in competition with the phytoplankton, depending on the C:N ratio of the substrate. As bacterial cells have a C:N ratio of about 5, utilization of particulate (PON) or dissolved (DON) organic nitrogen close to this ratio results in the regeneration of NH_4 N. At a carbon conversion efficiency of 25%, bacteria need a nitrogen supplement if the C:N ratio of the substrate exceeds 16. Thus, at C:N ratios above 15 they may compete with the phytoplankton for inorganic nitrogen, whereas at ratios below 15 they regenerate it. The flagellates, in turn (having a C:N ratio of about 5 and feeding on bacteria with a similar ratio), may be expected to regenerate much of the nitrogen as NH_4-N. Ciliates will do the same. A key factor determining the extent of nitrogen regeneration by the microbial loop is therefore the C:N ratio of the initial substrate (DON and PON).

Nitrogen may be lost to beach/surf-zone areas by export to adjacent coastal waters or by accumulation in the sediment. Losses to the sea may occur via active migration of fishes and zooplankton out of the surf zone or by the passive transport of DON, PON, and plankton out to sea, alongshore, or into estuaries. These exports may be significant where water turnover rates are high and surplus material is produced. Losses to sediments are unlikely to occur unless there is net deposition of sediment, as occurs on accreting shorelines. Elsewhere, turnover of the sediment by wave action and/or net erosion prevents any such accumulation in the long term.

Few beach systems have been studied in sufficient detail to compile even partial nitrogen budgets. Aspects of nitrogen cycling have, however, been studied on two South African beaches receiving high macrophyte inputs. Koop and Lucas (1983) showed that most nitrogen from stranded kelp was converted to microbial biomass in the short term and little returned immediately to the sea. Input was estimated at $4,380\,gN \cdot m^{-1} \cdot y^{-1}$, of which 4,123 g was initially converted to bacterial biomass, 88 g to invertebrate scavengers, and only 66 g returned directly to the sea. About $100\,gN \cdot m^{-1} \cdot y^{-1}$ was lost to the atmosphere in denitrification. Clearly these vast quantities of kelp-derived nitrogen cannot indefinitely be converted into interstitial bacterial biomass and must ultimately return to the sea. This could not, of course, be demonstrated by Koop and Lucas (1983), as their experiment lasted only eight days.

McLachlan and McGwynne (1986) addressed the question of nitrogen accumulation versus nitrogen export to sea. They studied a beach with high inputs of red algae totaling $14,338\,gN \cdot m^{-1} \cdot y^{-1}$. This figure was based on a turnover time of nine days for stranded wrack. By monitoring all forms of nitrogen in the beach (groundwater 19 to $146\,gN \cdot m^{-1}$, bacteria 2 to $3\,gN \cdot m^{-1}$, and meiofauna $0.1\,gN \cdot m^{-1}$, macrofauna being virtually absent), they were able to show that there was no net change in total nitrogen in the beach over a period of a year. Hence, the $14,000\,gN \cdot m^{-1} \cdot y^{-1}$ input must have been recycled. They further indicated that beaches in general could not form nutrient sinks. As most of the world's sandy beaches are either in equilibrium or undergoing erosion, they do not have the net sedimentation necessary to store nitrogen. Prograding beaches are rare, but only these can act as nutrient sinks — stranded material being trapped and rapidly buried in the sand. This is not to say that other beaches may not act as nutrient sinks in the short term, as chemical equilibria may shift with tides, storms, seasons, and changing inputs — so that the beach fluctuates between accumulation and exportation. As with carbon budgets (described previously), a relatively complete nitrogen budget has only been compiled for the surf-phytoplankton-driven beaches of the Eastern Cape (Cockcroft and McLachlan 1993) (Table 12.4 and Figure 12.10).

Table 12.4. Nitrogen budget for the Sundays River Beach surf zone (after Cockcroft & McLachlan 1993). Values in $gN \cdot m^{-1} \cdot y^{-1}$.

Process	Location/source	Dissolved and particulate organic–N (DON & PON)	Dissolved inorganic nitrogen (DIN)
Inputs	— Groundwater		780
	— Rainwater		300
	— Estuarine		600
		Total	1,680
Cycling	— Interstitial biota (intertidal)		590
	— Interstitial biota (subtidal)		4,610
	— Macrobenthos	547	364
	— Zooplankton	1,539	2,624
	— Fishes	336	323
	— Microbial loop		4,607
		Totals 2,422	13,120
Phytoplankton requirements			
	— Inner surf zone		10,100
	— Outer surf zone		3,100
		Total	13,200
Net excess of nitrogen = 1,600			

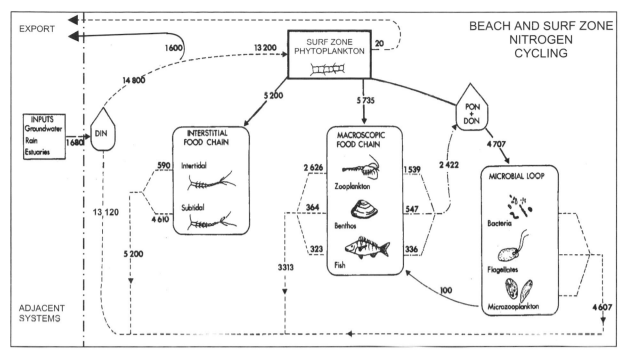

Figure 12.10. Nitrogen budget and recycling in the Sundays River Beach/surf-zone ecosystem (after Cockcroft and McLachlan 1993). Solid lines indicate grazing, broken lines recycling or nutrient flux. All values in $gN \cdot m^{-1} \cdot y^{-1}$.

The nitrogen requirements of surf diatoms and other phytoplankton were calculated from primary production estimates to be $13,200\,gN \cdot m^{-1} \cdot y^{-1}$ for the surf zone to the outer limit of surf circulation cells. Nitrogen inputs to this system include groundwater, rain, and export from estuaries, totaling some $1,680\,gN \cdot m^{-1}\,y^{-1}$. Nitrogen cycling within the ecosystem is accomplished by the components of the macroscopic food chain, interstitial biota, and microbial loop. The interstitial biota is estimated to recycle $5,200\,gN \cdot m^{-1} \cdot y^{-1}$, mainly as nitrate from the intertidal (11%) and subtidal (89%) sediments. Only very preliminary values are available for the microbial loop, where both bacteria and flagellates appear to be important mineralizers. These give a figure of $4,400\,gN \cdot m^{-1} \cdot y^{-1}$, mainly as ammonia.

Recycling by the macroscopic food chain is more complex, as many end products are formed. In addition to ammonia, nitrogen is excreted in dissolved organic form as amino acids and urea, and in particulate organic form as feces. The benthos, zooplankton, and fishes recycle $3,311\,gN \cdot m^{-1} \cdot y^{-1}$ as dissolved inorganic nitrogen (NH_4), and $2,422\,gN \cdot m^{-1} \cdot y^{-1}$ in organic form. The latter is available for reingestion by all faunal components, particularly those of the microbial loop. The importance of zooplankton in this regard may be seen in Table 12.4.

Total nitrogen recycled by the biota of all three food chains is thus estimated at $13,120\,gN \cdot m^{-1} \cdot y^{-1}$. When added to the inputs, a figure of $14,800\,gN \cdot m^{-1} \cdot y^{-1}$ is indicated as being available to the surf phytoplankton, a surplus of $1,600\,g$ over their requirements. In addition, $2,442\,gN \cdot m^{-1} \cdot y^{-1}$, excreted in organic form by the macrofauna, is again available to these food chains.

The ecosystem clearly must have losses to balance this surplus nitrogen. As this is not a prograding shore, there should be no burial and therefore no long-term accumulation in sediments. There is, however, a very small loss to bird predators taking their prey inland, but most of the loss occurs seaward by the mixing of PON and DON and by the emigration of fishes and zooplankton across the outer boundary of the system.

In this ecosystem, external inputs thus supply some 13% of phytoplankton inorganic nitrogen requirements, whereas recycling by the biota can provide 98% — of which the interstitial system contributes the most (40%), followed by the microbial loop (34%), and the macrofauna (26%). However, if organic excretory products are included, the macrofauna appears as more important — these products being sufficient to meet 24% of the estimated nitrogen requirements of the microbial loop. The surplus, 11% of primary production requirements, is exported mainly as PON.

The fact that the biota regenerate almost all of the nitrogen required for primary production in the surf zone supports the view that the beach and surf zone form a semi-closed ecosystem. Furthermore, far from being unproductive the system has been shown to have surplus production available for export. Clearly, all three food chains are important in this regard. A striking feature is the difference between the estimated role of the microbial loop in carbon flow and its role in nitrogen cycling. The microbial loop was estimated to account for 60% of carbon consumed but only 34% of nitrogen recycled. The reason for this probably lies in the $C:N$ ratios of the foods or substrates utilized by these different components. The microbes probably utilize much DOC (diatom exudates and mucus), which has high $C:N$ ratios.

12.9   Conclusions

Sandy beaches are marine systems whose food chains mainly begin and end in the sea. Four biotic components are present: primary producers, macrofauna, interstitial biota, and water-column microbes. In terms of interactions between physical beach states and these components, two distinct beach types occur: interface beaches and self-sustaining beach/surf-zone systems. The former have little or no surf zone, virtually no primary producers or microbial loop, an interstitial biota far more important than the macrofauna, dependence on marine inputs, and generally a low total biomass. The latter, on the other hand, have well-developed surf zones with significant primary production and all three food-chain systems. Such beaches may be self-sustaining, not dependent on marine inputs, and may have their energetics dominated by the microbial loop. Between these extremes, intermediate states occur. The relative importance of the three food chains and the occurrence of these two major beach types over a range of morphodynamic states typical of open ocean beaches are illustrated in a conceptual model in Figure 12.11.

Interface beaches process materials derived from the sea and recycle nutrients to the sea, whereas self-sustaining beaches tend to recycle materials within their own boundaries. No beach is an isolated entity, however, and all beaches exchange materials with other coastal ecosystems across both their seaward and landward boundaries. Nevertheless, if typical interface beaches are characteristically importers typical self-sustaining beaches may be exporters of material.

Sandy beaches are thus open marine ecosystems interacting with adjacent marine and terrestrial environments, the latter the subject of the next chapter. In the absence of biological structures, the processes of energy flow and nutrient cycling define important elements of the structure and function of the beach and surf-zone ecosystem, provide

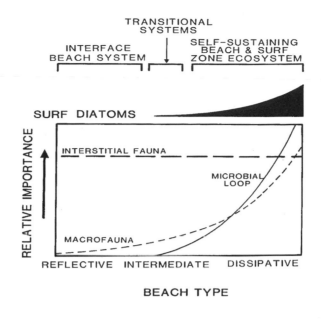

Figure 12.11. Conceptual model of the relative importance of the three food chains across the spectrum of beach morphodynamic types encountered on open sandy shores. The relative significance of surf-zone diatoms and the two distinct types of beach system are also indicated.

valuable avenues for research, and aid in the conceptualization of their dynamics and sensitivities. We have seen that the characteristics of the beach system depend on the interactions among sand particle size, wave climate, and tidal regime (Chapter 2). The super-parameter controlling beach ecology and directly or indirectly driving these processes is wave action. Just as rainfall is the most important climatic feature in terrestrial systems, so water movement is in the sea — and nowhere is this more apparent than on exposed sandy beaches.

Coastal Dune Ecosystems and Dune/Beach Interactions

13

13.1   Introduction

Whereas previous chapters have been primarily concerned with the marine beach and surf zone, this book would certainly be incomplete without some treatment of the coastal dunes, extensive along the world's sandy shorelines and backing most beaches (Figure 13.1). Indeed, on most shores beaches and dunes together form a linked system Tinley (1985) has termed the *littoral active zone* (see Chapter 15). This zone is characterized in many cases by extensive aeolian and wave-driven sand transport and sand exchanges between the dunes and the beach. Despite their obvious interactions, these are really quite distinct ecosystems — the beach/surf-zone being a marine wave-driven ecosystem and the dunefield a terrestrial wind-controlled system.

Coastal dunes are characterized by the exchange of sand with beaches and by the influence of aeolian forces. They are positioned between the landward edge of the backshore

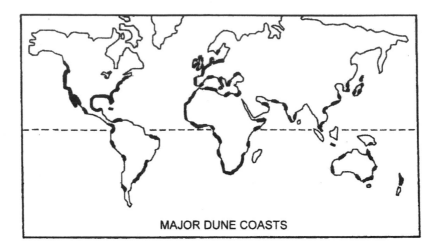

Figure 13.1. Distribution of the major dune coasts of the world (after Nordstrom *et al.* 1990). These are generally linked to extensive beaches.

MAJOR DUNE COASTS

and the landward limit of aeolian sand transport. The landward boundary may be abrupt or grade more gradually into other terrestrial ecosystems. A useful recent reference is Martinez and Psuty (2004).

13.2　The Physical Environment

Coastal dunes mostly consist of particles of silica and/or carbonate in the size range 125 to 300 μm (i.e., fine to medium sand). Generally, dune sand originates from the intertidal beach and is blown inland to form dunes. Dune sands therefore tend to be finer than beach sands. The amount of sand moved by the wind depends on three factors: the moisture content of the sand, the wind speed over the sand surface, and the grain size. When the sand water content exceeds 2%, very strong winds are necessary to move the particles, and therefore little sand is transported during rainy weather. Sand movement is proportional to the cube of the wind speed. For dune sands, movement is initiated at about 4 m·s^{-1} and increases dramatically at wind speeds above 10 m·s^{-1} (Figure 13.2). Over the particle size range of sand (60 to 1,000 μm), transport is greater for fine than for coarse sands at any wind speed.

Wind-blown sand that forms coastal dunes usually originates from the beach and travels in three ways: (1) in suspension as dust, (2) by bouncing or saltation, and (3) by creep, or rolling. Usually 75% of the transport is by saltation and 25% by creep, only sub-sand particles moving in suspension. Surface roughness also has an important effect on aeolian sand transport: the rougher the sand surface (due either to bed forms or to vegetation) the lower the rate of transport. Vegetation increases surface roughness to a considerable extent, tending to trap sand (Figure 13.3).

Dunes can form in two major ways: either vegetation traps wind-blown sand or the sand forms ripples and waves without vegetation, these gradually expanding to form more extensive and complex systems. After disturbance has initiated a dune, slip and windward faces develop once dune height exceeds 1 m. Sand is then transported up the windward face and down the slip face (Figure 13.4). Formation of the slip face creates a wind shadow, where sand transport ceases — the slip face thus slowly moving forward as sand traverses it. Slip faces usually have an angle of repose of about 34 degrees. The rate of advance of such dunes is highly variable, often ranging from 1 to 10 m·y^{-1}, depending on

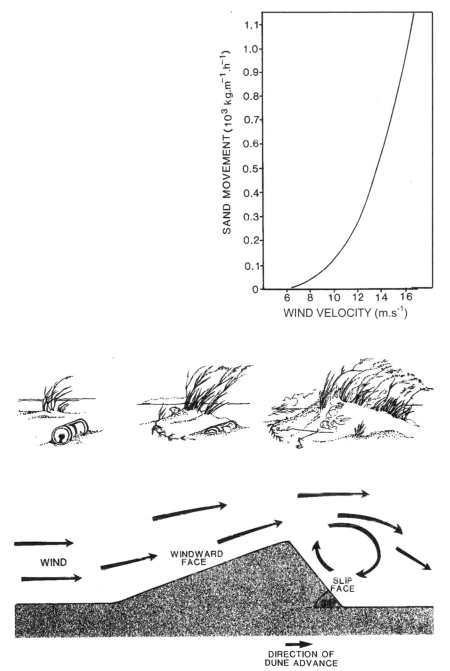

Figure 13.2. Aeolian sand transport rates as a function of wind speed, for average dune sand (after Bagnold 1941).

Figure 13.3. Sand can be trapped by vegetation or litter to initiate dune formation (after McGwynne and McLachlan 1992).

Figure 13.4. Diagrammatic section through a simple dune, showing slip and windward faces and airflow patterns.

dune height and wind regime. Different dune types are formed due to various wind regimes, the amount of sand available, and the vegetation that develops as the dunes grow.

13.3   Coastal Dune Formation by Vegetation

Many plant species are able to colonize supralittoral sands, despite initially poor nutrient conditions, lack of moisture, and sometimes very high temperatures. Such colonization may, on sheltered beaches, begin at or just above the strandline — aided by accumulations of wrack and tidal litter, which reduce the sand temperature and increase its moisture content. Strandline colonization is frequently by annual plants, the number of plants and the variety of species changing from year to year.

Higher up the shore, perennial grasses may be able to establish themselves, acting as a sand trap that may result in the establishment of a foredune a meter or two in height. This ability to bind sand depends on the development of extensive horizontal and vertical rhizome systems (Ranwell 1972). Even the dead shoots of such plants help to retain trapped sand. As the embryonic dune grows, either the shoots must keep pace with the sand surface or (more commonly) the rhizomes grow vertically upward toward the surface.

Although the formation of dunes is very largely dependent on colonization by plants, litter of all types may add to their development — particularly if these objects are not moved by the wind. Under natural conditions, such litter may be minimal and of little effect, but where there is pronounced human activity the amount of litter may be considerable.

The growth of a coastal dune is as hummocks or ridges (initially linear or curvilinear) parallel to the drift line. Most coastal dunes are relatively stable. However, certain dune types may erode after reaching maximum height and then move landward — in some cases remaining unstable for centuries before stability is imposed by the growth of vegetation. As a dune grows, it assumes one of two characteristic shapes. Nonvegetated dunes have a gently sloping windward side but a more abrupt slope to leeward, forming the slip face. Vegetated dunes, on the other hand, show the opposite tendency. As the windward slope of a vegetated dune only rarely displays complete vegetation cover, it is apparent that such a dune shape (with its steepest slope to windward) is inherently unstable.

The seaward development of the dune system is limited by the height of storm tides, which may undercut the seaward dune. Although sand is then eroded from the seaward face, strong onshore winds continue to deposit sand on the leeward slope — so that the dune effectively moves landward. This effect is enhanced if the dune grows to such a height that its vegetation no longer retains the sand. The entire seaward face and crest are then subject to severe wind erosion, the ridges moving away from the beach. Where prevailing winds are onshore, the highest dunes are likely to be encountered some distance inland from the beach and under these circumstances may reach 70 or even 80 m in height. However, where prevailing winds are offshore the greatest height is likely to be displayed by the coastal dune abutting the beach.

13.4 Dune Types

Many types and variations of dunes form along the coast (Figure 13.5), depending on prevailing winds, sand supply, climate, vegetation, and other factors. For the sake of convenience these multiple forms can be reduced to three main categories (Hesp 1991): foredunes, parabolic dunes, and transgressive dunefields. These are outlined in turn in the following.

Foredunes (Figure 13.6) are the commonest form of coastal dunes. They develop in regions of high energy where there are aggressive pioneer plants (e.g., *Ammophila*,

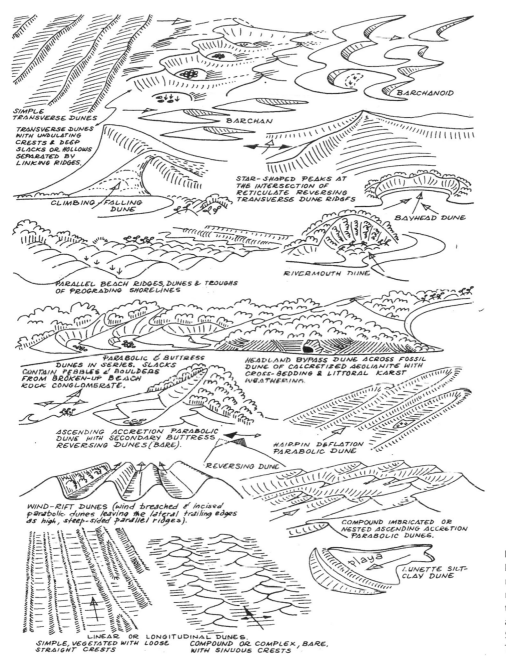

Figure 13.5. Diagrammatic representation of the range of coastal dune types encountered around the coasts of Southern Africa (after Tinley 1985).

Spinifex) or in areas of low wind speed or on sheltered beaches where limited sand transport results in sand being trapped by pioneers that prefer less sand inundation. Foredunes are often associated with reflective beaches. They may take the form of (1) incipient foredunes, which are dunes in the process of formation by pioneer plants, (2) established foredunes, which may include species of woody vegetation and reach reasonable height, or (3) relict foredunes, which are the remnants of earlier foredunes left behind by a prograding shoreline. They consist of hummocks, ridges, or both, usually in lines parallel to the shore.

Figure 13.6. Foredunes may consist of hummocks or ridges in one or more lines parallel to the shore (after McGwynne and McLachlan 1992).

Figure 13.7. Parabolic dunes and blowouts occur where strong winds blow sand inland and trailing ridges are held by vegetation (after McGwynne and McLachlan 1992).

Parabolic dunes (Figure 13.7) and blowouts usually occur in regions of medium to high wind speeds, where fairly large amounts of sand are being transported landward from the beach. Blowouts occur wherever the vegetation cover is destroyed. They are best developed on west coasts characterized by strong onshore winds and occasional storm-wave erosion. Parabolic dunes are formed by the downwind extension of blowouts or by the deflation of vegetated surfaces. They comprise a deflation zone and trailing dune ridges, and advance downwind. They are common on high wind-energy coasts, and can occur as compound, imbricate, long-walled, and cliff-top types. They are mostly associated with intermediate beaches.

Transgressive dunefields (Figure 13.8; see also Figure 13.17) occur where huge volumes of sand are transported back from beaches facing into strong wind regimes (i.e., dissipative beaches). Largely unvegetated, they generally consist of transverse dunes (i.e., with their axes running perpendicular to the main wind direction). However, barchanoid dunes, deflation plains, slacks, remnant knobs, and other dune types may occur in such dunefields. They are most characteristic of high-energy west and south Southern Hemisphere coasts, especially at the downdrift margins of longshore transport systems. They also occur in the Northern Hemisphere (for example, the Oregon coast).

Figure 13.8.
Transgressive dunefields develop where strong winds blow large amounts of sand inland from exposed, usually dissipative, beaches (after McGwynne and McLachlan 1992).

Dune and beach types are closely linked, as both are largely controlled by the wind regime, which in turn affects the wave climate. Simple foredunes tend to be associated with reflective beaches of low wave energy. Parabolic dunes are associated with intermediate beaches, and transgressive dunefields are most often associated with high-energy dissipative beaches, where there is the greatest potential for landward aeolian sand transport from the intertidal beach.

Three important elements of climate affect coastal dunes: moisture, wind, and sunlight (Tinley 1985). The most important of these is the wind regime, in terms of both the direction of the predominant winds and the wind speed — as these affect sand transport rates and consequently dune forms. On the other hand, moisture has an important impact on the development of vegetation, which in turn affects dune structure.

13.5   Edaphic Features

Dune soils are composed chiefly of quartz and carbonate. A small amount of heavy minerals may also occur. Distinct features of coastal dune sands are (1) alkalinity, due to the high carbonate content, (2) a tendency to become finer with increasing distance inland, (3) a tendency for the grains to be well rounded and positively skewed, (4) the input of large amounts of salt spray, and (5) stabilization by plant growth wherever rainfall is adequate (Ranwell 1972, Tinley 1985).

Dune soils evolve with time. This evolution, which is best seen on a shore-normal gradient across prograding shores, involves (1) an increase in organic matter with age, (2) a decrease in calcium carbonate as this is leached out of the sand, and (3) a corresponding decrease in pH (Figure 13.9). Organic content tends to increase continuously, but nitrogen levels stabilize after about 1,000 years. Humus accumulation causes soil acidity and dunes thus become more acidic as they age. However, they may display increasing alkalinity with depth. Dune sands are highly permeable, with good drainage. They are therefore edaphically dry. The presence of calcium carbonate may induce deficiencies in some elements, but at the same time it reduces the toxicity of sodium. Factors that perturb the processes leading to soil formation include the development of blowouts, the movement of slip faces over vegetation, and erosion due to waves and fires.

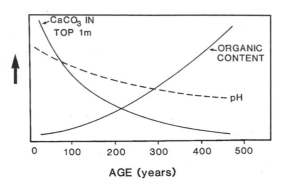

Figure 13.9. Evolution of dune soils: temporal changes in CaCO₃, pH, and organics.

13.6   Water

An important feature of coastal dunes is the position of the groundwater table, as this determines the deflation base or the lowest level to which wind can erode sand. It also has an important influence on the development of vegetation. Where the groundwater table is much more than a meter deep, it is not available to the majority of plants. Ranwell (1972) has distinguished four dune zones, based on the depth of the water table: (1) a semiaquatic habitat where the water table is less than about 0.5 m below the surface of the sand, (2) wet slacks, where the water table is about 1 m deep, (3) dry slacks, in which the table is 1 to 2 m deep, and (4) the open dune habitat, where the water table characteristically lies more than 2 m below the surface.

Old dune soils weigh only about half as much as equal volumes of young dune sand and have a moisture capacity of up to 33%, compared with only 7% for young dunes in most areas (because of their high organic content). The actual water content in young dune sand is commonly less than 1%, but minimum values increase with age as the organic content increases.

As the capillary rise of water (in even the finest sand) is no more than 40 cm, the water table below a dune 3 m or more in height cannot contribute directly to plant requirements, for the roots seldom penetrate below a depth of 1 m. Dune plants are thus dependent on rainfall and on condensation, either in the form of dew or as a nightly precipitation of water vapor rising upward through the interstitial spaces from the warmer wetter layers below. The situation is frequently different in the dune slacks or low-lying flat areas, where plant roots may have direct access to the water table.

In open sand in drier areas, moisture is generally about 4% by mass. Internal dew can, however, form when surface sands cool rapidly at night. The water table may rise after rain, and may even show weak tidal fluctuations near the sea. It may also be saline near the sea, and seawater can even intrude below the freshwater in areas of groundwater abstraction. Further, water table movements can be related to daily rhythms of evapotranspiration and sand movement.

The chemistry of dune groundwater is determined by the balance between humus input (which tends to make it more acidic) and the calcium carbonate content of the sand, which makes it more alkaline. Groundwater may be high in nutrients from various sources and wind may provide nutrients together with salt spray. Furthermore, material cast up on the drift line and subsequently decomposed can release nutrients that enter the water system.

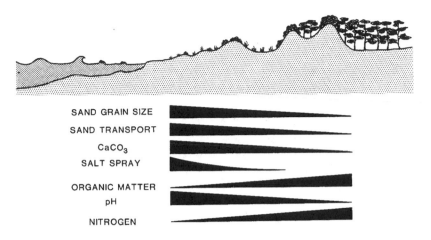

Figure 13.10. Abiotic gradients across a typical coastal dunefield (after McLachlan 1991).

Nutrient levels in dune groundwater are determined by a balance among five factors (Ranwell 1972): (1) inputs from weathering of minerals, (2) inputs from the atmosphere, (3) losses by leaching, (4) cycling by the flora and fauna, and (5) storage in the soil. The buildup of dune vegetation is accompanied by a general increase in the major nutrients, although much of this is tied up in plant biomass.

13.7   The Gradient Across Coastal Dunefields

The most distinct feature of coastal dunes is their connection with the beach and the impact of the maritime system on the dune climate. This takes the form of a distinct gradient across coastal dunefields from the sea inland. Next to the sea, conditions are harshest, due to onshore winds transporting much salt spray as well as moving sand particles. Airborne salt spray is an extremely important factor limiting plant growth. The levels of sodium chloride in the sand are frequently low, due to rapid leaching, but the impacts of salt spray on the vegetation are great.

Other features of the shore-normal gradient include a decrease in sand particle size landward, a decrease in temperature extremes and the amount of open sand, a decrease in pH, an increase in organic levels, and generally a decrease in rates of sand movement. Some of the major abiotic changes that would be expected across a typical dune gradient are illustrated in Figure 13.10. In many narrow foredune systems such a gradient will be truncated, whereas in wide transgressive dunefields it will be well developed.

13.8   Dune Vegetation

Vegetation is by far the most important component of the biota on sand dunes because it is directly involved in establishing the dune forms and creating the structure of the dune habitat. The primary colonization of aeolian sand along the coast is usually by microorganisms and fungi. These organisms promote aggregations of sand grains 1 to 2 mm in diameter. Fungi are most important in this regard, followed by bacteria and actinomycetes, and algae may also play a role. The presence of such aggregates reduces wind erosion, increases soil moisture, and increases the nutrient status of the sand. This is therefore a very important first stage in dune succession. Aggregation of sand grains also increases as dunes become more stable.

A common feature of dune vegetation in later stages of stabilization is the abundance of lichens and bryophytes. Pioneer bryophytes on open sand tend to display an upright growth pattern and are able to grow through small amounts of accumulating sand, whereas mosses (with a spreading growth habit) occur in more stable sand.

Distinct zones of vegetation can usually be distinguished at right-angles to the shore. These zones are most obvious on prograding shores, but may also be exhibited in dune habitats characterized by quite extensive sand movement. Although Doing (1985) has described six foredune vegetation zones around the world, four major zones may be more easily recognized (Tinley 1985). These zones can occur across typical coastal dune systems in any part of the world where rainfall is sufficiently high and the shoreline sufficiently stable or prograding — so as to allow succession to reach conclusion. The main features of these zones are as follows:

- *Zone 1, pioneers*. Closest to the sea, this zone is characterized by creeping grasses and succulent herbs with rhizomatous and stoloniferous growth. These plants fix sand above the drift line to form incipient foredunes. They can often establish themselves at the highest reach of the swash but are ephemeral and often destroyed by extreme storms. Many of these species die back after a few years. Dune pioneers occupy an extreme habitat with high sand transport rates, heavy salt loads, and temperature extremes. This represents the most physically controlled of the four zones. Plants occupying it are characterized by rapid growth to outpace sand accumulation, succulence to store water, cuticular protection against salt loading, and glands to exude salt. The European beach grass *Ammophila* is perhaps the most successful plant in this category, but other notable examples are *Cakile*, *Spinifex*, *Ipomoea*, *Scaevola*, *Honkenya*, and *Arctotheca*. On most shores, one or two species dominate this zone — often forming hummocks. Their seeds are dispersed by water and wind action.
- *Zone 2, shrubs*. This consists of a shrub community or dune heath, which may include a mixture of plants from zone 1 together with psammophytes from other habitats. There is usually moderate sand movement. The flora may include annuals, forbs, creepers, succulents, and shrubs with a wide distribution. The seeds of these forms are dispersed both by wind and by birds, and the community becomes increasingly invaded by bird-dispersed species as it ages. In the case of many narrow dunefield or foredune systems, only these first two zones may have space to develop.
- *Zone 3, thicket*. The scrub-thicket zone typically has a flat canopy due to wind pruning. It is characterized by little or no sand movement, in contrast to zone 1, and only develops where rainfall exceeds about 250 mm per year. This community may consist of dwarf trees and shrubs, with a compact canopy whose height increases with rainfall. There is usually little understory but a distinct litter layer. Height tends to increase in the lee of dunes. Seed dispersal is mainly by birds, and widely differing species of plants dominate in different areas.
- *Zone 4, forest*. Tall thicket or forest only develops in areas of high rainfall, behind the shelter of larger dunes. This tall thicket and/or forest includes a mixture of thicket species and true forest species, as well as other minor elements. It generally only occurs where rainfall exceeds 700 mm a year and where soils are mature. Canopy height increases with rainfall and shelter. There is a closed upper canopy, below which there may be a layer of smaller trees and shrubs, a field layer, and a ground layer. Seed dispersal is predominantly by birds.

Coastal sands are colonized and built into wooded dunes by the interaction of wind and plant growth across the zones in a serial succession. Wind, shelter, and rainfall are key factors. On dry coasts (less than 100 mm rainfall), only the first one or two zones may occur, whereas on wet coasts with rainfall exceeding 900 mm all four zones may be well developed, especially where the shore is prograding. This idealized zonation or succession is often broken up by blowouts or interrupted by estuaries (barrier dunes) or headlands. Although the intrinsic direction of succession is toward a mature closed, woody community, in reality it is multidirectional as changes take place in response to erosion, accretion, disturbance, and the remobilization of sand — so that recurring arrays of habitats and niches occur (Tinley 1985). The most complex dune areas occur where there is both strong wind disturbance and aggressive plant growth (Doing 1985).

To survive in coastal dunes, plants require specific adaptations to the unique conditions prevailing there (Hesp 1991). Stress factors and the corresponding adaptations of plants to cope with these are listed in Table 13.1. A major feature of dunes is that sand is edaphically dry and that dryness may be increased by the salt load. Adaptations to this include the presence of succulents and water savers, which restrict transpiration by having thick cuticles or the ability to close the stomata. Some succulents may fold or shed their leaves or reduce leaf area. Spreading roots can rapidly take up rainwater, and in some cases deep roots may go down to the water table. Finally, many plants are able to absorb dew.

Further adaptations and modifications are encountered in species colonizing zone 1, and to a lesser extent zone 2. Leaves may have a light color and high reflectance, as well as transpirational cooling to reduce heat load. Salt stress is prevented by protective cuticles and glands that excrete salt. Wind stress and sand accumulation rates may be counteracted by rapid growth in response to sand burial and low-growth forms with tough leaves. Pollination is usually accomplished by having the flowers situated below the leaves for protection, by making use of wind pollination, and by the anthers opening only in favorable weather. In many cases, germination is inhibited by high salt loads. Thus, marked changes in vegetation types and plant adaptations occur across coastal dunes from the sea inland (Figure 13.11). As with physical changes, these changes may be truncated in narrow foredune systems and not all coastal dunes will display the full range of vegetation features. There are also clear patterns of change in diversity and adaptations with latitude (Figure 13.12). Diversity is lowest and endemism highest in arid regions, and diversity highest in tropics/subtropics and areas with Mediterranean climates. Adaptations to stress are best developed in arid regions.

13.9   The Fauna

The fauna of coastal dunes is limited but may contain unique elements. Arthropods and vertebrates usually predominate, particularly insects, birds, and mammals. Arachnids are also common, and crustaceans may be important near the beach. Mollusks may be well represented on lime-rich soils, and frogs and other groups have also been recorded. However, insects are usually dominant, especially Hymenoptera, Coleoptera, and Diptera. Van Heerdt and Morzer Bruyns (1960) distinguished six groups of insects in their classic study of the yellow dune region of Terschelling. These were halobionts found only in coastal areas, halopsammophiles or salt-tolerant species that also occur elsewhere, psammophiles found everywhere on sandy soil (but not tolerant of places with a high salt

Table 13.1.   Stress factors and corresponding plant adaptations in coastal dunes (after Hesp 1991).

Stress factor	Adaptations	Examples
Salt spray	Salt-resistant or salt-preferring/ tolerance (e.g., hypertrophy)	*Cakile* spp. (salt-resistant) *Salsola* spp. (salt-preferring)
Sand burial	Increased seed, node, root, shoot, and rhizome/stolon development	*Spinifex sericeus*, *Ammophila* spp. *Chamaecrista* spp.
Swash inundation and ponding	Flooding resistance Fruit buoyancy	Some tolerance in *Cakile maritime* Swash tolerance in *Spinifex* spp. *Cakile* spp.; pan-tropical beach taxa
Dryness, high light intensity, high temperatures, wind exposure	Leaf roll Leaf orientation (phototropism) Leaf hairiness Leaf loss Epicuticular wax layer Succulence Mechanical resistance (sclerophylly) Accumulation of solutes Efficient water use Various root adaptations Aerodynamic growth forms	All Monocotyledonae, *Ammophila* spp. *Hydrocotyle bonariensis* *Spinifex* spp. *Ambrosia chamissonis* Most species *Carpobrotus* spp.; *Cakile* spp. Most species *Ammophila arenaria*; many species Most species Many species? 'Cushion' growth habits of some species
	C_4 and CAM photosynthetic pathways	*Spinifex* spp. & many tropical & subtropical species
	Increased heat tolerance Osmotic adaptation	*Ammophila arenaria* Several species
Sand salinity	Salt resistance NaCl accumulation Salt bladders Succulence Osmotic adaption	*Salsola kali* *Salsola kali* Chenopodiaceae; *Atriplex hastata* Dicotyledonae; *Carpobrotus* spp. *Cakile*; *Ammophila*
Nutrient deficiency	Restrict or accumulate inorganic ion Plant plasticity Nitrogen fixation by rhizosphere bacteria Uptake of nutrients from strand litter Phosphorus from endomycorrhizae fungi Redistribution of nutrients	*Suaeda maritime* *Festuca rubra* Some species *Cakile edentula* Some species *Carex* spp.
Various stresses e.g., high light intensity, sand burial, salinity, inundation, dryness, wind exposure	Hygroscopic leaf movement Variation of life cycle and flowering times Seed dispersal Seed morphology Germination strategies Plant morphology Stand density and morphology Reduced leaf size	Mosses; *Tortula princes* *Elymus mollis*; *A. arenaria* *Spinifex* spp. Various species *Uniola* spp. *Spinifex* spp. Various species Some species

content), ubiquists found everywhere, social ants, and a small number of flying insects found everywhere. They also distinguished two communities separated by the crest of the main dune ridge: a seaward and a landward community (the former dominated by halobionts).

Dune soils, especially in areas of high moisture content, may support rich interstitial biotas (van der Merwe and McLachlan 1991), which have unfortunately been little studied. These components consist of bacteria, fungi, and meiofaunal forms, particularly deposit feeding and plant parasitic nematodes.

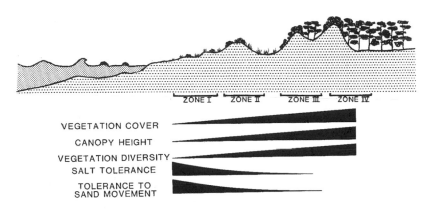

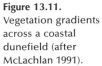

Figure 13.11. Vegetation gradients across a coastal dunefield (after McLachlan 1991).

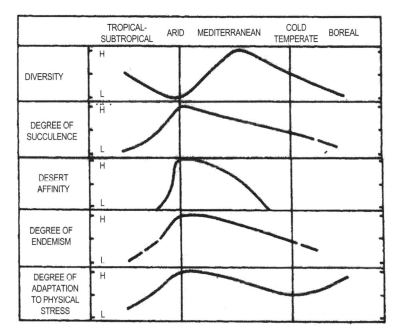

Figure 13.12. Global trends in coastal dune plants as a function of latitude (after Hesp 1991). H = high and L = low.

Vertebrates are usually represented by small mammals and birds, with reptiles and amphibians also present. Rodents may be the commonest mammals in the first vegetation zone, and birds (such as terns) nest at the beach/dune interface.

The dune fauna has limited diversity, and few species are unique or restricted to the dune habitat. Examples of unique species are the beach vole *Microtus breweri*, the American beach mouse *Peromyscus polionotus*, and Damara Terns *Sterna balaenarum*. However, coastal dune endemics are not abundant, limitations on endemism being imposed by the small extent of most coastal dunefields and their extensive sea and landward boundaries — which allow free exchange of the fauna with surrounding areas.

The dune fauna may be especially impoverished in arid areas but may include many more species where moisture is less limiting, and up to 370 species of arthropods have been recorded in single dune systems. The most exceptional recorded case, however, is of a fauna that includes more than 1,500 species of insects on two small dune islands — both less than 150 years old (Hoeseler 1988). Diversity is thus considerably higher than on the adjacent intertidal beaches. Species diversity responds primarily to changes in vegetation and habitat along the gradient perpendicular to the shore. Species diversity usually increases away from the sea and is positively related to the structural diversity of soil and vegetation, which usually increases in cover and complexity away from the sea as salt spray and sand movement rates decline.

In coastal dunes, the extremes associated with desert dunes are largely absent. Physiological specializations seem unlikely to be necessary, and the most important adaptations of animals are behavioral. A key adaptation is rhythmic activity. Near the beach and supralittoral zone, tidal rhythms may be evident — these often taking the form of activity during the low tide to avoid inundation (see also Chapter 6). However, for the true dune fauna the rhythms are primarily diel. Most species seem to be nocturnal, and there may even be separate day and night faunas.

Other distinguishing features of dune faunas are the ability to bury themselves or to burrow rapidly, and cryptic coloration. In low-lying areas, such as dune slacks and in the supralittoral zone, the ability to survive flooding (for example, by flotation) may also be important. Roberts (1984) described the adaptations of sand frogs to life in the dunes — adaptations that included pairing of males and females for at least five months before the egg deposition and having the entire larval development take place in the egg capsule. Apart from emerging during the wettest period of the year, dune frogs survive the arid summer in dry Australian dunes by staying underground in moist sand. Their survival under these conditions is apparently facilitated by the predictable winter rainfall.

Dune animals need to be tolerant of flying sand, heavy salt loads, and the absence of protective litter. Adaptations to extremes of temperature or aridity appear to be less important. Species associated with plants, rather than the sand surface, show few specializations and differ little from the fauna elsewhere. Typical faunal changes expected across a coastal dune gradient moving landward (McLachlan 1991) (Figure 13.13) include the following:

- Initial dominance by crustaceans near the beach, followed by their decline and eventual disappearance
- Increasing abundance and diversity of insects and vertebrates
- Increasing development of interstitial biota as soil develops and pH drops
- Increasing impact of the fauna on the vegetation due to grazing, seed dispersal by birds, and so on

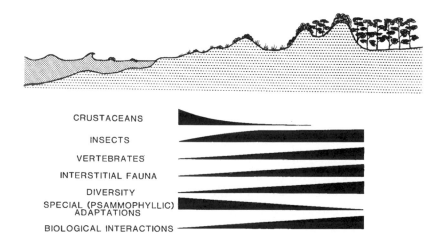

Figure 13.13. Faunal gradients across a coastal dunefield (after McLachlan 1991).

CRUSTACEANS

INSECTS

VERTEBRATES

INTERSTITIAL FAUNA

DIVERSITY

SPECIAL (PSAMMOPHYLLIC) ADAPTATIONS

BIOLOGICAL INTERACTIONS

- Decreasing specialization of the fauna for a psammophyllic lifestyle, including decreasing adaptations to salt spray and sand movement
- Escalating biological interactions

There may also be shifts from primarily r-strategist species near the beach to k-strategists inland, but this is not clear and more research on dune populations is required.

13.10   Food Chains

Dune food chains are driven by primary production of the dune flora, organic inputs from the sea, and inputs from land. A high proportion of plant biomass, both living and dead, may occur below ground — where it is more available to the interstitial biota than to the macrofauna. Indeed, a feature of coastal dunes appears to be the great importance of interstitial pathways. Moist conditions and a high proportion of buried organics favor the development of a rich interstitial community of bacteria, fungi, and meiofauna. In wet dune slacks, the biomass of the interstitial community may far exceed that of the macrofauna, and most consumption of carbon in dune slack ecosystems will be via the interstitial pathway (McLachlan et al. 1996b). Broadly, three main food chains may be recognized in coastal dunes (Figure 13.14):

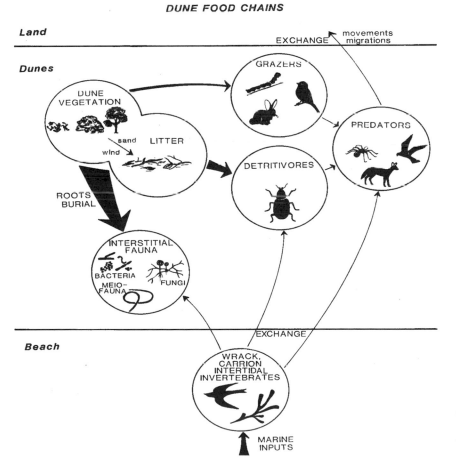

Figure 13.14. The three main food chains of a coastal dune system (after McLachlan 1991).

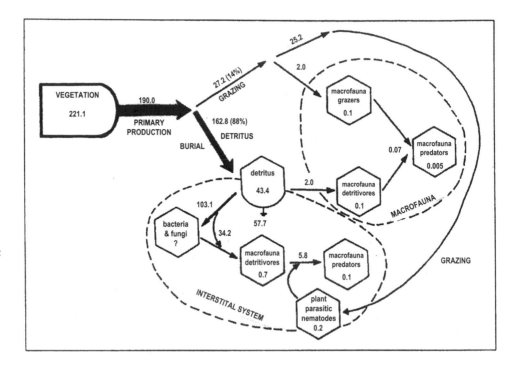

Figure 13.15. Circuit diagram showing energy flow and trophic relations in a coastal dune slack (after McLachlan *et al.* 1996b). Units are dry mass in g·m⁻² (biomass, inside modules) and g·m⁻²·year⁻¹ (flows).

- A grazing pathway consisting primarily of macroscopic herbivores (insects, mammals, and fruit-and-seed-eating birds)
- A detritus pathway consisting of macroscopic animals (insects) feeding on vegetation detritus (litter), which may accumulate in the lee of vegetation hummocks, at the base of slip faces, and elsewhere
- The interstitial biota, feeding on buried detritus, and in the case of plant parasitic nematodes on the roots themselves

There is very little published quantification of dune food chains. Mention is occasionally made of feeding studies, and some authors have examined the effects of animals (especially rabbits and gulls) on the flora (Ranwell 1972). The impact of seed eaters on dune seed reserves has received limited study. However, in most situations only a small proportion of primary production are grazed directly by herbivores. The only study to quantify energy flow in a coastal dune system investigated moist dune slacks in a transgressive dunefield (Figure 13.15). Because of fairly rapid dune movement, vegetation showed a clear five-year succession and high burial rates. Vegetation biomass was 221 g·m⁻², with 31% above ground. Detritus biomass was 40 g·m⁻², with 3% above ground. Plant production was 190 g·m⁻²·y⁻¹, and faunal biomass was mainly in interstitial meiofauna (1.0 g·m⁻²) — with macrofauna, arthropods, and vertebrates less important (0.2 g·m⁻²). It was estimated that 14% of plant production was grazed directly (93% by parasitic nematodes and 7% by macrofauna), and 86% entered the detritus pool. Of the detritus, 2% was consumed by macrofauna detritivores, 63% decomposed by the interstitial fauna, and 35% accumulated in the sand — being decomposed later when the slack was covered by advancing dunes. In the absence of other studies, it is difficult to know how typical this is of dune ecosystems. However, the contrast with the marine system (Chapter 12) is strong because dunes are characterized by a relatively high biomass and low turnover of macroscopic plants, whereas dissipative beaches may be driven by microscopic plants with small biomass but high turnover. In both cases macro-faunas are usually less important in energy flow than microfaunas.

Because of the relatively small scale and extended seaward and landward margins of most coastal dunefields, much exchange occurs across the boundaries. The extent of the sea and landward boundaries is large relative to the area of coastal dunefields, which range from a few meters to a few kilometers in width. Consequently, animals move across both boundaries to seek food and shelter — in the dunes, on the beach, or inland. Such systems are thus very open, and many of their food chains may start in the sea (e.g., with wrack cast ashore and subsequently eaten by dune animals) and end inland (e.g., in predators moving into the dunes to feed).

13.11   Dune/Beach Exchanges

In addition to climate and moisture, four materials are exchanged across the dune beach interface (Figure 13.16). These are sand, groundwater, salt spray, and living and dead organic material. Virtually all coastal dunefields receive sand from and/or supply sand to beaches. The direction and extent of these exchanges are controlled by the direction of prevailing winds and the alignment of the coast. The growth of dunes is thus dependent on the supply of sand to the intertidal zone. As most coasts display predominantly onshore winds, beaches are usually sand sources for the dunes. Headland bypass dunes, however, deliver sand from one beach to another.

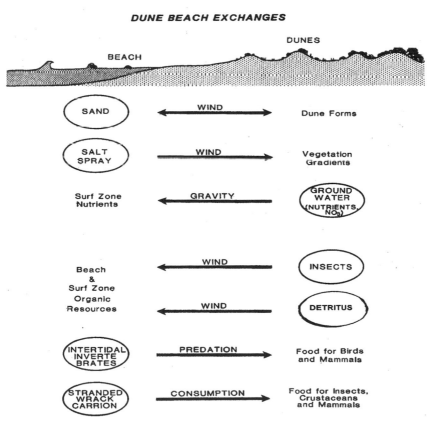

Figure 13.16.
Exchanges of materials between dunes and intertidal beaches (after McLachlan 1991).

Groundwater seeps from the land into the sea through dunes in unconfined aquifers. This groundwater may be discharged in large volumes in some places (typically 0.1 to $10\,m^3 \cdot m^{-1} \cdot d^{-1}$) and often contains very high nutrient levels, thereby adding significantly to the nutrient pool of the beach/surf-zone system.

Salt spray blown inland has an important effect on dune vegetation. It derives from the bursting of bubbles and generation of spray in the surf zone, followed by aeolian transport of the droplets inland. Because these droplets are relatively large, salt spray load increases exponentially with wind speed and decreases exponentially with distance from the beach. The bulk of the salt load is therefore dropped close to the sea, usually within 200 m of the intertidal beach — where it has a significant impact on dune vegetation and seedling germination. Salt spray, together with sand movement, is a major factor structuring vegetation communities and landward gradients in coastal dunes.

Four types of organic materials may be exchanged across the dune/beach interface: insects are often blown from the land toward the sea; carrion and wrack cast ashore may provide a source of food for dune animals, which move down to the beach to feed; larger animals in the dunes, such as birds and mammals, may be dependent on intertidal organisms for food; and organic detritus from the dunes may be blown into the beach system.

In addition to sand, which may be transported in massive volumes in high-energy situations, the absolute quantities of the previously cited materials transported across most dune/beach interfaces are fairly small. However, the impacts of these materials (groundwater nutrients, salt spray, and organic materials) may be significant to both ecosystems. Thus, although dunes and beaches are discrete (respectively terrestrial wind-controlled and marine wave-controlled) systems, and although the interactions between them are quantitatively small, they are essentially interdependent and interacting.

13.12   A Case Study of Dune/Beach Exchanges

The Alexandria dunefield system (Figure 13.17) in Algoa Bay, southern Africa, covers $120\,km^2$ and has been well studied — particularly in terms of dune/beach exchanges. For detailed references, see Brown and McLachlan (1990) and McLachlan et al. (1996b).

The sand of the Alexandria coast consists predominantly of quartzite (61%), with a lower proportion of calcite in the form of molluskan shell fragments (38%) and a small

Figure 13.17. The Alexandria dunefield. (Photo: W. Illeberger; by permission of the Geological Society of South Africa.)

amount of heavy minerals. The average grain size is about 0.2 mm in the dunes, as compared with 0.25 mm on the beach. The sand is well sorted, becoming finer further back in the dunefield, which averages 2 km in width.

Sand exchange. The coast is characterized by extensive longshore sand transport, driven by wind and waves. Sediment transport in the surf zone in the longshore direction totals 5 to $9 \times 10^5 \, m^3 \cdot y^{-1}$, whereas total longshore aeolian transport in the dunefield averages 4 to $5 \times 10^4 \, m^3 \cdot y^{-1}$. The average rate of sand transport in the dunefield is about $20 \, m^3 \cdot m^{-1} \cdot y^{-1}$, but this is not uniform throughout the width of the system, averaging $30 \, m^3 \cdot m^{-1} \cdot y^{-1}$ near the beach but only $10 \, m^3 \cdot m^{-1} \cdot y^{-1}$ further inland. Thus, the amount of sand moving past a fixed point in one year is an order of magnitude higher in the surf zone than in the dunes, a consequence of the greater transporting capacity of water.

There is also a strong net onshore movement of sand to the dunefield, which is accreting by some $375,000 \, m^3$ a year. Because of the configuration of the coast, little of this sand is transported alongshore and the dunefield largely functions as a sand sink. However, some sand is lost to the sea by wave erosion of the Woody Cape cliffs to the east, at a rate of some $45,000 \, m^{-3} \cdot y^{-1}$. The dunefield thus grows by about $330,000 \, m^3 \cdot y^{-1}$. This net gain in sand volume is accommodated in two ways: first, the dunefield creeps landward at an average rate of $25 \, cm \cdot y^{-1}$ along its 45-km length, and second it slowly becomes thicker by about $1.5 \, mm \cdot y^{-1}$.

This sand input to the dunes is by no means constant, and sand is supplied erratically, with three major pulses having occurred over the past 6,500 years. This pulsing sand supply could result from small variations in sea level, from climatic changes, or from destruction of vegetation by either human activity or bush fires.

Sand supply from the beach represents an obvious and important exchange between the two systems. The movement of sand exerts a direct and powerful control on the structure of the dunefield and on the development of its vegetation. It may be considered one of the primary factors controlling dune vegetation types and the structure of dune plant and animal communities.

Salt spray. Large amounts of salt spray are blown into the dunefield at all times of year. The prevalence of salt spray shows an inverse correlation with vegetation cover and plant species diversity. Cover increases from 3% in slacks near the beach to 50% on the precipitation ridge, and diversity increases from 12 to 20 species — despite there being more moisture near the beach. This confirms other studies on the effects of salt on the germination of seedlings and the growth of a number of foredune plants. In *Juncus*, *Arctotheca*, and *Gazania*, germination is markedly reduced in salinities above 2 to 3 ppt.

Groundwater. The Alexandria dunefield has a strong rainfall gradient, ranging from $400 \, mm \cdot y^{-1}$ at its western extremity to $800 \, mm \cdot y^{-1}$ at its eastern end. Groundwater flow rates are thus highly variable. Groundwater release occurs in the form of unconfined aquifers, discharging through the dunes to the beach and surf zone. Flow rates vary by more than an order of magnitude, but average about $1 \, m^3$ per meter of beach per day. Values in excess of $20 \, m^3 \cdot m^{-1} \cdot d^{-1}$ have been recorded in front of blind rivers leading into the dunefield. In general, however, the volumes of groundwater seepage are small in comparison to the volumes of saltwater flushed through the sand by wave action and the tides. Thus, dramatic declines in beach water salinity seldom occur. Nevertheless, this groundwater constitutes an important exchange because of its very high concentrations of

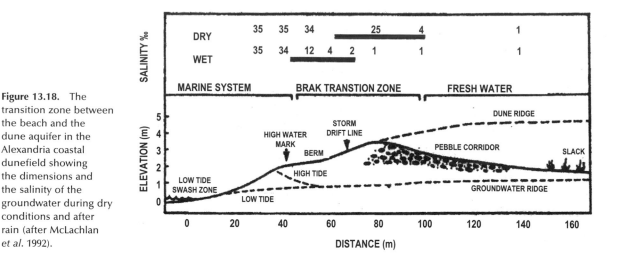

Figure 13.18. The transition zone between the beach and the dune aquifer in the Alexandria coastal dunefield showing the dimensions and the salinity of the groundwater during dry conditions and after rain (after McLachlan *et al.* 1992).

inorganic nutrients, particularly nitrates. With an average inorganic nitrogen content of 2.13 mg·ℓ^{-1}, the groundwater provides an annual input of some 777 gN·m^{-1}·y^{-1} (or 30 tons) along the entire stretch of the beach. This is sufficient to provide about 5% of the nitrogen requirements of the surf-zone phytoplankton, the sole primary producers in this beach/surf-zone ecosystem. Although the supply of freshwater is relatively unimportant with regard to beach salinities, where the fresh and saline groundwaters mix below the backshore a brackish zone is created and colonized by a unique brackish interstitial fauna worthy of study.

This transition zone (Figure 13.18) is 20 to 30 m wide at the dune/beach interface. Strong tidal pulses in the beach water table decay landward and are negligible at the landward end of the transition zone, which has a salinity gradient from 30% at the seaward end to 1 to 3% at the landward end. The transition zone supports a sparse interstitial fauna of nematodes — in contrast to the rich meiofauna in the beach, which is supplied by organic materials from the sea and a rich meiofauna also in the slack (fueled by dune vegetation).

Organic materials. Organic materials exchanged between the beach and the surf zone include stranded carrion from the sea (which is consumed by terrestrial carnivores) intertidal invertebrates being eaten by land predators, windblown insects carried seaward and fed on by marine animals, and litter blown from the dunefield into the sea. Such exchanges are episodic and difficult or impossible to quantify. Nearshore islands in Algoa Bay include penguin, gannet, and seal breeding colonies — with the result that relatively large vertebrates are cast ashore with some regularity. Although they may partially be consumed by beach animals, the main scavengers are terrestrial mammals — notably jackals, which patrol the beach nightly.

The chief terrestrial predators are birds. Their impact on the beach in question has been quantified, and they remove some 20% of the total population of intertidal invertebrates annually. As the birds are mostly residents of the dunes, their movements constitute an exchange of organic materials across the dune/beach interface. Some beach animals, such as supralittoral crustaceans, move into the dunes to feed — ghost crabs (Ocypodidae) providing possibly the most important example.

Another form of exchange, through insects, has proved impossible to quantify. When winds blow offshore, numbers of insects are blown from the dunes toward and frequently into the sea. They may include large numbers of flying ants, as well as moths, butterflies, and others. Many of those landing on the beach are consumed by birds, crabs, and whelks of the genus *Bullia*. In addition, insects have been recorded regularly from the stomachs of surf-zone fishes. Mullet, in particular, pick them up from the surface of the water while feeding on surf phytoplankton foam.

The final organic exchange between the dune and beach systems involves organic detritus, mainly fragments of vegetation blown seaward by the wind. Much of this material may be refractory and not available to marine animals. It nevertheless represents a potentially significant input. The energy flow through a dune slack in this system was outlined previously (Figure 13.15).

13.13   Conclusions

What are the primary factors that control the ecology of coastal dunes, and are these systems physically controlled or are they biologically structured? Noy-Meirs' (1980) autecological hypothesis suggests that in physically extreme environments, such as deserts, control is primarily physical and communities result from the independent responses of each population to their physical environment. Can this model be used to describe coastal dunes?

On coastal dunes the picture is complex, as an entire spectrum may be encountered — from physically controlled situations such as transgressive dunefields with high rates of sand movement and salt spray to systems with considerable biological structure (for example, forested dunes). The main feature of coastal dunefields is the gradient from the sea landward and in a conceptual sense they may be considered as grading from physically controlled situations near the beach through to a biologically structured configuration landward.

Coastal dunes are not unique, except for their relationship to the sea, and have many features in common with terrestrial ecosystems in general. Their floras exhibit fairly high degrees of endemism but moderate diversity, whereas coastal dune faunas contain few unique species but display relatively high diversity. Of the factors impinging on them, wind is perhaps the super-parameter. By its strength, frequency, and prevailing direction — especially in terms of the land/sea interface — wind controls sand movement, microclimate, salt spray load, and the form of the vegetation gradient across the dunes.

Although they are terrestrial ecosystems distinct from their adjacent marine beaches, coastal dunes are coupled to beaches by exchanges of sand, groundwater, salt spray, and organic materials and the maritime influence is strong. Indeed, that is what distinguishes coastal dunes from other dunes. The close coupling between coastal dunes and beaches, particularly with regard to sediment exchange and storage, creates an interdependence that underlies all concepts and strategies when it comes to conservation and management of sandy coastlines (see Chapters 14 and 15).

Human Impacts

14

14.1   Introduction

A variety of anthropogenic and other external factors impact sandy beaches on various temporal and spatial scales. Pollution, recreational activity, expanding human populations, development, structures that change the landscape and modify geological processes, and global change all affect beaches. In addition, natural processes such as storms, hurricanes, and tsunamis can have dramatic impacts. In general, beaches are most sensitive to pollution under low-energy or sheltered conditions because pollutants are not easily dispersed, whereas it is the more exposed beaches that are most sensitive to disruption of their sand budgets because of the large volumes of sand transported under high-energy conditions. This chapter reviews all of these impacts. It is largely based on Chapter 11 in the first edition, with additional material from Brown and McLachlan (2002), Brown *et al.* (2006), and Nordstrom (2000).

14.2   Pollution

Sandy beaches (like other types of shore) suffer pollution or contamination from a number of sources, including sewage, industrial effluents, and oil spills. There is nothing to suggest that organisms from different marine habitats differ in their tolerances of pollution or that there are marked differences based on phylogenetic considerations. Nevertheless, conditions within the habitat — as well as the behavior of the organisms — may have a marked influence on the ecological consequences of pollution. Sediments in general are pollution traps, and the finer the sediment particles the more efficient the trap. Thus, although a polluted rocky shore begins to recover as soon as pollution ceases, sandy and muddy shores may retain the polluting material for some time. A sandy beach community may in fact experience the effects of pollution for months or even years after the actual event. The classic example of this phenomenon is the pollution of sandy beaches by crude oil.

Crude Oil Pollution

Oil reaches beaches from a variety of sources, the most spectacular pollution arising from tanker accidents — so that the *Torrey Canyon* and the *Amoco Cadiz* have become household names. Accidents to other vessels are actually far more common than are tanker disasters, and these also often result in oil pollution (although not on such a vast scale). Oil slicks also originate from inadvertent spillages and from the cleaning of oil tanks at sea. Spillages at oil terminals are frequent, and the areas around such terminals may suffer chronic pollution, particularly as terminals are often sited in sheltered bays and lagoons from which oil escapes only slowly. Pollution may also result from offshore drilling, from runoff from the land, and from natural seepages. Indeed, for the oceans as a whole land-based pollution may be the most important source of petroleum hydrocarbons. Most attention has, however, been paid to single oil pollution events resulting from accidents to shipping.

On high-energy relatively coarse, sandy beaches the degree of faunistic damage observed following a spill is usually small because there is normally relatively little macrofauna present to begin with. However, the supralittoral macrofauna (insects and crustaceans) may be most susceptible because oil tends to accumulate at the top of the shore and cleanup activities (which are focused in this area) may also impact the fauna (de la Huz *et al.* 2005). The meiofauna is affected to a greater or lesser degree but is likely to recover within a year. Low-energy sheltered sands not only show a much higher initial mortality but oil trapped within the sediment may continue to affect the ecosystem for many years. Most species may be eliminated by the spill, but a few — notably polychaete worms belonging to the Capitellidae and the Cirratulidae — may increase in numbers or invade the area if not previously present, and soon dominate the system. Such a pollution-tolerant fauna is eventually joined by other species as oil concentrations diminish, and both biomass and species composition then undergo a series of fluctuations that decreases in amplitude over a number of years until relative stability is regained (Sanders *et al.* 1980, Southward 1982). The largest longest-lived species are in general the slowest to return, and hydrocarbons can still be detected in the tissues of the animals years after a spill.

The sandy-beach ecosystem may also be affected by the oiling and subsequent death of birds, some of which feed intertidally and possibly nest among the dunes. Others, such as cormorants, dive for food in the surf zone and are thus especially vulnerable to oiling. Oiled birds are later the focus of public attention following a spill. The numbers of birds killed by oil in European waters are at most a tenth of the natural mortality rate (Dunnet 1982), and thus the ecological effect is likely to be local and short lived. Death from oil pollution may, however, become a significant factor in bird populations already under stress from other causes (Brown 1985).

As far as the invertebrate fauna is concerned, some species are far more at risk than are others, depending on the zone they occupy, their method of feeding, and their behavior. For example, crabs that molt in their burrows may be particularly affected, as they then expose their soft and permeable post-molt integuments directly to the oiled sediment. Animals with delicate filter-feeding mechanisms (such as bivalve mollusks) seldom survive smothering with oil, whereas crustaceans (such as *Eurydice*, which display a tidal cycle of migration into the overlying water) may no longer do so after contact with crude oil.

Most work on marine organisms has, however, been done on the effects of dissolved fractions of oil, although the conditions under which such experiments are performed seldom reflect conditions in the field. Low-molecular-weight hydrocarbons tend to be the most immediately toxic and are the most soluble (Patin 1982). They are also the first to disappear from the system, largely by evaporation, so that oil spilled at sea has inevitably lost much of its short-chain highly toxic component before reaching the shore. Its composition on stranding depends on the original composition of the oil (which is variable), the length of time the slick has been at sea, and the conditions it has experienced — including wave action and temperature. Some of the oil deposited on the sand will be washed up the shore by waves and tides, whereas large amounts may become buried and so enter the sand column. This oil has a double effect, in that it not only affects the fauna directly but changes the physical characteristics of the sediment — clogging interstitial spaces and reducing water flow and possibly the oxygen supply (McLachlan and Harty 1981). Figure 14.1 gives some indication of the complexity of the physical parameters involved in oil pollution, and Figure 14.2 the persistence of different fractions.

Many of the biochemical, physiological, and behavioral changes that may be observed in animals after exposure to crude oil can be explained as reactions at the cellular level

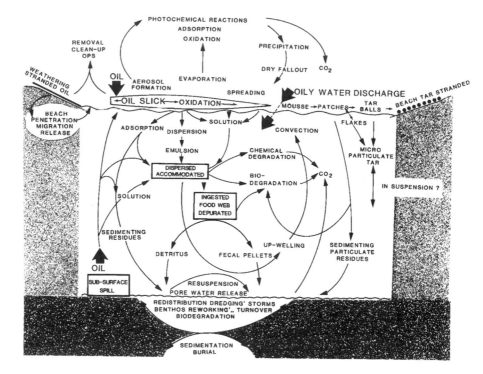

Figure 14.1. General pathways and mechanisms of the fate of crude oil polluting a sandy ecosystem (after Whittle *et al.* 1982).

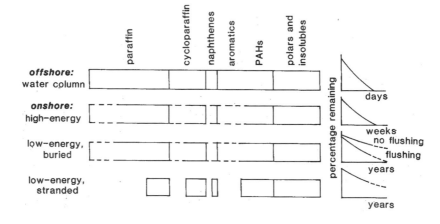

Figure 14.2. Persistence of oil fractions as a function of location and environmental conditions (after Vandermeulen 1982).

to changes in membrane permeability. Interference with membrane function is associated with the fact that hydrocarbons accumulate selectively in lipids and may alter lipid structure. This concept is also germane to the study of pollution effects on embryos and larvae, as hydrocarbons accumulate in the lipid component of yolky eggs, interfering with subsequent lipid mobilization and metabolism during development. In addition to effects in lipids, there are direct effects on enzyme systems, including respiratory enzymes.

Microsomal mixed-function oxygenases, capable of transforming aromatic hydrocarbons into more water-soluble metabolites, are widespread — although their presence in mollusks is uncertain. Mixed-function oxygenase activity is induced by exposure to oil (Neff and Anderson 1981), and such induction may be important in the development of resistance to oil pollution (Brown 1985). It may be noted, however, that hydrocarbon

Figure 14.3. How not to deal with oil pollution. Heavily oiled detritus being collected from a sandy beach and buried in a pit just below high-water mark, thus ensuring pollution of the beach for many months and possibly for years. (Photo: A. C. Brown.)

metabolites are often more toxic than the parent compounds and may add to, rather than diminish, the pollution of the beach — at least in the short term.

Oil Dispersants

Oil dispersants and emulsifiers are commonly used to disperse oil slicks at sea and are often sprayed too near the coast, despite regulations now in force in most countries. In some cases, they have even been applied to the shore itself — with disastrous consequences for the biota. Patin (1982) summarized much of the published data on dispersant toxicity. However, such toxicities by themselves are of limited value, as dispersants normally occur in the field only in the presence of oil. The literature on the toxicities of oil/dispersant mixtures is difficult to interpret and has in general little relevance to field conditions (Brown 1985). In principle, every mixture of oil and dispersant is more toxic than is the oil itself. The more effective the dispersant the more toxic it is to the biota. On the other hand, the more effective the less that needs to be used and thus a balance between toxicity and effectiveness needs to be achieved.

The markedly synergistic effects of oil and dispersant are probably due to the fact that dispersants, like hydrocarbons, tend to act on lipids — including the lipids of the cell membrane. The primary mechanism by which dispersants harm organisms may be the decreased surface tension at the tissue/water interface, accompanied by impairment of osmotic and ionic regulation, but the synergistic effect may also be due to enhanced biological accessibility and the penetrating power of hydrocarbons in the presence of dispersants. It should also be noted that spraying dispersants on or near a beach not only affects the biota directly but may increase the retention of oil by the sand (Southward 1982). Retention of oil on the shoreline, in any form, is not desirable (Figure 14.3).

Sewage and Organic Enrichment

Sewage is the oldest form of marine pollution of any consequence and the earliest marine antipollution laws were concerned with its disposal. This was mainly on aesthetic grounds, and only much more recently has it become apparent that the discharge of raw sewage to sea may damage the ecosystem or represent a health hazard to man. Raw sewage continues to be discharged by many countries, although there has been a

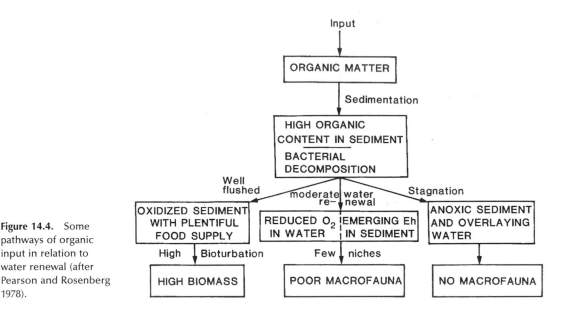

Figure 14.4. Some pathways of organic input in relation to water renewal (after Pearson and Rosenberg 1978).

tendency to increase the length of discharge pipes in an attempt to keep sewage away from the shore. Such measures are not always effective. Some countries have adopted partial treatment, often consisting of macerating the sewage and adding chlorine. A better method is to allow the solid component (sludge) to settle and to dispose of it on land or by incineration — only the fluid component being discharged to sea. The contamination of bathing beaches with sewage is a health problem that cannot be ignored. It is clear that any discharge of pathogens into coastal waters must be considered adverse from a health point of view and should be avoided, although in practice this ideal is not attained by any country in the world.

Unlike crude oil or factory effluents, sewage generally contains few substances that are directly toxic, although in some cases there may be a buildup of metals such as lead. The chief effects of sewage are a smothering of organisms near the discharge point by sludge, if this has not previously been removed, and an enrichment of the seawater leading to an increase in primary productivity and bacterial biomass. Organic enrichment of the beach itself may lead to a lowering of oxygen tensions and an upward encroachment of the anoxic black layers (see Chapter 3). Dissolved organic matter may be absorbed directly by meiofaunal species and even by some members of the macrofauna.

Pearson and Rosenberg (1978) reviewed the field of organic pollution in general, being mainly concerned with macrobenthic succession and gradients of enrichment, species composition, and biomass. Their representation of the importance of water movement in determining the effect of organic pollution is depicted in Figure 14.4.

Factory Effluents

Factories should ideally be sited where their effluents can enter deep water with strong currents. In practice, they tend to be built on estuaries or on shallow sheltered bays or inlets where sand or mud predominates. Thus, sandy beaches come in for more than their fair share of factory pollution, despite the increasing length of effluent pipes. Small factories and home industries seldom discharge into an effluent pipe at all, and too often rely on the sewers or on stormwater drains — which may open onto the beach above the high-water mark. Different countries have different regulations in this regard while some have no control measures at all.

Factory pollution is a long-term continuous process and necessitates the ongoing monitoring of possible damage to the ecosystem. Ideally, one should have a good knowledge of the communities to be monitored before the factory begins discharging, and most countries now require impact assessment studies before permission to discharge is given. The actual impact of the effluent is not always as predicted, of course, but at least the assessment implies a pre-pollution study of the ecosystem, thus facilitating the monitoring of change.

The difficulties of assessing pollution damage without such a pre-pollution study are formidable. One approach has been to compare the polluted site with that of a nearby unpolluted area, but in fact no two sites (even if completely unpolluted) offer identical conditions, so that two reference sites may differ in their faunas as much or more than a reference site and a polluted site. Sometimes there is a gradient of pollution that can be correlated with a gradient in biomass or in species diversity, but even here the conclusions drawn may be fallacious. Brown (1983b) studied the density of meiofauna along the length of a sandy beach subject to pollution from a marine oil refinery. This effluent was added to a small stream that escaped onto the beach through a stormwater drain. Meiofaunal numbers were consistently and significantly depressed around the outfall, and could be correlated with depressed oxygen tensions, elevated phosphate, and other indices of pollution. After some years the effluent was rerouted, thus promising a rare opportunity to study the recovery of the sandy-beach system. Meiofaunal densities did not recover, however, and it was eventually shown that the depression in numbers was actually due to reduced salinities attributable to the freshwater stream to which the effluent had previously been added and which continued to run down the beach. It had nothing to do with pollution!

It is, in fact, still impossible to make accurate predictions as to the ecological effects of a factory effluent because so many factors are involved, including the proximity of other discharges, current patterns, and type of shore. For this reason, there has been a move away from legislation governing effluent quality toward regulations aimed at ensuring acceptable seawater quality. What is acceptable depends on the type of shore and the use to which it is put. As far as the relevant ecosystems are concerned, the concept of assimilative capacity has gained ground. This was originally defined as the amount of material that could be contained within a body of water without producing an unacceptable biological impact. This has since been redefined in a number of ways (Figure 14.5). It is, for example, accepted that all animals have some capacity to withstand contamination. This is to be expected, as many pollutants released by man are either naturally present or are similar to naturally occurring compounds.

Thermal Pollution

Factory effluents are commonly warmer than the seawater into which they discharge, and therefore potentially add thermal pollution to the chemical pollution they produce. Such thermal pollution is, however, usually small or even insignificant — and it is only with the advent of nuclear power stations (which use vast quantities of water for cooling

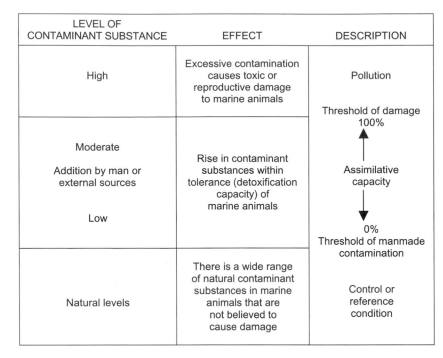

Figure 14.5. The concept of assimilative capacity (after Bascom 1984). Many chemicals that are toxic at high concentrations occur normally in the sea and can be tolerated by the marine fauna over a wide range of lower concentrations. No locations are known where open coastal waters have no contaminants.

purposes and discharge it back into the sea at temperatures higher by 10° C or more) that the subject has evoked serious study. Infaunal sandy-beach animals may of course, be buffered to some extent against short-term changes in temperature in a way that rocky shore animals are not. However, they are in no way protected against the long-term temperature changes that accompany the continuous operation of generating plants.

Subtle ecological changes should be expected in areas affected by heated effluents, including alterations in species composition, in the size of individuals, in population densities, in growth rates, and in breeding behavior. At Hunterston, in Scotland, no catastrophic changes were apparent after the power station became operational. However, over a 10-year period the bivalve *Tellina tenuis* showed a considerable drop in population density, although it had a faster growth rate in the warmer water. *Urothoe* commenced breeding earlier in the year and juveniles grew for a longer period, attaining a 28% greater size than individuals from unaffected areas (Barnett 1971). Another possible consequence of thermal pollution in some areas is the invasion of warmwater species, to the detriment of the original fauna.

Indicators of Pollution

Pollution often results in replacement of the fauna of a beach by related but more tolerant forms. The greater the pollution stress the more different the replacing species are likely to be (Figure 14.6), making indicator species useful. However, species identifications are not always essential. Defeo and Lercari (2004) showed that surveys would be able to discriminate between disturbed sandy-beach sites using different taxonomic levels, including class and phylum.

As stated previously, field studies aimed at monitoring the effects of pollution by considering whole communities are fraught with difficulties because of the considerable temporal and spatial fluctuations that occur in the composition of communities under normal

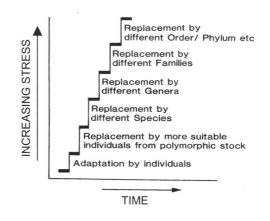

Figure 14.6.
Relationship between environmental stress (including stress due to pollution) and taxonomic variability (after Pearson and Rosenberg 1978).

circumstances, no community ever being completely static. An approach that ignores these complexities is to concentrate on indicator species, such as the polychaete worm *Capitella*, which increases in abundance in response to certain types of organic pollution (Figure 14.6). Alternatively, it may be useful to investigate just those species known to be most sensitive to toxic substances, the first to disappear under pollution stress (Gray and Pearson 1982). Among the sand-beach meiofauna, for example, copepods can be considered such an indicator group. A possibly promising technique in this regard is to assess the proportion of nematode worms to copepods, the former being markedly more resistant to most forms of pollution than the latter (Raffelli and Mason 1981) (see Chapter 9). This ratio has been used in practice by some workers, whereas others have severely criticized it (Gee *et al.* 1985). The concept may be worth pursuing further, although its value still needs to be resolved. Wenner (1988) summarized previous work on the use of invertebrates as indicators of beach pollution and made a convincing case for the use of the sand crab *Emerita* in this regard, where this animal is common. However, faced with apparently insurmountable complexities and uncertainties many pollution biologists have fled the field altogether in favor of the laboratory, and put their trust in toxicity testing.

Toxicity Studies

The assessment of the toxicity of single pollutants, of mixtures of pollutants, and of factory effluents may be seen as complementary to field monitoring and is essential to the drawing up of permissible levels of discharge. The difficulties of assessing the toxicity of pollutants to sandy-beach animals are very much greater than those attached to the use of rocky shore or pelagic organisms. This is because the experimental animals should clearly not be subjected to stresses additional to that imposed by the pollutant. This implies that the animal must have access to its natural substratum in the laboratory, and this in turn complicates the experiment, makes it difficult to observe the animal within the sand, and makes it difficult to assess accurately the level of pollution to which it is being subjected. It is thus hardly surprising that sandy-beach animals have been little used in assessing toxicity or that the results gained have frequently been of doubtful value.

The simplest and most obvious way of assessing toxicity is to place animals in a known concentration of pollutant and note the time of death of each individual. One can then

Table 14.1. Short-term effects of some pollutants on adult females of the sandy-beach whelk *Bullia digitalis* (after Brown 1982c). (All metals added as chlorides.)

Pollutant	Highest conc. without effect	Conc. at which burrowing ceases	Conc. causing irreversible stress	96-hour LC_{50}	Effect on oxygen uptake
Cadmium	0.1 ppm	0.5 ppm	0.7–0.8 ppm	0.9 ppm	30% increase at 0.5 ppm 20% decrease at 0.75 ppm
Zinc	1 ppm	2 ppm	3 ppm	3 ppm	40% decrease at 2 ppm
Lead	0.5 ppm	1 ppm	—	—	Slight but significant decrease at 5 ppm
Copper	0.1 ppm	0.2 ppm	0.35 ppm	0.5 ppm	70% decrease at 0.3 ppm
Mercury	0.5 ppm	2 ppm	7 ppm	—	20% decrease at 5 ppm
Selenium	1 ppm	4 ppm	7 ppm	—	15% decrease at 5 ppm
Phenol	10 ppm	50 ppm	1 ppt	—	100% increase at 50 ppm 80% decrease at 1 ppt
Ammonium nitrate	50 ppm	60 ppm	550 ppm for 15 h	About 300 ppm	No effect up to 1 ppt

calculate the concentration that will kill 50% of the animals in a given length of time (usually 48 or 96 hours). This value is the LC_{50}. Permissible levels of discharge are still frequently based on LC_{50} values, and its defenders say that it at least ranks pollutants in terms of toxicity. Even this is by no means always the case, however, and in fact different life-history stages of a species may show very different susceptibilities — whereas sublethal effects, at lower concentrations than the LC_{50}, may be just as damaging to the ecosystem over a period of time.

It is safe to say that there is no aspect of the behavior and physiology of sandy-beach animals that is unaffected by pollution. Frequently the first sign of disturbance, at very low levels of pollution, is reduced feeding activity or its complete cessation. This may be the result of impaired chemoreception, whereas in filter-feeders many pollutants inhibit the cilia used in feeding. Pollutants such as hydrocarbons may also markedly depress absorption efficiencies at low concentrations, such effects often being associated with histopathological changes. Respiration is also commonly affected at low concentrations. This may in part reflect changes in activity but is in particular due to chemical interactions with membranes, oxygen-carrying pigments, or respiratory enzymes. The rate of oxygen uptake is commonly depressed, but may actually be increased at very low levels of pollution, sometimes due to the uncoupling of oxidative phosphorylation (see Table 14.1).

At somewhat higher pollution levels than those affecting feeding, and associated with stressful situations in general, the normal responses of animals to a variety of factors may cease or be reversed. This is particularly apparent in sandy-beach, as opposed to rocky shore, animals because of their greater mobility and clearcut responses. Thus, the burrowing response of the whelk *Bullia* is dramatically reversed at quite low levels of stress, already-buried individuals emerging from the sand, lying on their backs, and spreading their feet — a response that encourages transport by waves and currents (Figure 14.7). *Eurydice longicornis* also displays a reversal of the burrowing response at a certain level of pollution and no longer shows the marked positive rheotaxis typical of the species. It is possible that the reversal of such responses, which normally help the animals maintain position on the shore, has survival value in that it encourages their transport to other, possibly less stressful, areas.

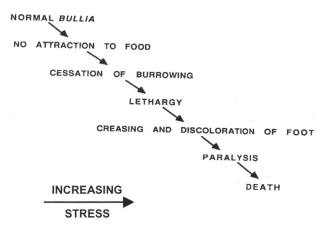

Figure 14.7. Sequence of events usually observed when an adult *Bullia* is subjected to pollution or some other form of stress (after Brown 1986).

The reactions to stress mentioned previously are generally reversible once the stress is removed, but at higher pollution levels (or if the pollutant acts for a longer period of time) irreversible changes begin to occur, leading ultimately to death. The onset of irreversible change might theoretically be the best point to assess, as it can be argued that it is the point at which the animal is effectively lost to the population. It is not always the easiest to assess, however.

Many workers have tried to discover and develop an appropriate physiological indicator of stress, including pollution stress, for marine animals. A promising indicator of this nature is the rate of nitrogen excretion, for stress frequently results in an increased utilization of protein for energy production, with a resultant increase in the excretion of ammonia or other nitrogenous waste products. In addition, invertebrates (such as mollusks) tend to leak amino acids and this leakage may increase under stress.

An important aim in all science must be the ability to predict. Thus, toxicity studies should move away from the purely empirical (in which each pollution source or body or water is tested separately) to a scheme whereby the toxicity of a pollutant or mixture of pollutants can be predicted with some certainty from purely chemical parameters. As far as organic pollutants are concerned, many attempts have been made to correlate toxicity with a physical property of the pollutant. So far, the most promising of these properties has been the molecular valence connectivity index (*mc*-index), which takes into account the number of valence electrons, the degree of separation of the atoms, and the degree of branching of the molecule. Not only does the *mc*-index correlate well with LC_{50} values (Koch 1983) but also with at least some sublethal responses, including cessation of the burrowing response in the sandy-beach whelk *Bullia* (Figure 14.8).

Scientists frequently lose sight of the fact that combating pollution is a practical matter rather than an academic discipline, and that however refined monitoring and pollution-testing techniques may become the real aim should be to reduce or prevent the pollution in the first place. In this regard, the most difficult task is not trying to come to terms with the complexity of pollution effects on marine ecosystems but rather persuading the appropriate authorities to pass adequate legislation or to take effective action.

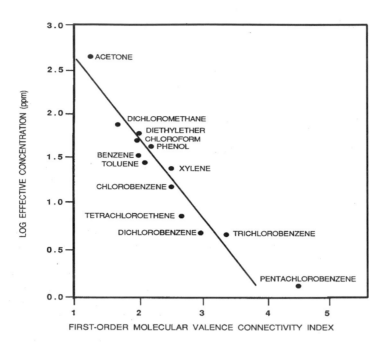

Figure 14.8.
Correlation of first-order molecular valence connectivity indices of organic pollutants and the concentrations at which the burrowing response of the sandy-beach whelk *Bullia digitalis* is reversed (after Brown 1986).

14.3   Recreational Activities

Off-road Vehicles

A variety of vehicles connected with recreation may invade a sandy shore, causing different types and degrees of negative impact. Some recreational vehicles — such as motorcycles, 4-wheel drive vehicles, and vehicles with large wide tires — driven up and down dunes, often at considerable speed, cause displacement of sand and destroy dune vegetation. This activity tends to focus mainly on the backshore near the dune/beach interface, where vehicles can drive above the tide but below the dune vegetation. This can be extremely damaging in view of the fragile nature of coastal dune ecosystems. In addition, shorebirds are easily disturbed, frightened off their nests — and any nests, eggs, and young on the backshore may even be destroyed. Indeed, shorebirds may be more affected than any other component of the fauna by vehicle use on beaches. Furthermore, vehicle ruts on beaches may act as traps for fledgling plovers and turtle hatchlings (Hosier *et al.* 1981).

Both off-road vehicles and more conventional vehicles may be driven along the beach itself. This often causes little impact along the wet foreshore where the sand is wet and firm, although this is not true of all beaches. On some New Zealand beaches, vast numbers of sand dollars (*Echinodiscus*) dominate the foreshore and can be crushed by vehicles. Higher up the slope, where the sand is drier and less compact, vehicles are liable to plough deeper tracks and crush semiterrestrial invertebrates such as isopods, talitrid amphipods, and ocypodid crabs. These can be damaged on the surface or in their burrows. Wolcott and Wolcott (1984) considered the negative effects of off-road vehicles on populations of the crab *Ocypode*. The deep-burrowing oniscid isopod *Tylos* is also affected (Figure 14.9).

Greatest damage by vehicles and even pedestrians is on dune vegetation. Several studies have shown how vehicles damage dune vegetation, both above and below the surface (Godfrey and Godfrey 1981, Anders and Leatherman 1987, Judd *et al.* 1989,

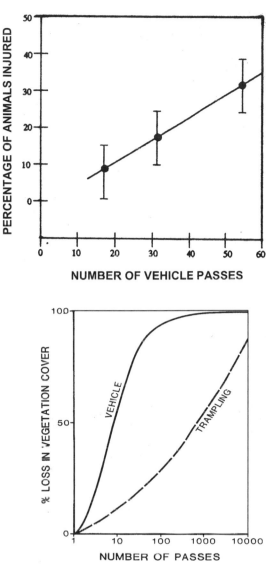

Figure 14.9. Regression of percentage injury to the deep-burrowing semiterrestrial oniscid isopod *Tylos capensis* against number of off-road vehicle passes (after van der Merwe and van der Merwe 1991). Y = 0.75X − 4.6, $r^2 = 0.99$, and 95% confidence limits are shown.

Figure 14.10. Generalized model of the effects of off-road vehicles and human trampling on dune vegetation cover. The curves shift to the left in arid conditions.

Rickard *et al.* 1994). Damage to vegetation exposes bare sand and initiates the formation of blowouts, especially if following the prevailing wind direction. It also changes the microclimate and affects the fauna. Studies that investigated this on sites representing pioneer and climax dune vegetation communities have found that vegetation height and cover decreased in all treatments (more so the heavier the intensity of traffic), but effects are greater in climax shrub communities than among the pioneers. Vehicles describing curved paths do more damage than those driven in straight lines because they tend to gouge deeper tracks on corners. High-intensity treatments result in strong immediate impacts (Figure 14.10), followed by further gradual decline in vegetation over a period of a few months — and pioneer dune communities display faster recovery than shrub or climax communities due to their faster growth rates. These studies indicate that dune vegetation has no tolerance for vehicles. Greatest damage is done in the first few passes and recovery can take years. The presence of vehicles on a beach also detracts from the wilderness quality of the area and may spoil the recreational experience for many beach users.

Trampling

Recreation on the coast tends to focus on beaches and is a rapidly growing area of the tourism industry. If not controlled, however, it can reduce the physical stability of beach and dune surfaces. In an experiment with barefooted human trampling, Moffet *et al.* (1996) showed that trampling could damage delicate crustaceans and juvenile bivalves on the lower shore. Supralittoral species, such as talitrid amphipods, may be most susceptible to the effects of trampling and can disappear altogether from heavily used urban or resort beaches (Weslawski *et al.* 2000, Veloso *et al.* 2005) but may later recolonize them once the peak tourist season has passed (Fanini *et al.* 2005). In most cases, however, pedestrian activity on the beach may have only slight impacts on invertebrates relative to the physical effects of changes in wave climate. Nevertheless, these activities may frighten avifaunas and change their use patterns (Burger *et al.* 1995). However, it is dune vegetation that is most sensitive to trampling, which can compact and change the bulk density, organic matter, and moisture content of soil; reduce the cover and height of vegetation; decrease production of biomass; reduce the number of flowering species; cause disappearance of vulnerable species; introduce weeds and exotic plants; interfere in natural vegetation succession; cause loss of biodiversity; and disrupt fauna (Andersen 1995). These effects are usually greatest near access points, such as car parks.

Trampling by domestic stock and their grazing of dunes also has impacts. It can increase available light at the ground and decrease the store of organic matter and nutrients due to decreased input of vegetation litter and accelerated rates of decomposition (Kooijman and de Haan 1995). These activities can also increase aeolian sand transport, leading to greater dune mobility. A general model of the effects of vehicles and trampling on dune vegetation is shown in Figure 14.10.

Beach Cleaning

Many beaches in popular areas are regularly cleaned of wrack during the holiday season, and in some cases throughout the year. This is because decomposing wrack, with associated odors and flies, is considered a problem for beach users. In other cases, wrack may be harvested for use as fertilizer, animal feed, and production of alginates and agar. Colombini and Chelazzi (2003) have reviewed the importance of beach wrack and effects of its removal on beach ecology. Cleaning commonly takes the form of clearing the beach not only of debris left behind by visitors but of kelp and seagrass wrack and other stranded biota. This deprives the ecosystem of valuable nutrient input — impacting wrack-dependent species, including semiterrestrial forms such as talitrid amphipods, oniscid isopods, and ocypodid crabs. Mobile beach-cleaning machines are employed on some tourist beaches. These suck up and sieve the sand, capturing not only debris but any small animals, such as talitrids, near the surface. Talitrid populations can be effectively eliminated by this process. These mobile machines can also crush more deeply buried invertebrates in their burrows. Removal of wrack can seriously affect beach ecosystem trophic dynamics, and Dugan *et al.* (2003) have shown how removal of wrack affects shorebirds that feed on the associated invertebrates.

Bait and Food Collecting

The collection of invertebrates for use as bait is common on beaches stable enough to support the burrows of prawns, worms, and other invertebrates. Indeed, 15 species of beach clam are harvested extensively around the world in recreational, artisanal, and commercial fisheries — and in several cases overexploitation has led to the collapse of the fishery (McLachlan *et al.* 1996a). Nevertheless, although populations are often drastically reduced by these activities they are seldom eliminated, as they reach a level at which the effort of collecting fails to justify the reward. Recovery usually begins as soon as collecting ceases (see Chapter 8). The exploited clam *Mesodesma* in Uruguay recovered rapidly after the beach was closed to the fishery for 32 months (Defeo 1996). In northern KwaZulu-Natal, South Africa, ghost crabs (*Ocypode*) and mole crabs (*Hippa* and *Emerita*) are harvested in a subsistence fishery that appears to be sustainable (Kyle *et al.* 1997). In addition to the removal of beach animals, collecting typically involves digging, the use of suction devices, trampling, and other disturbances. The results of these physical disturbances may be more deleterious to the habitat than the actual removal of target animals (Wynberg and Branch 1997).

Fishing

Surf zones are important nursery areas for fish and are home to a number of resident species (see Chapter 10). Both recreational fishing and commercial seining have significantly depleted populations of the latter in many areas, thus impacting the surf-zone system and reducing predation in the intertidal zone as the tide rises. Recreational fishing commonly involves the use of off-road vehicles, whereas seining often results in a bycatch of sand crabs and other animals of noncommercial value, which may be dispatched on the spot or left to die on the beach.

Direct human fishing pressure along sandy coastlines is likely to increase, and this may lead to ecosystem degradation. Although much of the pressure can be controlled, small-scale artisanal fisheries are frequently not amenable to such management, and present a growing regional-scale threat to sandy-beach macrofauna populations. This is significant in the context of growing human populations and rising food demand. The 25% predicted increase in global human fish consumption by the year 2030 (FAO 2002) should be partially met through aquaculture in developed nations but in many of the poorest states will be met through increased take by artisanal fishers. In the mid 1990s about 200 million people worldwide depended directly on artisanal fisheries, and their catch accounted for about 50% of all human fish consumption (Pauly 1997).

Other Recreational Activities

The recreational value of sandy beaches is considerable and increasing. Recreational activities such as swimming, wading, surfing, running, dog walking, picnicking, ball games, horse riding, sand sailing, and wave kiting must all have some impact, although this has not been quantified. In general, recreational activities decrease sand stability and increase its mobility. However, impact on the intertidal beach is usually slight, and surf-zone invertebrates are little affected. The experiment of Jaramillo *et al.* (1996), in which a fenced-off strip of Chilean beach was compared with an adjacent area open to the public and heavily used, indicated no significant effect of recreation on the crustacean infauna — probably because sand movements due to changes in wave climate overshadowed physical effects of human disturbance. However, fish may be frightened into deeper water by bathers, whereas shorebirds (such as sanderlings) are reluctant to come onto the beach to feed, possibly resulting in nutritional stress or causing them to migrate to less populated beaches. On the Florida coast, increasing human presence near sanderlings was found to lead to decreased foraging times of the birds during the day and increased nocturnal foraging (Burger and Gochfeld 1991).

Ecotourism

Tourism associated with sandy beaches occurs in various forms. Ecotourism refers to the recreational use of pristine undeveloped areas, whereas nature-based tourism refers to developed areas where there is a recreational component associated with nature. Both ecotourism and recreation in developed areas with dense human populations encourage appreciation of coastal environments, but may have severe impacts on coastal bird populations. Rare birds may attract more attention than common species, adding to their vulnerability, but bird colonies are more vulnerable to disturbance than are isolated individuals (Burger *et al.* 1995). In the USA, least terns have been severely impacted by coastal development and ecotourism, and piping plovers are widely threatened. Piping plovers in less disturbed areas spent more time foraging and less time being vigilant than birds at other sites. The presence of people caused stress for breeding adults and chicks, possibly accounting for decreased reproductive success (Burger 1991). Ecotourism affects bird behavior, reproductive success, and abundance of both breeding and migratory birds in New Jersey (Burger *et al.* 1995), and frequent human intrusion leads to avian habituation and learning. The exclusion of people from sensitive habitats has had beneficial effects for some species.

Litter

Litter left behind on the beach and in the dunes by human visitors is an escalating problem. Nonbiodegradable plastic materials have become the main items of litter, affecting surf-zone animals as well as those higher up the slope. Moore *et al.* (2001) studied the composition and distribution of beach debris in Orange County, California, and Claerboudt (2004) quantified litter on beaches in the Gulf of Oman — showing that plastic debris topped the list and that most debris was of local origin. In some regions of the world, litter is simply left, being allowed to accumulate on the backshore or to be washed out to sea. In others, it may be collected but then buried above the high-water mark or among the dunes, where it tends to resurface. Only in countries with a serious commitment to environmental conservation is beach litter removed to landfill or incinerators. In addition to environmental impacts, litter also detracts from the aesthetic value of a beach.

14.4 Global Warming

Global warming, due to the release of greenhouse gases (and in particular carbon dioxide) together with the destruction of forests, has been underway for at least the last 150 years (since the start of the industrial era) but has only attracted serious attention in the past few decades. There is now widespread agreement that the greenhouse effect poses real and substantial problems for the environment. In addition to temperature change, sea level rise is implicated as the sea warms and polar ice and glaciers melt. Rising sea levels will promote increased erosion along sandy shores. In addition, global warming may be expected to cause increased storminess, at least in some regions, as well as changes in rainfall patterns. Increased storminess will result in erosion, retreat of beaches, and dune scarping with vegetation loss.

Slow long-term trends in accretion or erosion are only apparent over periods of decades or even centuries. On some beaches, retention of newly available sand leads to accretion and the coastline may slowly advance seaward. More common worldwide, however, is long-term beach erosion — with a loss of sediment, diminishing beach volumes and consequent retreat of the coastline. A number of factors, some mentioned previously, may be involved in such chronic erosion. Excessive precipitation and flooding behind a beach also favor erosion, in that the escape of this water to sea carries sand with it and causes a rise in groundwater. Past sea level changes must have caused accretion and erosion cycles. Coastal emergence leads to coastline advance, whereas a rise in sea level (as in many areas in recent decades) results in recession and loss of beach sand to the sea floor. Long-term climatic changes, including changes in rainfall patterns as well as increases in the frequency and/or intensity of storms, have significant effects on beach dynamics. Because sand is transported to sea during storms, returning slowly in calmer weather, increasing storminess will change this balance and lead to continuing erosion.

Predicted rises in sea level due to thermal expansion of seawater and the melting of polar ice and glaciers pose more serious problems. Average sea level rise is predicted to be up to 90 cm by the year 2100. This process will be extremely slow, so that highly adaptable sandy-shore biota will not be at direct risk from it. The most significant problem will be loss of habitat, especially if sea level rise is accompanied by increased storminess. The observed tendency for beach erosion, which is more common worldwide than long-term accretion, will be enhanced — while beaches on which the sand budget is at present balanced will also suffer a reversal of this tendency. Some narrow beaches may disappear completely, whereas others (lacking dune systems) will become severely restricted. Sandy shores that currently incorporate extensive dune systems storing reserve sand should suffer the least, the habitat remaining essentially unchanged, though moving landward. Erosion may be mitigated by deliberate beach nourishment, which should become more widespread. Where there are fixed structures or resistant geological features, such as cliffs, landward migration of the shoreline will be halted.

Changes in sea temperature can have severe effects on marine populations. Events such as El Niño on some South American beaches give an indication of changes that might be expected from rapid global warming. On Peruvian beaches, the abundance of many species plummeted during El Niño events, but this was followed by recovery when conditions returned to normal (Arntz et al. 1988, Tarazona and Parendes 1992). Subtidal areas that had been anoxic saw an increase in abundance and diversity, and extension of vertical distribution in many species during El Niño events. These changes were largely related to changes in productivity, indicating that effects of elevated temperature on beach biota may be indirect.

In comparison with El Niño events, sea temperature changes due to global warming should be gradual, allowing marine populations time to adjust and acclimate. Temperature rise for the oceans as a whole is likely to be only about 1 to 2° C in the next decades, although semienclosed marine lagoons and shallow bays may mirror the atmospheric temperature rise. This is a small change compared with that experienced by beaches close to cooling water discharges from nuclear power stations or those subject to El Niño.

Moreover, aquatic sandy-shore animals are in many regions adapted to rapid changes in temperature and in areas of upwelling these can be extreme, the temperatures changing by up to 10° C in an hour or so.

Sandy-shore animals seldom experience temperatures close to their upper tolerance limits. Further, sandy-beach animals are capable of burrowing and of escaping below the sand if conditions at the surface become hostile. However, if the global temperature rise were to be added to natural warmwater events (such as El Niño in some regions) this combination may have some limited impact. Some redistribution of species could then occur, animals from the tropics and subtropics tending to invade slightly higher latitudes.

14.5   Direct Human Pressure

The estimate of the global human population is 7 billion people by the year 2020 and up to 75% of these will live within 60 km of the coast (United Nations 1998). Future pressures on sandy shores cannot be predicted by multiplying present pressures by the ratio of future coastal populations to existing populations because the increase in tourism is economically driven rather than population driven. In developed and in some developing countries, measures to preserve the coast advance continually and are increasingly the result of well-informed legislation. Many countries in Africa have been involved in strife in the last few decades, and political instability is the norm rather than the exception. In these circumstances, conservation is seldom taken seriously, and what legislation there might be is not enforced. Further, sandy shores generally come low down on any list of conservation priorities.

Figure 14.11 presents a generalized representation of past trends in the intensity of development on sandy coasts based on activities in first-world countries. The changes have evolved from accidental actions to direct modification in response to changes in population pressures, income, leisure time, and technological advances. Other locations have gone through similar phases of landscape conversion, but the phases may have been at different dates or have had different durations. A site-specific curve would show great short-term fluctuations. There is a change in the slope of the curve starting about two centuries ago, corresponding to the advent of steam power that enabled man to make

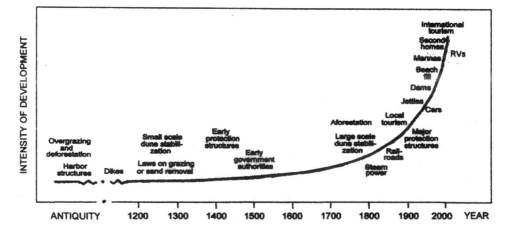

Figure 14.11. A generalized model of trends in development on soft coasts (after Nordstrom 2000).

large modifications to the landscape. The slope of the curve has steepened with increases in the availability of machinery, with development of the internal combustion engine, and with the growth of tourism (Nordstrom 2000).

14.6   Altering the Landscape

Changes to beaches and dunes, due to both planned and unplanned human actions, are listed in Table 14.2. Many of these activities have been covered in the foregoing sections. This section focuses on activities that modify the geomorphology. The most destructive alterations reduce sediment supply and can completely replace local landforms and habitat with buildings, transportation facilities, or engineering structures such as groins, revetments, and breakwaters. Some changes may retain the shape of the landform and provide habitat, although the surface characteristics and biotic interactions may differ.

Military Activities

Military activities may be destructive to landforms and biota at specific sites in the short term, but they may also provide the opportunity for long-term evolution of natural processes in intervening areas. Military use of dunes for maneuvers or firing ranges can prevent urbanization or agricultural uses while they are in use and allow for natural development of dunes after this use ceases.

Table 14.2. Ways that habitat and landforms on sandy coasts are altered by human actions (after Nordstrom 2000, Brown & McLachlan 2002, Brown *et al.* 2006).

Eliminating habitats for alternative uses	***Altering landform mobility***
Constructing buildings, roads, promenades	Shore protection/navigation structures
Alternative surfaces (golf courses, land fills)	Shore perpendicular structures (groins, jetties)
Mining for minerals, construction aggregate	Shore parallel structures (seawalls, revetments)
	Offshore structures (breakwaters, artificial reefs)
Altering habitats through use	Constructing marinas, buildings
Recreation, swimming, horse riding, trampling	Introducing different sediment
Off-road vehicles	Stabilizing landforms (fences, vegetation)
Fishing, harvesting (food, bait, wood, fruits, etc.)	Altering vegetation, exotics/plantations
Grazing	Controlling vegetation (mowing, grazing, fires)
Extraction of oil and gas	Clearing the beach of litter
Extraction and recharge of water	
Military uses and buildings	***Sand nourishment and restoration***
	Providing protection to human structures
Reshaping the geomorphology	Creating larger recreation platforms
Removing sand that inundates facilities	Creating new habitat
Breaching barriers to control flooding	Burying unwanted or unused structures
Dredging channels to maintain inlets	Managing sediment budgets (bypassing)
Widening beaches to allow more visitors	
Removing obstacles to beach access	***Altering external conditions***
Grading dunes to provide sea views	Damming or mining streams
Enhancing natural landscapes for recreation	Pollution, oil spillage, fertilizers, herbicides
Altering environments for wildlife	Burial of waste, underground seepage
	Chemical and thermal effluent
Altering faunal viability or use patterns	Radioactive pollution
Introducing pets or feral animals	Organic enrichment from sewage
Nature-based tourism	Diverting runoff

Reshaping

Reshaping landforms using earth-moving equipment is common, but its effects are poorly studied. Beaches and dunes may be graded or reshaped for several reasons: to create a barrier to prevent wave overwash and flooding; to uncover human facilities inundated by wave overwash or wind drift; to create or maintain channels to facilitate navigation, relieve flooding, or alter circulation patterns; to enhance beach recreation by creating wider, higher, or flatter platforms; to eliminate dunes that restrict access or obscure views of the sea; and to create new landforms to enhance environmental quality (Nordstrom 2000).

Disrupting Sediment Transport

The most extensive alterations in beach landforms are associated with shore protection and navigation projects that disrupt sediment transport. On a highly developed shore, protection methods in beach environments have changed in type and frequency of implementation, revealing an early preference for groins, followed by a period of construction of shore-parallel structures, to a period of beach nourishment that is currently favored (Figure 14.12). Groins are shore-perpendicular structures designed to trap sand moving alongshore (Figure 14.13). By blocking longshore transport of sediment (see Chapter 2), they create differences in beach widths, with accretion on the updrift sides and erosion downdrift. Use of groins has tended to decrease relative to shore-parallel structures and beach nourishment (in part due to accelerated erosion on their downdrift sides), but new groins are still being constructed, often to designs that allow for some bypass of sediment.

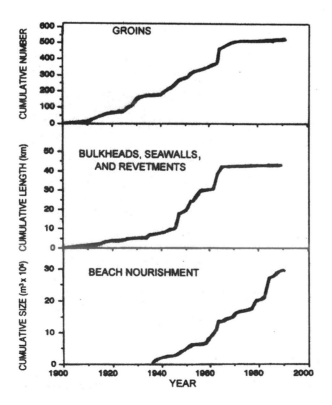

Figure 14.12.
Cumulative trends in shore protection projects in New Jersey (after Nordstrom 1994, 2000).

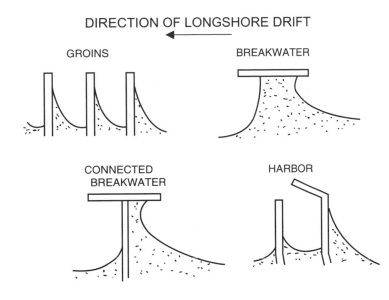

Figure 14.13. Effects of hard engineering structures on littoral sand transport showing accretion and erosion updrift and downdrift, respectively.

Breakwaters are shore-parallel structures built offshore to reduce wave energy or hold beach sand in place (Figure 14.13). They are more common where wave energics and tidal ranges are low and their size and construction costs can be kept low. Shore-parallel structures built landward of the beach provide a barrier between the dynamic beach and sites of human development. They truncate the landward portion of the beach that would be reworked by storms and restrict or prevent exchanges of sediment and biota between the beach and dune. These structures prevent the beach from migrating landward in response to sea level rise or reductions in the sediment budget. They also allow human structures to survive closer to the water than would normally be the case. Unlike groins that rearrange sediment being transported by natural processes, shore-parallel structures are designed to resist natural processes and are less environmentally friendly than groins. They can compress, truncate, or fragment across-shore gradients and have serious effects on beach habitats and zonation of the fauna. In California, for example (where 10% of the coast is armored with hard structures), Dugan and Hubbard (2006) showed significant beach narrowing, loss of upper intertidal zones, and reduction in wrack accumulations and shorebird diversity and abundance on beaches armored with concrete seawalls as compared to beaches backed by natural bluffs. Rising sea level will exacerbate these impacts.

Marinas and port facilities can act as artificial headlands that break up the orientation of the shoreline, change refraction and diffraction patterns, trap sediment, deflect sediment offshore, and starve adjacent downdrift beaches of longshore sand transport. New beaches may be intentionally created adjacent to these same marinas, but they are likely to be managed for beach-related activities, and provisions have to be made to ensure that recreational use does not preclude development of natural features (Nordstrom 2000).

Buildings block sediment transport, but they can also constrict and accelerate wind flow between them, increasing the likelihood of local scour. Actions taken by shore-front owners — such as constructing sand barriers, removing sand, and planting exotic species — can result in greater changes to landforms than the passive effects of the buildings.

Sand fences, including manufactured forms and natural brush, are used to trap sand moved by wind to create or repair dune ridges or prevent inundation of landward developments (see Figure 15.4). Sand fences are inexpensive and easy to construct, and their deployment usually occurs at the highly dynamic boundary between the beach and dune. Dune surfaces have been stabilized in many ways, such as spreading straw, brush cuttings, matting, bitumen spray, or even old tires. These efforts at stabilization often address threats that are more imagined than real, resulting in unnecessary environmental losses. Aforestation (planting trees) has been conducted on dunes to protect crops from sea winds, establish a forest industry, and stabilize dunes. Ecological problems — including loss of flora and fauna, changes to soil characteristics, lowering of the water table, and seeding into adjacent unforested areas — have led to attempts to remove some areas of woodland to restore dune habitat that has greater conservation value (see papers in Carter *et al.* 1992). Dunes may be destabilized by removing surface vegetation. Introducing more mobility into coastal landforms has been a recent phenomenon, and it has normally been conducted at small scale in order to reintroduce sediment to the beach (see Chapter 15, case studies).

A means of changing the morphology and mobility of landforms is through accidental or intentional introduction of exotic vegetation. Plant invasions can decrease biodiversity, interfere with successional processes, change the appearance of the landscape, and reduce the value of land for conservation and recreation. Exotics that owe their introduction or spread to human actions include Japanese sedge (*Carex cobomugi*) introduced to coastal dunes in the USA, pine trees (e.g., *Pinus nigra* ssp *laricio* and *P. contorta*) used to stabilize dune systems throughout the world, bitou bush (*Chrysanthemoides monilifera*) introduced to Australia from South Africa, *Acacia cyclops* introduced to South Africa from Australia, the Australian *Casuarina equisetifolia* used to stabilize mobile dunes in Mexico and elsewhere, and European beach grass (*Ammophila arenaria*) introduced to stabilize dunes on the Pacific coast of the USA, South Africa, and Australia. Stabilizing dunes with exotic vegetation is a problem that is often disguised as a beneficial green alternative.

Beach Nourishment

Beach nourishment is a soft engineering alternative to hard structures on the shore. It is used to (1) protect buildings and infrastructure from wave attack, (2) improve beaches for recreation, (3) create new natural environments, (4) eliminate detrimental effects of shore protection structures by burying them, and (5) retain sediment volumes during sea level rise. Bypassing operations move sediment from one side of a barrier (e.g., a groin) to the other by pumping sludge in a pipeline or by trucking. Such operations remove sediment from accreting beaches and deposit it on nearby eroding beaches and are local-scale sand management operations that can also be considered forms of beach nourishment. Beach nourishment is practiced extensively in developed countries and is the preferred strategy for combating erosion in some countries (Hamm *et al.* 2002). Hard engineering is still important in countries where flooding and sea level rise threaten public infrastructure and private settlements (Turner *et al.* 1998). In cases where removal of structures and retreat landward is not possible, beach nourishment may be the best alternative to increase the resilience of coastal systems without loss of ecological functions or capacity to provide recreational and other services.

Beaches can be made more stable by introducing coarser sediments, such as gravel, but there are differences in habitats associated with the different types of gravel used in nourishment operations. Adding coarser sediments to sandy beaches is likely to occur more frequently as suitable sources of sandy beach-fill are exhausted and technologies for transporting gravel improve. It does, however, change beach morphodynamics and may result in the loss of some species.

Fill sediments can be deposited in various positions, such as deep-water offshore, in the nearshore, on the subaerial beach, or on the dunes. Nourishment of the upper beach has been the most common practice. However, nourishment on the upper beach can result in a widened and often overly steep beach with a morphology and sediment composition that may be physically different from the original beach in terms of sand compaction, shear resistance, moisture content, and the size, sorting, and shape of sediments. These changes can have negative impacts on the macrobenthos (Rakocinski *et al.* 1996, Peterson *et al.* 2000, Greene 2002), although few studies have been sufficiently thorough to provide detailed insights on this (Peterson and Bishop 2005). Nevertheless, beach nourishment has proved effective in combating erosion (Hamm *et al.* 2002), and if done properly may also enhance the habitat of selected species of biota. Beach nourishment is best done as a series of small additions of sand at one- to two-year intervals rather than as a single major episode.

Beach nourishment has had a positive impact on some plant and animal species — including seabeach amaranth (*Amaranthus pumilus*), piping plover (*Charadrius melodus*), and several species of sea turtles — by creating habitat (Crain *et al.* 1995, National Research Council 1995). Nourishment also provides a source of sand for dune building and space that allows well-vegetated dunes to be rapidly established (Freestone and Nordstrom 2001). However, beach nourishment operations will not improve conditions if active recreational uses take precedence. Environmental costs may be incurred due to (1) removal of habitat and death of biota in the borrow area; (2) increased turbidity and sedimentation in both the borrow and nourished areas; (3) disruption of mobile species that use the beach or borrow area for foraging, nesting, nursing, and breeding; (4) increases in undesirable species; (5) changes in wave action and beach morphology (e.g., from dissipative to reflective); (6) changes in grain size characteristics; (7) higher salinity levels in aerosols associated with placement by spraying; and (8) change in community structure and evolutionary trajectories resulting from new conditions in the borrow and nourished areas (National Research Council 1995).

Detrimental environmental effects of nourishment are often considered temporary, but many indirect or complex impacts are unknown (Gibson *et al.* 1997) and prediction of the long-term cumulative implications of large-scale projects is difficult (Peterson and Lipcius 2003). The longest recovery times are required for biota when there is a poor match between the grain size characteristics of the fill materials and original substrate (Rakocinski *et al.* 1996, Peterson *et al.* 2000). The considerable time required to conduct quality studies of nourishment and the difficulty of applying results from one location to another will result in many unanticipated and unwanted environmental impacts. However, nourishment projects will continue to occur in locations where protection and recreation projects are economically justified, and this will be a fruitful area for future research in sandy-beach ecology.

Dune Nourishment and Enhancement

Nourishment of the upper beach can alter aeolian transport, dune growth, and vegetation change by increasing beach width. This increases the likelihood of aeolian transport and decreases the likelihood of marine erosion of the foredune. Dunes may also be nourished directly (Mendelssohn *et al.* 1991). The sediments used to nourish dunes often come from the same sources as those used to nourish beaches, causing the substrate to resemble the backbeach rather than the better-sorted finer-grained sand of the dunes. Direct fill creates a substrate not found in either natural dunes or dunes created by accretion at sand fences or vegetation plantings. Fill materials borrowed from estuarine environments may contain seeds and rhizomes of marsh plants, leading to establishment of salt-tolerant plants such as *Phragmites* and *Spartina* in dunes where these species were previously uncommon.

Mining

Mining activities have severe effects on impacted habitats. In the case of beaches, it is usually sand mining and excavation that is of concern. However, in addition to removal of sand itself mining may take place for precious stones, such as diamonds, or for various minerals. Extensive sections of sandy beach along the Atlantic coast of South Africa and Namibia are mined for diamonds. Mining may be undertaken on the beach itself or in the surf zone beyond. In all cases, heavy vehicles and machinery are involved on the beach. Strip mining along intertidal sandy beaches effectively destroys the ecosystem. Many animals, including the meiofauna, may eventually return as the beach reestablishes its former characteristics. However, some species (such as semiterrestrial crustaceans) may fail to do so. Offshore mining can be equally disruptive, as the material is usually pumped ashore and the tailings are left on the beach, altering the beach profile and its particle-size structure. At Elizabeth Bay, in Namibia, the dumping of coarse tailings from diamond mining resulted in the beach becoming more reflective — with a consequent loss of fauna (McLachlan 1996). Mining in the dunes destroys the vegetation and may disrupt sand transport, in addition to adversely affecting shorebirds. Further, tailings or topsoil runoff from mining behind the dunes may pollute the beach.

Dams

Natural reduction in sediment supply, driven by global climate change through the Holocene, established a very long-term pattern of beach erosion. In the past this was partly balanced by deforestation for cultivation, grazing, and construction of settlements — which resulted in delivery of vast quantities of sediment to coasts by streams. However, the damming of rivers — which was widespread in the twentieth century — dramatically reversed this trend, resulting in increased rates of erosion. The damming of rivers deprives estuaries and the shoreline of natural fluvial input of sediment. In 1950, there were 5,270 large dams in the world, whereas there are currently more than 36,000 (World Resources Institute 1998). Construction of dams has been a significant cause of coastal erosion in many locations throughout the world. It is difficult, however, to determine the magnitude of this impact or to distinguish between the effects of the dams and effects of associated activities such as quarrying, land reclamation, urbanization, aforestation, and agricultural use (Nordstrom 2000, Shesma *et al.* 2002).

Groundwater Level Changes

In addition to pollution of groundwater, water abstraction from dunes commonly results in a lowering of the water table. One such activity is the drawing off of water for domestic or agricultural purposes. This can cause salinization of the groundwater — which may be intensified by the hardening of surfaces — so that surface water from rain is diverted to stormwater drains instead of sinking into the soil. Lowering of the water table can have serious adverse effects on the dune ecosystem, which in turn may affect the intertidal beach. Whereas flooding or raising the water table hastens erosion, extraction of groundwater can lead to subsidence, local increases in sea level rise, and increased rates of beach loss (Nicholls and Leatherman 1996) or accretion. Artificial recharge with water from other sources, such as occurs in The Netherlands, may add nutrients and disturb the circulation of ground water — leading to unnatural water table fluctuations. Recharge can also occur from watering lawns. Watering and wastewater disposal above coastal bluffs can add weight to weaken cliff materials (by solution in some cases) and lubricate surfaces along which slides develop.

14.7   Natural Impacts

The human activities covered in this chapter pose serious threats to sandy coasts on many scales, but some natural processes can also disrupt beaches — though such disturbances are seldom severe or long lasting. Storms are an important part of a natural cycle molding the morphodynamics of sandy beaches, and represent a natural hazard faced by sandy-shore animals. Sand and some animals are washed out to sea, whereas others may be stranded upshore, where they die of exposure. Such episodic events, where extreme, often result in greater mortality than does predation. Some animals regain the shore if not carried too far out to sea, in that the ability to survive storms by behavioral means is a key feature of sandy-shore animals. However, these mechanisms do not always give adequate protection, especially if significant sand erosion occurs. In compensation, interspecific competition is minimized, as few macrofaunal species can tolerate these conditions. In extreme cases, so much sand may be eroded from the beach that bedrock becomes exposed. This prevents burial and disrupts the laminar flow of the swash, making colonization by swash-riding species impossible. Some beaches are ephemeral and only occur seasonally, sand deposited during relatively calm periods being totally removed during the months when storms are prevalent. In such cases, there may be some colonization by bacteria and meiofauna as sand is deposited. However, except for highly mobile species (such as benthoplanktonic crustaceans) macrofauna usually has not enough time to establish itself. More common are beaches simply too inhospitable to accommodate macrofauna for much of the year. Some macrofauna species may remain offshore until storms have flattened the beach and then colonize it during calmer weather. The most severe natural episodic events, hurricanes and tsunamis, may completely destroy the beach and dune systems — especially where their sediment stores have been reduced by human activities and they are not free to respond naturally.

14.8   Human Influence on the Evolution of Beaches

A wide range of beach and dune types are managed by humans, and Nordstrom (2000) gives a full account of this. Dune growth can be prevented in many locations by beach

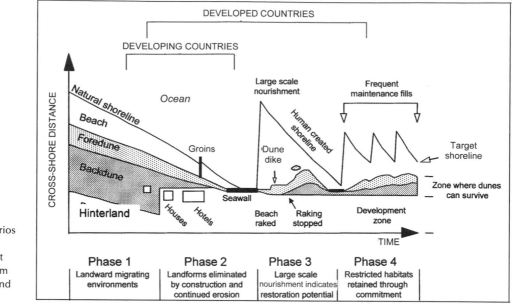

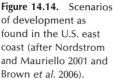

Figure 14.14. Scenarios of development as found in the U.S. east coast (after Nordstrom and Mauriello 2001 and Brown *et al.* 2006).

cleaning or mechanical grading to maintain beaches for recreational use. Other locations may have some dunes, but these may be restricted, and they may not be built by aeolian processes, colonized by indigenous coastal species, or free to migrate inland. Coastal erosion and attempts to retain a fixed shoreline can result in loss or truncation of beaches and dunes. Beach nourishment operations can then replace lost sediment. However, nourishment is usually intended to protect buildings and provide recreation space, not to restore natural systems. If dunes are rebuilt, they are often small and linear because they are designed to form a dike against wave attack, and not to take up recreational beach space or interfere with views of the sea. Dunes along a developed coast are usually more restricted than natural dunes, and are often located where only the dynamic seaward section of a natural dune would occur. Accordingly, the species normally found on the back-dune environment can only exist on modified dunes in developed areas if growing conditions are enhanced by providing a relatively stable environment protected from inundation by sand, water, and salt spray. This can be done by restoring the fronting beach or maintaining sacrificial foredunes.

The coastal environments identified previously represent stages that can evolve, or be retained in a given condition, by human efforts. A conceptual model of changes in relative dimensions of shoreline environments with time has been developed (Figure 14.14), based on the situation on the east coast of the USA. This is typical of many low-lying shores and includes a vision of what is possible in the future. Most of the changes depicted in Figure 14.14 have occurred in well-developed areas, such as portions of the coast of the USA and The Netherlands. However, many locations in other countries are still in earlier phases. A view along the coast of many countries reveals a series of segments that are now at different states in this temporal continuum. All countries with coasts have segments in phase 2, but only a few developed countries have shore segments in phase 4.

Phase 1 depicts natural conditions on a soft shore. Here, dimensions of the natural beach and dune are initially unhindered by humans, but surfaces become increasingly altered by trampling, harvesting, or grazing. Many locations that are currently migrating under near-natural conditions are being modified by nonintensive tourist activities such as nature-based tourism in remote areas and controlled tourism in protected areas. Many locations subject to nature-based tourism are evolving. Increasing numbers of visitors, supported by increasing investment in support infrastructure, will eventually lead to elimination of much of the natural resource base and conversion to intensive development of phase 2. Seashore-protected areas may not undergo this phase of intensive development. However, some such areas may be threatened by erosion and they may require artificial nourishment, just as the developed areas do in later phases (phase 4).

Phase 2 represents the conversion of landward portions of the dune system to cultural environments, with increasing intensities of development through time. Here, increasing levels of investments make larger-scale shore protection projects more feasible. The end of phase 2 is marked by the loss of beach and dune habitat that is one consequence of using static protection structures without beach fill. Although small-scale beach nourishment may be conducted within this phase, such operations provide only short-term relief of erosion problems. Many segments of shoreline in developing countries are within phase 2, and local authorities now face decisions on whether large-scale shore protection projects are feasible and whether hard structures or beach nourishments are the appropriate solution.

Phase 3 represents the initial commitment to large-scale beach nourishment projects. The construction of a wide beach creates the potential for full environmental restoration, but the effect varies according to subsequent management actions. The new sand may be used purely as a recreation platform and graded and raked, preventing flood protection or sand drift, and then used to create the linear sand dike often associated with a developed coast. Suspension of beach raking can lead to a wider dune with greater topographical and biological diversity (Freestone and Nordstrom 2001), but loss of these new environments can occur if administrative delays prevent nourishment from occurring in a timely fashion (end of phase 3).

Fencing and vegetation plantings can be used to build dunes in a sand-deficient environment, but inevitably beach nourishment may be required to maintain well-vegetated dunes (Mendelssohn *et al.* 1991). Phase 4 represents the need for nourishment to retain dune integrity, given erosion and competition for space for recreation and construction of facilities. Beach and dune environments are prevented from migrating landward. They must therefore be maintained on the seaward side of private properties and public promenades and thoroughfares. A significant aspect of the management approach in phase 4 is that not only the beach but the restored dune landscape is considered a resource and is protected by nourishment. Preservation of existing landforms and habitats is best done by small frequent nourishment operations. Design studies for projects in New Jersey (USA) include nourishment every three years, which should be adequate to retain the natural environments that are able to form to seaward of cultural features. Restored dunes in developed areas are not as wide as similar landforms in the undeveloped enclaves that remain in phase 1, but judicious use of sand fences and vegetation plantings can recreate the types of habitats lost, if not the area and spatial relationships (Freestone and Nordstrom 2001, Nordstrom *et al.* 2002). These and other aspects of coastal zone management are dealt with in Chapter 15.

14.9   Conclusions

Few generalizations can be made about the future of sandy shore ecosystems as a whole. Table 14.3 presents a summary of some of the main short-term factors impacting sandy shores, and their geographical extent, severity of impact, and type of beach likely to be most sensitive or affected. Impacts on sand budgets are most severe and most likely to have the greatest implications for sandy-beach ecology.

If short-term impacts/threats on sandy beaches are grouped in the categories of pollution, engineering structures, and human activities (Figure 14.15), it is on sheltered shores that the former is likely to have greatest impacts because pollutants (which mostly affect beaches) are unlikely to be rapidly dispersed under low-energy conditions. Human activities (including recreation), on the other hand, tend to have most impact on dunes and dune vegetation — and effects are more severe under more dynamic or exposed conditions. Hard engineering structures, especially those that modify sediment transport on

Table 14.3.   Summary of short-term factors currently impacting sandy shores (after Brown & McLachlan 2002).

Factor	Extent	Type of beach most affected	Importance (max. = 10)
Storms	Widespread	Exposed	7
Disruption of sand transport	Near hard structures	Exposed	6
Pollution	Localized/widespread	Sheltered	6
Trampling	Localized	Vegetated dunes	5
Recreation/tourism	Localized (increasing)	Resorts	4
Human litter	Localized (increasing)	Resorts	4
Beach nourishment	Fairly localized	Urban/resort	4
Beach cleaning	Localized	Urban/resort	4
Mining	Localized	Various	3
Groundwater changes	Widespread	Arid areas	3
Bait collecting	Widespread	Beaches with rich fauna	3
Fishing	Widespread	Beaches with rich fauna	3

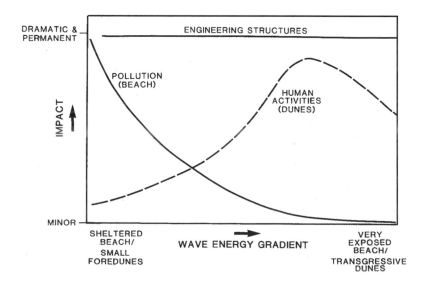

Figure 14.15.   Relative importance of three types of human impact on sandy shores along an exposure gradient. Pollution chiefly affects the subaerial beach; trampling, off-road vehicles, and other activities mainly impact the dunes; and coastal engineering structures permanently alter sand budgets for the entire shore system.

the shore, have severe effects in all shore types, although the volumes of sediment affected will be greatest on exposed shores.

The chief long-term threat facing sandy shores, virtually worldwide, is loss of habitat resulting from increasing erosion attendant on sea level rise, changes in storm patterns associated with global warming, and expanding human population with associated construction and use of human facilities. On a regional basis, sandy-shore ecosystems may experience increasing pressures from artisanal fisheries in developing nations, especially as other food sources become scarce. Temperature rise by the year 2050 is likely to have relatively subtle effects on the biota, but projected changes in sea level and current patterns may significantly alter sand transport — enhancing erosion and loss of habitat in many areas, but possibly favoring accretion in others. Human pressures on the coast are expected to increase, in some cases dramatically, but increased pressure on sandy shores may be mitigated by improved legislation and management resulting from better understanding of sandy-shore processes.

The world is increasingly crowded, and human impact on the coast in all of these forms (short and long term) plus natural episodic events driven by climate change are likely to increase. It is essential that this is balanced by increasingly well-informed sophisticated and effectively implemented coastal zone management and conservation, the topic of the final chapter. We now have sufficient understanding of the structure, function, key processes, and sensitivities of sandy-shore environments on all scales to manage them as unique systems suitable for multiple use and sustainable development while maintaining high environmental quality. However, this knowledge will not be translated into sound management unless there is commitment and resources are provided at a local level to enable implementation of the practices covered in the next chapter.

Coastal Zone Management

15

15.1   Introduction

In view of expanding human populations, rapid development in many parts of the world, and increasing demands for coastal recreation, it is not surprising that most effort that goes into coastal zone management (CZM) worldwide is focused on soft coasts, particularly beaches and dunes. Where such systems are pristine or provide habitat for threatened species, the main purpose of management may be conservation of the habitat and/or the target species. Where there is greater human pressure and development, coastal zone management may be mainly concerned with managing for multiple uses, limiting the impacts of development, ensuring sustainability, and maintaining key processes such as sediment transport.

In general, sandy beaches are fairly resilient to most activities other than major engineering structures (Chapter 14), and are well suited to nonvehicular recreation — but coastal dunes, because of their sensitive vegetation, are much more prone to damage and require some measure of conservation in almost all cases. This is a central feature of coastal zone management of soft coasts: providing access for recreation on beaches while limiting damage to the dunes behind them.

This chapter begins by emphasizing the importance of managing the sandy coast and its sediment budget as a unit. We reconsider the threats covered in Chapter 14 and then outline the main principles of coastal zone management relevant to sandy coasts. We finish with some case studies and conclusions. Useful references for further information on coastal zone management include Clark (1996) and Nordstrom (2000).

15.2   The Littoral Active Zone

The area 200 m vertically above and below the shoreline covers a fifth of the earth's surface, a quarter of primary production, 60% of the human population, more than two-thirds of the world's large cities, and 90% of fisheries. Whereas international law limits the seaward extension of the coastal administrative area to 12 nautical miles, the landward boundary is highly variable. Within this broad strip of coastline, inshore water, and hinterland lies a narrow strip, the coast, including the ecosystems we have been considering: beaches, surf zones, and dunes. The coast is characterized by dynamic features, productive ecosystems with high biodiversity, and unique environments and habitats. Although there is no simple definition of the coast, all definitions will include some land and some sea elements.

Three properties of soft coasts highly relevant to coastal zone management are their malleability, their temporary stabilization by plant growth in dunes, and their sensitivity to disturbance. They do, however, have a durability conferred by their ability to recover from episodic events and to return to equilibrium. Hard coasts consist of rocky shores and sea cliffs as well as boulder beaches. Soft coasts consist of beaches, dunes, surf zones, and estuaries, which exchange sediment and form a single geomorphic system of sand storage and transport. Movement, exchange, and storage of sediment in this system — mostly driven by water movement — are the most important processes. In this, beaches play a central role. The main sand sources to the coast are rivers carrying sediment (mostly silica from erosion of land), carbonates provided by biogenic sources in the sea, and weathering of sea cliffs. Losses of sand or reduced input can be caused by sea level rise, dams on rivers, and storms (Figure 15.1).

Sandy shores consist of three entities — surf zones, beaches, and dunes — which are linked by the interchange of material, particularly sand (Dyer 1986, Komar 1988). Together, they comprise a single geomorphic system, termed the littoral active zone (LAZ) (Figure 15.1) (Tinley 1985). This is the part of the coast characterized by wave- and wind-driven sand transport that lies between the outer limit of wave effects on bottom stability (usually between 5 and 20 m in depth) and the landward limit of aeolian sand transport (i.e., the landward edge of the active dunes). Although this area constitutes a single geomorphic system, it consists of two distinct ecological systems: a marine beach/surf-zone ecosystem populated by marine biota and strongly influenced by wave energy and a terrestrial dune system inhabited by terrestrial plants and animals and

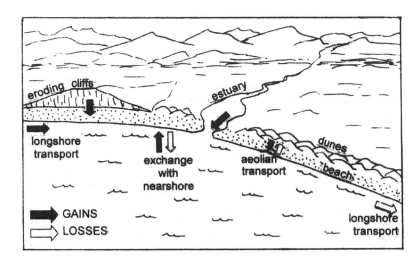

Figure 15.1. Components of the sand budget in the littoral active zone indicating gains and losses from beaches (after Komar 1988). Sand storage areas are dunes, beaches, and the nearshore.

strongly influenced by wind energy. In managing sandy coastlines, it is imperative that this contrast be borne in mind, although the two should be managed as a unit.

In simple terms, managing a sandy coast means managing the sand budget (sand storage transport) and thus it is essential to understand the coupling of dune and beach systems, their exchanges of sand, and their interdependence. The role of dunes in sand storage is especially important. Plants and animals per se are of secondary importance to sand budgets in sandy-shore management, although of course they must be taken into account if the ecosystem status of the two systems is to be maintained or reestablished. It is essential, therefore, to develop insight into the sand budget and methods of conserving it, so as to allow normal exchanges of sand not only between beaches and dunes but alongshore. Management must also appreciate that most soft coasts are in the process of retreating and must allow for this by fixing recession lines.

Secondary concerns involve recognizing the ecosystem status of the component systems and appreciating its sensitivity to human impacts (Figure 15.2). The most sensitive areas are the backshore and the foredunes. Generally, the beach and surf zone are more resilient and suited to recreational activities, whereas the dunes are best demarcated as conservation areas. In terms of impact, major engineering structures and other large scale disturbances — which disrupt sand movement and change wave and wind climates — have the most severe effects on all beach types. As far as pollution is concerned, however, sheltered coasts are much more sensitive than exposed shores, whereas human impact on the foredunes is greater on exposed coasts.

The littoral active zone can be tough and effective as long as it is allowed to change and respond within its natural boundaries, particularly where dune plants and sand reservoirs are left intact. The protection of the backshore-frontal/active-dune zone is therefore of paramount importance. It can serve as the natural buffer against storms. Further, sand budgets in general (and longshore sand transport in particular) are easily disrupted and often need to be managed. In addition to the general issue of sand budgets, other sensitive features of the littoral active zone are as follows.

- Dune vegetation, especially in the foredunes
- Nesting turtles

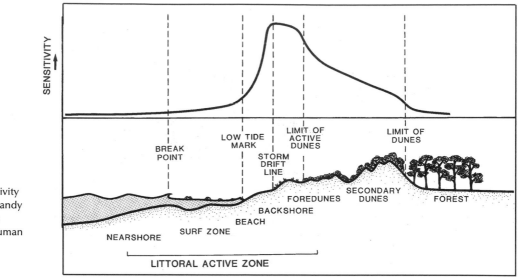

Figure 15.2. Sensitivity curve for a typical sandy coastline. The areas most sensitive to human impact are the backshore and foredunes.

- Supralittoral fauna
- The water table
- Fishery areas
- Rare species
- Archeological sites
- Dynamic and fragile habitats
- High wilderness quality

Most of these sensitive features are concentrated around the dune/beach interface and the foredunes. This needs to be borne in mind when managing or conserving sandy coasts.

15.3   Summary of Threats

Sandwiched between rising sea level on one side and increasing human activity and development on the other, sandy coasts are under great pressure. Problems and threats facing sandy shores were covered in Chapter 14. These are as follows.

- *Pollution*, which is usually localized and has greatest impact on sheltered shores.
- *Human recreational activities*, including off-road vehicles, trampling, angling, and so on, which generally have greatest impacts on foredune vegetation and the backshore, in that intertidal sandy beaches are fairly resilient to these activities.
- *Development* associated with expanding human populations, towns, harbors, and resorts often results in structures and services intruding into the littoral active zone — where the damage may include direct loss of habitat, disruption of sand transport, and elimination of sand storage.
- *Global change* causing sea level rise and increased storminess will tend to hasten erosion and raise storm damage to sandy coasts, causing them to retreat inland. This is specially problem where the littoral active zone is not free to move inland because of constraints imposed by manmade structures.
- *Natural disasters*, such as tsunamis and hurricanes, cause tremendous damage to the littoral active zone — often flattening dunes and eroding their stored sand to the

surf zone. If sufficiently intact, the littoral active zone can cope with this, slowly regaining its sand and reestablishing damaged dunes after the episode. It is when the littoral active zone has been damaged or reduced, or is constrained by development and not free to respond in the natural way, that serious problems arise.

Bearing these problems and issues in mind, let's consider the principles of coastal zone management appropriate to dealing with sandy coasts.

15.4 Principles of Coastal Zone Management

Keep the LAZ Intact

A number of general principles form the core of any strategy of managing sandy coasts for sustainable use and development. The central paradigm in this context is the unity of the littoral active zone. Understanding the main features of the littoral active zone and the linkages within them is paramount when managing sandy coasts. The littoral active zone must be managed as a unit. Dunes and beaches cannot be managed in isolation. Unfortunately, this is often not appreciated by coastal authorities, who tend to manage these two systems in isolation. Various other principles aid in identifying the correct approach to managing the littoral active zone.

Setbacks

It is critical, wherever possible, to leave the entire littoral active zone intact and locate all development landward of it. A setback line should therefore be established landward of the littoral active zone and all development kept landward of this line. Because of sea level rise and possible coastal retreat in many areas, it is advisable to locate the setback line well landward of the littoral active zone, thereby providing space into which the zone can retreat if necessary. The setback or recession line is, in effect, a line of retreat that anticipates erosion, storms, and sea level rise. Thus, along eroding beaches setbacks should be set further inland than along stable shores. Examples of setbacks in various countries are shown in Figure 15.3. In most cases, this is 50 to 100 m from the shoreline. Setbacks can also follow topographic/morphological features of the coast.

Ideally, no structures should be allowed seaward of the setback line, in order to create an adequate buffer zone. Had this been more fully adopted in the case of the Aceh tsunami and hurricane Katrina, an intact LAZ might have reduced the damage and loss of life that ensued. In many countries, the shoreline is public land that cannot be developed and this provides a form of setback. Any activity, especially development, inside the setback area should be subject to the strictest scrutiny, environmental impact assessment, and monitoring.

Zoning

Ribbon development, which occurs in strips along the coast, is unwise because it seriously impairs the function of the littoral active zone and reduces the aesthetic/wilderness quality of an area. The correct approach is to concentrate development in nodes separated by protected areas. Zoning along the shore has a custodial purpose, allowing the coasts to be subdivided into sections allocated to (and managed for) different purposes

COUNTRY	DISTANCE INLAND FROM SHORELINE
Ecuador	- 8 m.
Philippines	---- 20 m.
Mexico	---- 20 m.
Brazil	----- 33 m.
New Zealand	------ 30 m
Oregon (USA)	------------ Permanent vegetation line (variable)
Colombia	------------------ 50 m.
Costa Rica	------------------ 50 m.
Indonesia	------------------ 50 m.
Venezuela	------------------ 50 m.
Chile	-------------------- 80 m.
France	---------------------- 100 m.
Norway	---------------------- 100 m.
Sweden	---------------------- 100 m.
Spain	---------------------- 100 to 200 m.
Uruguay	----------------------------- 250 m.
Indonesia	----------------------------------- 400 m.
Greece	--------------------------------------- 500 m.
Denmark	--------------------------------------- 1 – 3 km

Figure 15.3. Setbacks adopted by some countries (after Clark 1996). Distances are horizontally from the high-tide level.

(e.g., conservation of unique habitats, fishing, water sports, and development). Zoning also has a regulatory purpose in aiding planning — defining what types of development may be permitted in different areas (e.g., resorts, navigation, aquaculture). Zones can have both spatial and temporal components and can be enforced by controlling access. Zoning can also serve to minimize clashes and interference between different user groups. Indeed, zoning can be especially important where there is potential for conflict between different activities/users of the littoral active zone. To work well, zoning should match marine and terrestrial areas in a consistent fashion and avoid uses that would cause clashes between the two.

It is useful to have distinct boundaries between zones in the form of natural topographical or morphological features. Further, adjacent zones should not clash. For example, a conservation area should not be sited adjacent to an intensive use area. Ideally, conservation zones should include some areas that are well protected, located between buffer zones, which in turn are surrounded by multiple-use zones. Public education is an important part of coastal zone management in general, and of zoning in particular, and the public should participate in the planning process leading to zoning. Zoning can be done in a variety of ways, and zones can include the following uses.

- Recreation in different categories (e.g., bathing, boating)
- Conservation of nesting areas, dunes, beaches, or the entire LAZ
- Intense development, preferably behind the setback line
- Ports and shipping
- Defense and military purposes
- Buffer purposes (i.e., intermediate between conservation and intensive use)

Where a zone is identified for development, this does not mean that development should automatically proceed. It should always be preceded by a thorough environmental impact study.

Environmental Impact Assessment

Ideally, development should be kept out of the littoral active zone, but this is not always possible. In most countries, no major development can proceed — especially in an area as sensitive as the littoral active zone — without thorough environmental study in the form of an environmental impact assessment (EIA). Such assessment must not only include ecological issues but should consider social and economic factors. It requires a thorough study of what the environmental, social, and economic effects, costs, and benefits of a proposed development are likely to be, as well as possible alternatives and, should the development proceed, mitigating measures. This is essential for wise decision making and planning in coastal zone management. In the case of soft coasts, this is absolutely critical where there is any plan for development in the littoral active zone, and no development or planning should proceed without it. The environmental impact assessment process should include the following.

- Description of the proposed project
- Description of the site: physical, biological, human
- Identification of resources at risk
- Screening, to distinguish between major and minor impacts/issues
- Public scoping, involving all interested and affected parties, to assess all major impacts, consult with stakeholders, and provide recommendations
- Identification of alternative sites and project options
- Assessment of magnitude of likely positive and negative environmental, social, and economic effects during both construction and operation phases of the project or development
- Identification of suitable mitigation measures
- Preparation of the EIA report, including impact assessment, environmental management, and mitigation recommendations and baseline monitoring

This is a lengthy and consultative process that may have more than one iteration. It can identify cause-and-effect relationships, predict impacts, and guide authorities to wise planning and decision making. The purpose of environmental impact assessment is not to stop development but to guide it for long-term sustainability. It is essential that the environmental impact assessment process start as early as possible, well before development. To this effect, it can be done in stages: first a preliminary review to ascertain if there are likely to be impacts, then an environmental examination, and finally a full environmental impact assessment if significant impacts are expected.

Much of this involves good common sense rather than sophisticated techniques. Methods that can be used to identify impacts include checklists and matrices (Table 15.1). Checklists and matrices basically consist of listing all environmental, human, social, and economic issues and attributes of the affected areas and indicating what impact each aspect of project development will have on them. Developments on sandy coasts for which environmental impact assessment is absolutely essential are as follows.

- Industrial plants
- Infrastructure, such as ports, highways, and airports
- Any form of mining or sand removal
- Any alteration of water resources (e.g., dams, recovering groundwater)
- Major recreational or tourism facilities (e.g., hotels, resorts)

Table 15.1. Example of a simple matrix used to evaluate impacts of construction of a small harbor on a sandy coastline. X = negative impact; + = positive impact.

Activity or impact source	Attribute	Physical				Biological					Human			
		Beach sediment	Dune sediment	Surf water	Groundwater	Beach fauna	Seabirds	Surf zone vegetation	Dune vegetation	Breeding birds	Recreational value	Wilderness quality	Existing structures	Jobs
During construction	Dredging	X		X		X	X	X			X	X		
	Construction of structures	X	X	X		X	X	X			X	X	X	
	Wastes			X	X						X	X		
	Land clearing		X						X	X	X	X	X	
	Disturbance & noise						X			X	X	X		
	Fresh water abstraction				X									
During operation	Sewage			X	X									
	Other pollution			X	X									
	Boats & shipping						+				x +	X		+
	Erosion & accretion	X		X		X	X				X		X	
	Tourism										+	X		+

- Release or treatment of wastes, sewage, chemicals, and hazardous materials
- Large-scale and/or intense forestry, fishing, or agriculture
- Housing and settlements
- Power, desalination, or other plants

It is beyond the scope of this book to go into more detail on the environmental impact assessment process. The environmental impact assessment process is normally followed by the development of an environmental management plan.

Dealing with Erosion

Most of the world's coastlines are in a state of erosion or retreat as a consequence of sea level rise and other natural processes. Erosion is often exacerbated when a development takes place, with or without an environmental impact assessment. This has negative consequences for the sand budget and the result is erosion of the beach. When dealing with sand loss and erosion of the shore, coastal zone managers have four options (Komar 1998), which are listed in order of increasing level of action/interference.

1. No action
2. Retreat and relocation inland
3. Sand nourishment or soft solutions
4. Stabilization structures or hard solutions

Where an area is not developed, as in a wilderness state or with minimal development, options 1 and 2 can be best. This involves simply allowing natural processes of coastal retreat to proceed, the beach and dune system moving landward while maintaining a dynamic equilibrium. Where there is a generous setback line, it may provide space for this. If development of the area is limited, this may be considerably more cost effective than options 3 or 4. No action and retreat is the most natural approach, and is the preferred option whenever possible. However, many coasts are highly developed and removal of infrastructure, structures, and buildings is not feasible. In this event, the coastal zone manager has to resort to options 3 or 4 in the case of an eroding shoreline.

Options 3 and 4 involve direct and expensive actions and are widely practiced on developed coasts. The soft solution (option 3) is to add sand to replace that which is lost and thereby replenish or build the shore, whereas the hard solution (option 4) is to construct hard engineering structures to anchor the shoreline and protect it from wave attack. Beach nourishment, the soft solution, is preferred to hard structures as it is more natural and less disruptive to ecological processes. Sand can be brought by truck or pumped as sludge from the borrow area and placed on the beach, in the nearshore or even on the dunes, depending on the particular situation. If trucking is the mode of transport of sand, the backshore or dunes are the most convenient site for deposition. If sand is pumped as a sludge, usually from a subtidal borrow area, the beach or surf zone may be the best area for deposition. Essentially, nourishment attempts to maintain the natural condition, but it is expensive (costing up to $5 million per kilometer) and needs to be maintained. Generally, several small nourishments at intervals of months or years are better than single major nourishment activities. Perhaps the most critical aspect of nourishment is the particle size of the borrowed sand: too fine and it will tend to be rapidly washed away by waves and currents; too coarse and it may change the beach morphodynamic state and its ecology. In addition, the borrow area should not be located in the littoral active zone because that would simply relocate the erosion problem from one sensitive site to another.

Although hard solutions may be more permanent, they also cause more disruption to natural processes. Indeed, that is how they work. In many cases, it is hard structures in the littoral active zone that cause erosion problems in the first place, usually by disrupting longshore sand transport. However, in some cases the situation is such that hard structures are the only solution to protect a developed coast from serious erosion. There are three categories of such structures (Figure 14.13): (1) seawalls/revetments on the shore parallel to the shoreline, (2) groins perpendicular to the shore, and (3) offshore breakwaters. Seawalls are usually hard concrete structures and revetments are made of stones or artificial concrete "boulders" called dolos. They are expensive, but if done properly can last very long. Groins, often in series along the shore, protrude across the beach and surf zone, blocking longshore sand transport and trapping sand. They are usually half the width of the surf zone and trap sand to form pocket beaches. The spacing and length of groins determine the nature of the artificial shoreline. Breakwaters are shore-parallel offshore structures that break wave energy to create sheltered conditions on the shore that favor sand deposition. They also disrupt longshore sand transport to trap sand and can be used in conjuction with groins.

Clearly, different situations require different responses. Faced with an eroding shoreline the coastal zone manager needs to consider the extent of development, the nature and speed of erosion, the desirability of keeping the shore natural, and the costs and benefits of different options and their long-term implications. From an ecological point of view, no action or retreat is most desirable and hard structures least desirable, but the decision in each case will depend on the weighing of the previously cited factors.

Restoration and Rehabilitation

Damaged habitats or ecosystems in the littoral active zone often need repair from destruction or erosion caused by storms or human activity. This is particularly the case for dunes where vegetation has been removed or damaged. Restoration is the process of returning an area to its previous (pristine) condition as far as possible, whereas rehabilitation involves repairing an area, but not necessarily to its original or natural condition. Although beach nourishment is also a form of restoration or rehabilitation, depending on the procedures adopted and the end result, most commonly it is the dunes and dune vegetation that need repair.

During extreme storms and episodic events, coastal dunes may be eroded and their sand stores removed and deposited in the surf zone, thereby dissipating wave energy and helping to protect the coast from further attack. This ability of dunes to store and yield sand to the beach and surf zone is what makes them such effective elements of natural coastal defense. Following an episodic erosion event, it is the trapping of wind-blown sand by vegetation that slowly rebuilds the dune and replenishes its sand store. This dynamic equilibrium between rapid storm erosion and slow recovery keeps the frontal dune fluid. It alternately yields and recovers sand, whereas back dunes may be more stable and permanent.

To maintain this natural buffer function, coastal dunes need to be left intact, and thus sand removal should never be allowed under any circumstances. Further, the vegetation

Figure 15.4. Slat fencing in New Jersey. (Photo: K. Nordstrom.)

Figure 15.5. Sandy beaches and surf zones are ideal recreation playgrounds. Their recreational value is, however, diminished once the carrying capacity is exceeded. (Photo: Robin Ford.)

should be protected from trampling, vehicles, and so on. In addition, stabilizing the dune and trapping sand can be hastened, especially where human impacts have contributed to sand loss, by using brush or slat fencing and spreading mesh, mulch, or even bitumen spray — followed by planting. Fencing should have about 50% porosity, be aligned parallel to the shore, and if possible be aligned perpendicular to the prevailing winds (Figure 15.4). Fences 1 m high usually fill up with sand within a year and can then be followed by additional fences. Planting should preferably be with indigenous species naturally occurring in the area and appropriate to the particular dune zone (see Chapter 13). People can be kept off the dunes by providing raised boardwalks that allow access to the beach from behind the dunes (see section on access).

Recreational Carrying Capacity

Recreational activities are increasing dramatically in developed countries as people enjoy more leisure time and expect higher standards of living. Clearly, much of this activity is (and will increasingly be) focused on soft coasts, including sandy shores. Open coast beaches are generally quite resilient, even to relatively heavy human traffic (Figure 15.5). They are thus ideal playgrounds and suitable areas for focusing recreational activities.

However, the backshore and dune areas (and particularly foredunes) are sensitive, even to barefoot human traffic. Pioneer plants maintain a tenuous foothold in foredune systems, in a balance of stabilizing (vegetation) and destabilizing (wind, storm erosion, salt spray) forces — and even moderate human activity can tip this balance.

Recreational carrying capacity is defined as the level of recreational use an area can sustain without an unacceptable degree of deterioration of the character and quality of the resource or of the recreation experience (Pigram 1983). Within this broad definition, four categories are recognized: economic, physical, ecological, and social carrying capacities (Brotherton 1973). Economic carrying capacity is seldom used in practice, as it only applies where the benefits derived from different uses of a site are being compared. In most cases, soft coastline resources are not simultaneously available for recreational and commercial use.

Physical carrying capacity is essentially a design concept, based on the number of use units (people, cars, boats, vehicles) that can physically be accommodated in a certain area. It needs to be taken into account when planning the types and levels of uses to which a piece of coast will be subjected, especially in designing and locating car parks. However, the most important forms of carrying capacity to be considered for sandy shores are ecological and social.

Ecological carrying capacity is the maximum use the biota or physical processes of an area can withstand before becoming unacceptably or irreversibly damaged. As any use of an ecosystem induces change, the decision as to what level of use will cause unacceptable change is to a large extent subjective. It will depend on the types of activity occurring on the site, the nature of animal and plant communities, and the amount of change that can be tolerated without detracting from the enjoyment and outdoor experience of the visitor. Management and protection of resources may thus be seen as means of ensuring a high-quality recreation experience and not as ends in themselves. The most sensitive elements of the sandy coast are the fauna (birds, invertebrates) and vegetation of the backshore and foredunes, and it is usually their tolerance to recreational activity that determines the ecological carrying capacity of a sandy shore.

Social carrying capacity is the maximum level of recreational use, above which there is a decline in the quality of the recreational experience from the point of view of the participants. This may vary widely on different beach types. On an urban beach a high density of users may be tolerated, even enjoyed, whereas on a wilderness beach much lower numbers can be supported if the wilderness experience is not to be compromised. Factors that influence the experience, and thus social carrying capacity, include the scale and variety of the landscape, vegetation cover that provides exclusion, and the facilities available. In addition, carrying capacity must be expected to vary with different population groups, which may have different backgrounds and different expectations. It must also be appreciated that even after individual satisfaction has begun to decline due to overcrowding the aggregate satisfaction obtained by the total number of users of a beach may still increase — although the nature of the recreation experience will change.

Sandy coasts are subject to both mass tourism (high numbers of people seeking a social vacation experience) and ecotourism (fewer visitors seeking a quality environment). Sandy coasts need to be managed for both. Because of their great sensitivity, even to trampling by barefoot humans (see Chapter 14), use of dunes is usually limited by ecological carrying capacity. However, for beaches (which are more resilient) social carrying capacity is more relevant and determines the quality of the recreation experience. Therefore, although sensitive dunes may have almost zero carrying capacity beaches may have a carrying capacity ranging from as low as 10 persons per kilometer on a beach where a wilderness experience is expected to more than 1,000 persons per kilometer or an urban or resort beach, where users are looking for an outdoor social experience.

Controlling Access

Human activity and recreation can most easily be managed, and carrying capacity accommodated, by controlling access to and through the littoral active zone. The frequency and size of car parks along the littoral active zone are a major determinant of the number of recreationists likely to use a beach. By locating car parks behind the littoral active zone and by having raised walkways (Figure 15.6) from the car parks over the dunes to the beach, coastal zone managers can both protect the dunes to ensure that their ecological carrying capacity is not exceeded and control the number of people on the beach within its social carrying capacity to ensure a quality recreational experience. Not only must access be controlled in most cases but activities on the beach may need monitoring. In particular, an authority that controls access is most likely also to be responsible for public safety on the beach. Beaches can be dangerous, and numerous drownings occur every year. Short (1999) discusses beach hazards in detail. These include rip currents, large plunging breakers, troughs, and other deep spots in the surf. Greatest care needs to be exercised on exposed intermediate and dissipative beaches, and lifeguards are necessary on popular beaches in high season.

Figure 15.6. Access to the beach, especially over foredunes, must be strictly controlled. (Photo: Planning Office, City of Cape Town, with permission.)

Pollution

Ideally, no pollutants should be allowed in the littoral active zone. However, where the release of pollutants of any form is unavoidable this should be well away from recreation and conservation areas. Further, it should be preceded by treatment to reduce toxicity, and discharge should be as far offshore as practicable. Location of any facility releasing potential pollutants into the littoral active zone should always be preceded by EIA.

Planning, Permits, and Regulations

Planning, permits, and regulations are all important techniques that aid the coastal zone manager. Although it would be ideal to have free access to the beach for everyone and no regulations to govern activities or behavior, this is seldom feasible. Planning is required to manage the littoral active zone effectively for harmonious multiple use. Some developments and activities may need to be controlled by permits (for example, speed boating or fishing), and regulations are usually required to manage such activities.

15.5   Management, Planning, and Implementation

Using the techniques and principles outlined in the previous section, coastal zone management on a sandy shore generally aims for the following.

- An integrated approach, addressing all resources and considering all interests
- Coordination across all sectors for the terrestrial and marine parts of the coast
- Sustainable multiple use that does not compromise the future and controls the use of renewable resources
- Conservation of biodiversity, especially in the dunes, and protection of valuable species
- Protection against natural hazards (such as hurricanes, floods, and tsunamis) by keeping the LAZ as intact as possible
- Optimal long-term sustainable use of the sandy coasts and maintenance of the most natural environment possible
- Restoration of damaged ecosystems and protection of sensitive habitats
- Control of pollution in all forms
- Guidance of planning and development in a harmonious and safe way
- Raising public awareness of the uniqueness, sensitivity, and value of the sandy coast and encouraging participation
- Resolution of conflicts

Developing an integrated CZM program generally involves four phases: policy formulation, strategy planning (or preliminary planning and exploring options), program development (or developing master plans), and implementation, with the last often the most difficult. The idea is to finish with a comprehensive master plan that can feasibly be implemented. It is generally best to plan for and consider the widest possible area of coastline but to manage a smaller area of shoreline within this. Although the land and sea are both always included in such a program, most control is exerted through managing access to the shore from the land. Thus, the shore is often conserved by managing the adjacent resources on the land. The purpose of an environmental management plan is to ensure that development and conservation can coexist and to guide development.

15.6   Case Studies

U.S. Barrier Islands

Barrier islands, which are a characteristic feature of some coastlines, require particular and special attention as far as planners and conservationists are concerned. These are unstable systems, which bear the brunt of wave attack and are hence of major importance in protecting the mainland, especially during storms. They also frequently act as refuges for plant and animal communities, including a number of rare species, and are therefore of considerable importance in their own right. The long-term survival of barrier islands, as well as barrier beaches, is dependent on maintenance of their dune systems, for these store sand and replace that lost during storms. Any activity that may deprive barrier islands of sand must be viewed with disapproval. Barrier islands, built of beaches and dunes, are typical of the United States east coast and are dynamic systems that have seen much human disturbance. This account of human impacts on them over more than 100 years is taken from Nordstrom (1994, 2000) and is depicted in Figure 15.7.

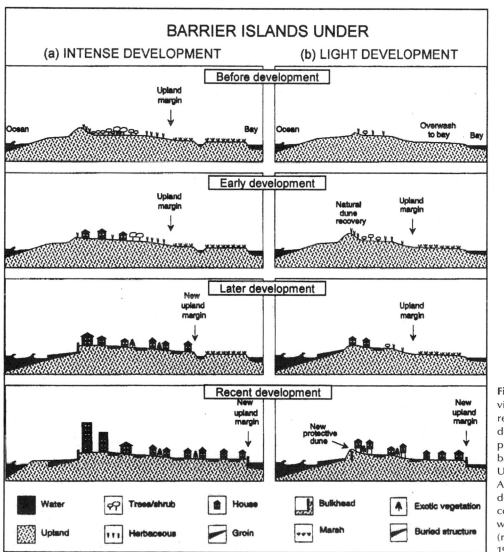

Figure 15.7. Profile view of phases representing the development and protection of coastal barriers in New Jersey, USA, through time. Areas with intense development (left) are contrasted with areas with less development (right) (after Nordstrom 1994, 2000).

Beaches along the New Jersey shoreline had some limited use as a recreational resource as early as 1790, but only a few cabins, homesteads, and boarding houses existed on the islands prior to the mid nineteenth century. The barrier islands were low and narrow, and they were backed by salt marshes and fronted by foredunes that formed ridges in relatively stable portions of the barriers, and hummocks in more dynamic areas. The higher portions of the barriers had lush growth of cedar, holly, and other trees. New Jersey barriers were highly mobile prior to human development, especially adjacent to inlets.

Railroad lines, built to support real estate speculation ventures, extended along much of the New Jersey shoreline by 1886. Recreational business districts then clustered near railroad stations on the higher portions of the barriers. Subsequent modifications included grading dunes to a flatter form to facilitate construction of buildings and roads and destruction of the natural vegetation. Expansion of settlements was rapid thereafter. The permanent population of Atlantic City was over 5,000 by 1880 and land values grew from almost nothing in 1854 to $50 million by 1900. The barrier island resort industry in New Jersey developed rapidly in the early twentieth century, due primarily to increased use of cars. New communities appeared, in some cases linked by roads rather than railroads. Growth then extended outward from these locations, both alongshore on the higher portions of the barriers and landward onto the marsh surfaces. Filling of the marshes occurred on several barriers between 1880 and 1910. Settlement expansion involved transformation of the physical environment into a cultural one. Shore protection measures were first employed during settlement expansion, although the intensity of development was probably too limited early in this stage for public works to be cost effective. The widespread use of hard protection structures developed during settlement expansion after 1900. Dredging channels into the marshes to accommodate boats occurred in the larger settlements between 1905 and 1913. More extensive conversion of the marshes to lagoon developments for private housing and their associated boat docks occurred on many of the barriers after 1950. Most of these projects placed materials dredged from the new waterways onto the marsh surfaces to provide a platform for houses and roads. Construction of lagoon housing is now severely restricted by regulations governing use of wetlands, but marshland now only remains in isolated enclaves on the landward sides of the developed portions of barriers.

Many inlets that existed in 1886 and all inlets that formed since that time were closed artificially by the U.S. Army Corps of Engineers or kept from reopening after natural closure to eliminate undesirable shoreline fluctuations, facilitate land transportation, or increase the flow through nearby controlled inlets. Navigation improvement projects at inlets were undertaken to maintain navigation channels and static shorelines. Five of the New Jersey inlets existing today are stabilized by jetties constructed since 1911, and two inlets without jetties are maintained by dredging.

The proximity of the New Jersey shoreline to the large urban population centers of New York City and Philadelphia provided the stimulus for further intensification of development. Restrictions on the number of units per building in New Jersey have limited high-rise constructions to only a few locations. At Atlantic City, for example, the buildings are so large and so close to the beach they profoundly affect wind processes and sand transport. Other communities may have a high density of multiple-unit low-rise structures or detached houses that are fronted by a narrow dune, a bulkhead, or both. The level of development portrayed in Figure 15.7 has been sufficiently intensive to justify use of large-scale nourishment operations to protect buildings and infrastructure. Human actions along a coast with lower levels of development depend on less costly public expenditures, such as dune building and raising buildings and bulkheads for backup protection.

The greatest change in the dimensions of the New Jersey barriers due to development is the dramatic increase in the width of dry surface area through filling of the marshes. The maximum elevations of isolated dunes in both developed areas and undeveloped areas are generally higher than they were in 1886, but they do not extend as far inland. Dunes no longer exist in many areas that are now protected by bulkheads and seawalls. In numerous locations, dunes exist but are no higher than 1.5 m above the elevation of the backshore and no wider than 15 m. Despite this limitation to their human utility value as a form of flood protection, the dunes form a more continuous barrier than they did prior to development and are more likely to restrict barrier island mobility. Several communities in New Jersey are protected today by a combination of soft and hard solutions (beach nourishment, groins, bulkheads, and dunes), and many communities employ several of these methods of protection. Over 43 km of the 205-km-long shoreline of New Jersey is protected by shore-parallel structures, and there are over 300 groins.

Shallow offshore areas have generally lost sand since the beginning of development in the mid nineteenth century, but the back beach and dunes have undergone less erosion since development, and several barriers even have accretional trends. As a result, there has been little onshore migration of the beach over the past century, but the subtidal area has probably deepened. Major storms have periodically eroded these beaches and destroyed or damaged buildings, support infrastructure, and protection structures, but post-storm reconstruction efforts have rebuilt these facilities, usually to larger proportions. In most cases, the seaward line of human structures has not retreated. The result of human activity on the New Jersey shore has been an increase in the number of buildings subject to storm hazards, accompanied by a decrease in the role of natural processes of sand transport and storage on the islands or the likelihood that they will migrate. More recently, massive nourishment projects have provided the potential for natural processes to return in some areas and more management options are now available. Thus, although New Jersey led the way in the degradation of a sandy coast it now has the potential to lead in restoration (Nordstrom and Mauriello 2001).

Impact of Hard Structures on Longshore Sand Transport

Coastal engineering structures built out into the water from the shore (such as groins) block the natural littoral drift of sand prevailing along most coasts. This deprives beaches of sand and initiates erosion on the downdrift side of the structure, while sand deposits updrift where the beach advances seaward. Littoral drift is not a constant phenomenon at any given site. It varies enormously with wave action and the direction of wave attack, and there is commonly even a reversal of drift direction under different conditions (notably during storms). The summation of all individual sediment transport events over a year is the net longshore transport, and it is this value that is important in determining the effects of coastal structures on erosion and deposition, rather than any single transport event.

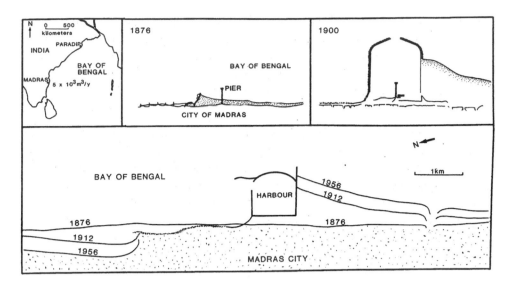

Figure 15.8. Coastline changes associated with the development of Madras Harbor, India, from 1876 to 1950 (after Komar 1998).

Among the several examples cited by Komar (1998), the development of the Port of Madras in India is of particular interest because it was constructed on the open shore in an area of strong longshore drift and because resulting changes have been documented since its construction was sanctioned in 1875 (Figure 15.8). Due to strong northward drift, sand deposited rapidly on the beach to the south of the breakwater while it was still under construction, and erosion to the north dictated the placement of groins. The original harbor entrance faced east (out to sea), but accumulation of sand on the south side built the shoreline seaward until sand began to drift around the eastern end, causing shoaling of the harbor entrance. This became so serious that by 1920 the entrance had to be moved to the north side of the harbor and an outer quay constructed in an attempt to deflect sand away from the new entrance. The entire shoreline has changed for many kilometers on either side of the harbor — severe erosion having taken place to the north (despite attempts to halt it) and dramatic deposition to the south, the shoreline advancing out to sea.

In another example, jetties constructed in 1935 (with the object of stabilizing the inlet of Ocean City, Maryland, USA) blocked the north-to-south transport of sand, resulting in a seaward advance of the shoreline to the north. This was not considered a disadvantage. However, the barrier island Assateague, to the south, suffered such severe erosion that the shore retreated some 450 m over a period of 20 years. By 1961, the south beach had separated from the inner end of the jetty (leaving a gap of almost 240 m of open water), and the following year a storm opened a breach over 1 km wide. Attempts to restore the beach by dredging and filling have apparently only been successful in the short term.

Literally hundreds of other well-documented examples could be given from all over the world. One remedy is, of course, nourishment by transferring sand from the updrift side of the construction to the downdrift side. Bypassing the obstacle in this way can be done by transporting the sand by truck or dredging and pumping the sand, mixed with water as a sludge, through a pipe to the downdrift area.

It might be thought from what has been said previously that structures built on coasts with no significant net longshore sand transport would cause neither erosion nor deposition of sand. This is by no means always the case, however, and there are several examples of quite unpredicted erosion and loss of property resulting from construction in such areas. It is not that erosion fails to occur on a coast with zero net drift but that following construction of a barrier a new equilibrium is established, an equilibrium that may in fact involve drastic changes to the shoreline. On the other hand, where there is a net littoral drift no balanced sediment budget can ever be established without bypassing, and the processes of erosion and deposition continue indefinitely.

Clearly, the ability to predict the effects of marine constructions on shorelines is of paramount importance in planning and management. Computer-based modeling of nearshore processes and shoreline changes have considerable potential for such prediction. The details of computer-simulation models lie outside the scope of this book and the reader is referred to Komar (1998) for an introduction to the techniques and applications involved.

Stabilization and Destabilization of Dunes

An obvious modification of the sand budget involves changing the sand inputs to, or outputs from, dunefields by stabilizing or destabilizing them. One may tend to think of human disturbance only in terms of destabilization consequent on vegetation damage, but in fact deliberate stabilization may have equally serious effects by trapping sand needed to nourish beaches naturally.

The south coast of southern Africa is exposed to predominantly southwesterly wind and swell and has large spiral bays characterized by considerable sand transport. Typically, sand is transported eastward within a bay, mainly in the surf zone by wave action. Toward the eastern end of such a bay, more exposed beaches and an increasingly onshore wind direction result in much aeolian transport of sand off the beach into a transgressive dunefield. The dunefield then acts as a shunt, transporting sand across the headland to feed the beaches at the sheltered western end of the next bay, thereby maintaining the process of littoral sand transport and forming an important link in an extensive coastal conveyor belt of sand. These dunefields, termed headland bypass dunefields (Tinley 1985), are usually large, sparsely vegetated, and transport sand in the form of advancing dune ridges transverse to the wind direction. In the past, such extensive areas of mobile sand were seen as wastelands with no inherent value. Indeed, in one case the advancing sand was seen as a threat to the city of Port Elizabeth. This led to efforts to stabilize the bypass dunefields using exotic vegetation, without any appreciation of the role the dunes played in littoral sand transport. Major stabilization efforts were carried out at the turn of the last century and into the mid 1900s, with great success. In three cases, stabilization of such dunefields for development purposes stopped them in their tracks. Only later was it realized that this cut off the sand supply to beaches used for recreation, thereby creating severe erosion problems. In one case, where the dunefield has not been developed, an experimental destabilization program was successfully carried out to remove vegetation, reactivate part of the bypass dunefield, and get sand blowing back on beaches at the western end of the next bay (Figure 15.9). Had the structure and linkages within the littoral active zone been appreciated in the first place, this would never have

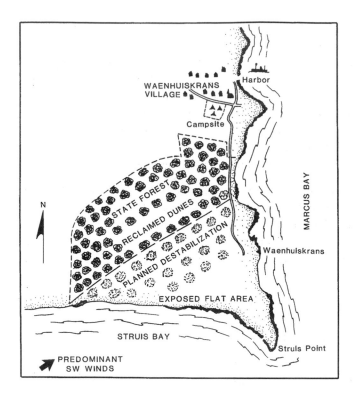

Figure 15.9. The headland bypass dunefield at Struis Bay, South Africa, showing the extent of the area that was initially stabilized with vegetation and then partially destabilized to get sand back onto eroding beaches in Marcus Bay (after Swart and Reyneke 1988).

happened. Stabilization of moving sand is no longer seen as essential to dune conservation unless rare species or habitats are threatened by moving sand.

15.7   Conservation

Increasing human impact on the coast and shrinking natural habitats argue for conservation. On land, many unique and sensitive systems are protected as reserves. The same is now happening in the sea in the form of marine protected areas (MPAs). MPAs can be established to protect endangered species or sensitive habitats, maintain biodiversity, maintain spawning biomass of exploited species, or provide educational opportunities. Usually a major goal is to protect target species, habitat diversity, and ecological processes. Several beaches in different countries are conserved in order to protect turtle nesting areas or to reduce angling pressure and allow resident fish stocks to recover. Although the World Conservation Union has recommended that 20% of the world's coasts be protected, this may not be feasible in many countries. However, there are different categories of protection. For example, an area can be managed for sustainable exploitation of beach clams, for limited recreational use in a wilderness setting, or for minimal use. Sites for protection can be selected on the basis of plant and/or animal diversity, productivity, representativeness of a biogeographic region, pristine condition of habitats, or extent of human disturbance. Siting an MPA next to a terrestrial reserve is ideal, as it enables combined management and control of access. The issue of size and spacing of reserves has been much debated, but in the case of beaches it is perhaps more important to cover a range of exposure categories/beach types. In general, though (as outlined previously) beaches are in less need of protection than dunes. They are tolerant of many recreational activities and are only seriously threatened by engineering structures. When a sandy shore is designated an MPA, management should cover the entire littoral active zone.

15.8   Conclusions

Sandy beaches and dunes form part of the coast's natural defense system against storm waves. The beach and its surf zone act as the wave dissipating apron, whereas the dunes constitute a reserve of sand that can be released during extreme wave onslaught and regained during calmer weather. The buffer effect of the dunes increases with dune height. Thus, any activity that tends to lower the dunes or erode or devegetate them leads to increased vulnerability to high seas and increased damage during storms. It must be stressed that because dune systems are so fragile even the slightest man-induced alterations to them may result in significant adverse effects. The dangers are especially great in view of rising sea levels and retreating coastlines, coupled with the fact that soft shores in general (and sandy beaches in particular) are the focus of much recreational activity — to which, it may be argued, the public has some right. The ideal formula to be observed is to use beaches for recreation while conserving dunes. Key principles for managing the sandy shore are as follows.

- Manage the littoral active zone as a unit
- Employ setback lines to allow retreat and to protect the LAZ and keep it intact
- Divide the coast into zones to be managed for different uses
- Undertake thorough EIA before any significant developments
- Use soft in preference to hard solutions to deal with erosion
- Manage development, recreation, and conservation for sustainability and compatibility of uses within the carrying capacity

Ocean sandy beaches are among the most beautiful landscapes on earth and harbor diverse faunas. Demand for access to their scenic vistas and recreational potential will surely increase with time. This must be matched by increasingly sophisticated coastal zone management and improved public awareness of the structure, function, and sensitivities of sandy shores. We hope this book will contribute to raised awareness of sandy beaches and that those who read it, when next they visit the sandy beach, will be able to identify with the words of T.S. Eliot:

> *We shall not cease from exploration*
> *And the end of all our exploring*
> *Will be to arrive where we started*
> *And know the place for the first time.*

Glossary

16

Accreting Accumulating sand.

Aeolian Wind-blown.

Allopatry(ic) When populations or species do not co-occur.

Anadromous Fishes spending part of their life in the sea and part in fresh water, where they spawn.

Aquifer The water body in a geological formation, such as a dunefield. Unconfined aquifers have open boundaries.

Autecology The ecology of a population of a single species.

Backshore The top of the beach above the high-water mark.

Backwash The swash running back down the face of the beach after a wave uprush.

Bar An accumulation of sand as a sandbank in the low intertidal or subtidal.

Barrier island A sandy island fronting the coast and usually running as a strip parallel to it.

Berm An accumulation of sand as a platform on the backshore, usually extending from the high-water mark back to the start of the dunes.

Batoid Belonging to an order of flattened cartilaginous fishes, including skates and rays.

BI The beach index, a measure of beach state incorporating sand particle size, beach face slope, and tide range — ranging from microtidal reflective (values < 1) to macrotidal ultradissipative (values > 3).

Bioturbation The disturbance of sediment with concomitant increasing turbidity through the action of animals.

Bore A broken wave as it crosses the surf zone.

Breaker A breaking wave.

Beach face The intertidal slope of the beach between the high- and low-tide marks.

Blowout A wind-eroded funnel developed in a dune, usually as a result of vegetation damage.

Bypass To go past or around. Pumping can bypass sand across a harbor from a beach on one side to the other. Dunes can bypass sand over a headland from one bay to the next.

Canopy The top layer of vegetation.

Carrying capacity The amount of activity or users a system can accommodate or support without damage.

Cusp A rhythmic feature on a beach face, usually reflective beaches, undulations spaced at 5- to 50-m intervals.

Deflation Wind erosion down to a base level.

Demersal Living near the bottom.

Denitrification Conversion of oxidized forms of nitrogen, such as nitrate or nitrite, to ammonia or nitrogen gas.

DFV Dimensionless fall velocity, an index of beach type incorporating measures of wave energy and sand mobility.

Dilatancy The tendency of an unsaturated sand to compact on agitation as water is driven out, the opposite of thixotropy.

Edaphic Concerning the soil.

Effluent line The area in the intertidal beach where the water table meets the surface and below which water, or effluent, drains out of the sand down the beach face.

Energy flow The transfer of energy, expressed as calories of units of carbon, through food chains from primary producers through herbivores to carnivores and other consumers.

Epipsammon Refers to organisms on the sand surface.

Exposure Refers to the extent to which a beach experiences wave action as opposed to shelter.

Fetch The distance over water across which a wave-generating wind blows.

Flatfishes Flattened bottom-dwelling fishes, including soles.

Geotaxis Orientational response to gravity.

Guild A collection of species with similar resource requirements.

Heteroparity Living more than one year.

High energy A beach exposed to strong wave action.

Infragravity With a period longer than normal gravity waves (i.e., usually 1 to 5 minutes).

Interstitial Between the sand grains.

Isotherm A line joining regions of similar temperature.

Iteroparous(ity) Breeding more than once in a lifetime.

K-strategist A species characterized by slow growth, longevity, and low fecundity (i.e., high investment in a few propagules).

Lecitrophy Dependence on yolk in the egg for nourishment.

Littoral Pertaining to the shore.

Liquefaction Becoming liquid, fluid.

Metapopulations Spatially segregated populations of benthic adults interconnected through larval dispersal.

Mineralization Release of nitrogen and phosphorous by breakdown of organic materials.

Morphodynamics The morphology and dynamics, referring to the structure, sediment movement, and water circulation of a beach and surf zone.

Neuropodium The lateral/ventral appendages of polychaete worms.

Nitrification Oxidation of ammonia to nitrite and nitrate.

Nourishment (beach) Adding sand to restore a beach.

Phototaxis Orientational response to light.

Pioneer A plant that colonizes bare sand and starts dune vegetation succession close to the beach.

P/B The ratio of production or growth to biomass (see also *Turnover*).

Planktotrophy Dependence on the plankton for nourishment.

Pocket beach A small isolated beach between headlands, outcrops, or other topographic features.

Precipitation ridge An advancing slip face where a dune system moves landward, usually over vegetation such as woodland, burying the area in front of it.

Prograding Advancing or accreting, gaining sand.

Psammophile A species that occurs in sand.

R-strategist A species characterized by rapid growth, large numbers of propagules, and short life span.

Redox Reduced/oxygenated status or condition.

Reduced layers Layers of sediment darkened due to the buildup of sulfides under reducing or anoxic conditions.

Rheotaxis Orientational response to flow.

Rip current A seaward-flowing return current that drains water from inside the surf zone out through the beakers.

Rip head The zone of discharge of a rip current outside the breakers.

RPD The redox potential discontinuity zone, a transition layer in sediments between surface-oxygenated layers and deeper reduced layers, often characterized by gray sand.

RTR Relative tide range, an index of beach type obtained by dividing tide range by wave height.

Saltation Bouncing along the surface.

Scope for growth The amount of consumed energy available for growth after catering for maintenance or respiration.

Seine A net operated by hand from a beach, usually with floats to keep one line up and weights to keep the other down, and drawn to the beach at two ends, often with a bag in the center.

Semelparous(ity) Breeding only once in a lifetime.

Shoaling Moving into shallower water.

Slip face The advancing face of a dune down which wind-blown sand cascades.

Succession Progressive replacement of one type of community by another in sequence toward a climax.

Swash The final stage of a wave after it has broken, crossed the surf zone as a bore, and collapsed. Swash runs up the beach face and falls back as backwash.

Swept prism The section of the sand body of a beach and surf zone that is mobilized (i.e., can gain sand during calms and lose sand during storms).

Sympatry Co-occurring.

Synecology The study of the ecology of groups of organisms (for example, a community) in relation to the environment.

Taxis Orientational response to an environmental cue.

Thixotropy The capacity of sand to liquify upon agitation due to high water content enabling the grains to float.

Transgressive dunes Mobile dune systems, usually with little vegetation, that advance inland.

Trophic Related to feeding.

Trough A hollow or deeper area between sand ridges or bars, usually in the surf zone.

Turnover The rate at which biomass is converted into growth (see also *P/B*).

Upwelling The ascent of deep, cold, nutrient-rich water near the coast to replace surface water blown offshore.

References

References

17

Aarnio, K. (2001). The role of meiofauna in benthic food webs of the northern Baltic Sea. *Acta Acad. Aboensis* 61B:1–37.

Alexander, R. R., Stanton, R. J., and Dodd, J. R. (1993). Influence of sediment grain size on the burrowing of bivalves: Correlation with distribution and stratigraphic persistence of selected neogene clams. *Palaios* 8:289–303.

Alheit, J., and Naylor, E. (1976). Behavioural basis of intertidal zonation in *Eurydice pulchra* Leach. *J. Exp. Mar. Biol. Ecol.* 23:135–144.

Anderson, J. G., and Meadows, P. S. (1979). Microenvironments in marine sediments. *Proc. Roy. Soc. Edinb.* 76B:1–16.

Anderson, U. V. (1995). Resistance of Danish coastal vegetation types to human trampling. *Biol. Conserv.* 71:223–230.

Anders, F. J., and Leatherman, S. P. (1987). Effects of off-road vehicles on coastal foredunes at Fire Island, New York, USA. *Environ. Mgmt.* 11:45–52.

Ansell, A. D. (1982). Experimental studies of a benthic predator-prey relationship. I. Feeding, growth, and egg-collar production in long-term cultures of the gastropod drill *Polinices alderi* (Forbes) feeding on the bivalve *Tellina tenuis* (da Costa). *J. Exp. Mar. Biol. Ecol.* 56:235–255.

Ansell, A. D. (1983). The biology of the genus *Donax*. In: *Sandy beaches as ecosystems*, A. McLachlan and T. Erasmus (eds.), pp 607–635. The Hague: Dr W. Junk.

Ansell, A. D., and Trevallion, A. (1969). Behavioural adaptations of intertidal molluscs from a tropical sandy beach. *J. Exp. Mar. Biol. Ecol.* 4:9–35.

Ansell, A. D., and Trueman, E. R. (1968). The dynamics of burrowing in the sand dwelling anemone *Peachia hastata*. *J. Exp. Mar. Biol. Ecol.* 2:124–127.

Ansell, A. D., Sivadas, P., Narayanan, B., and Trevallion, A. (1972). The ecology of two sandy beaches in South West India. II. Observations on the populations of *Donax incarnatus* and *Donax spiculum*. *Mar. Biol.* 17:318–322.

Ansell, A. D., McLusky, D. S., Stirling, A., and Trevallion, A. (1978). Production and energy flow in the macrobenthos of two sandy beaches in South West India. *Proc. Roy. Soc. Edinb.* 76B:269–296.

Armonies, W., and Reise, K. (2000). Faunal diversity across a sandy shore. *Mar. Ecol. Prog. Ser.* 196:49–57.

Arntz, W. E., Brey, T., Tarazona, J., and Robles, A. (1987). Changes in the structure of a shallow sandy beach community in Peru during an El Niño event. In: *The Benguela and comparable ecosytems*, A. I. L. Payne, J. A. Gulland, and K. H. Brink (eds.), *S. Afr. J. Mar. Sci.* 5:645–658.

Arntz, W. E., Valdina, E., and Zeballos, J. (1988). Impact of El Niño 1982–83 on the commercially exploited invertebrates (moviscos) of the Peruvian shore. *Meeresforsch.* 32:3–22.

Avellanal, M. H., Jaramillo, E., Naylor, E., and Kennedy, F. (2000). Orientation of *Phalerisidia maculata* Kulzer (Coleoptera, Tenebrionidae) in sandy beaches of the Chilean coast. *J. Exp. Mar. Biol. Ecol.* 247:153–167.

Bagnold, R. A. (1941). *The physics of blown sand and desert dunes.* London: Methuen.

Bally, R. (1983). The respiration of the marine isopod *Excirloana natalensis* (Flabellifera; Cirolanidae) from an exposed sandy beach. *Comp. Biochem. Physiol.* 75A:625–629.

Barnett, P. R. O. (1971). Some changes in intertidal sand communities due to thermal pollution. *Proc. Roy. Soc. Lond.* 177B:353–364.

Barros, F., Underwood, A. J., and Lindegarth, M. (2002). A preliminary analysis of the structure of benthic assemblages of surf zones on two morphodynamic types of beach. *J. Mar. Biol. Ass. U.K.* 82:353–357.

Bascom, W. (1964). *Waves and beaches.* New York: Doubleday & Co.

Bascom, W. (1984). The concept of assimilative capacity. *Southern California Coastal Water Research Project: Biennial Report*, 1984:171–177.

Branch, G. M. (1984). Competition between marine organisms: Ecological and evolutionary implications. *Oceanogr. Mar. Biol. Ann. Rev.* 22:429–493.

Branch, G. M., and Branch, M. (1984). *The living shores of southern Africa* (2d ed.). Cape Town: Struik.

Brazeiro, A. (1999). Community patterns in sandy beaches of Chile: Richness, composition, distribution and abundance of species. *Rev. Chilena Hist. Nat.* 72:93–105.

Brazeiro, A. (2001). The relationship between species richness and morphodynamics in sandy beaches: Which are the underlying factors? *Mar. Ecol. Prog. Ser.* 224:35–44.

Brazeiro, A. (2005). Geomorphology induces life history changes in invertebrates in sandy beaches: The case of the mole crab *Emerita brasiliensis* in Chile. *J. Mar. Biol. Ass. U.K.* 85:113–120.

Brazeiro, A., and Defeo, O. (1996). Macroinfauna zonation in microtidal sandy beaches: Is it possible to identify patterns in such variable environments? *Estuar. Coast. Shelf Sci.* 42:523–536.

Bregazzi, P. K., and Naylor, E. (1972). The locomotor activity of *Talitrus saltator* (Montagu) (Crustacea: Amphipoda). *J. Exp. Biol.* 57:375–391.

Brotherton, J. D. (1973). The concept of carrying capacity and countryside recreation areas. *Recreation News Suppl.* 19.

Brown, A. C. (1973). The ecology of the sandy beaches of the Cape Peninsula, South Africa. Part 4: Observations on two intertidal Isopoda, *Eurydice longicornis* (Studer) and *Exosphaeroma truncatitelson* Barnard. *Trans. Roy. Soc. S. Afr.* 40:381–404.

Brown, A. C. (1981). An estimate of the cost of free existence in the sandy-beach whelk *Bullia digitalis* (Dillwyn). *J. Exp. Mar. Biol. Ecol.* 49:51–56.

Brown, A. C. (1982a). Towards an activity budget for the sandy-beach whelk *Bullia digitalis* (Dillwyn). *Malacologia* 22:681–683.

Brown, A. C. (1982b). The biology of the sandy-beach whelks of the genus *Bullia* (Nassariidae). *Oceanogr. Mar. Biol. Ann. Rev.* 20:309–361.

Brown, A. C. (1982c). Pollution and the sandy beach whelk *Bullia. Trans. Roy. Soc. S. Afr.* 44:555–562.

Brown, A. C. (1983a). The ecophysiology of sandy beach animals: A partial review. In: *Sandy beaches as ecosystems*, A. McLachlan and T. Erasmus (eds.), pp 575–605, The Hague: Dr W. Junk.

Brown, A. C. (1983b). The effects of fresh water and of pollution from a marine oil refinery on the fauna of a sandy beach. In: *Sandy beaches as ecosystems*, A. McLachlan and T. Erasmus (eds.), pp 297–301, The Hague: Dr W. Junk.

Brown, A. C. (1985). *The effects of crude oil pollution on marine organisms.* South African National Scientific Programmes Report No. 99. Pretoria.

Brown, A. C. (1986). Molecular connectivity indices of organic pollutants: Correlation with cessation of burrowing in the whelk *Bullia digitalis* (Dillwyn). *Trans. Roy. Soc. S. Afr.* 46:109–114.

Brown, A. C. (1996). Behavioural plasticity as a key factor in the survival and evolution of the macrofauna on exposed sandy beaches. *Rev. Chilena Hist. Nat.* 69:469–474.

Brown, A. C., and da Silva, F. M. (1983). Acute metabolic rate:temperature relationships for intact and homogenised *Bullia digitalis* (Gastropoda, Nassariidae). *Trans. Roy. Soc. S. Afr.* 45:91–96.

Brown, A. C., and McLachlan, A. (1990). *Ecology of sandy shores*. Amsterdam: Elsevier.

Brown, A. C., and Odendaal, F. J. (1994). The biology of the oniscid Isopoda of the genus *Tylos*. *Adv. Mar. Biol.* 30:89–153.

Brown, A. C., and McLachlan, A. (2002). Sandy shore ecosystems and the threats facing them: Some predictions for the year 2025. *Environ. Conserv.* 29:62–77.

Brown, A. C., and Trueman, E. R. (1991). Burrowing of sandy beach molluscs in relation to penetrability of substratum. *J. Mollusc. Stud.* 57:134–136.

Brown, A. C., da Silva, F. M., and Orren, M. J. (1985). Haemocyanin and protein concentrations in the blood of the sandy beach whelk *Bullia digitalis* (Dillwyn). *J. Mollusc. Stud.* 51:99.

Brown, A. C., Stenton-Dozey, J. M. E., and Trueman, E. R. (1989). Sandy beach bivalves and gastropods: A comparison between *Donax serra* and *Bullia digitalis*. *Adv. Mar. Biol.* 25:179–247.

Brown, A. C., Nordstrom, K., McLachlan, A., Jackson, N., and Sherman, D. (2006). The future of sandy shores. In: *The waters, our future, prospects for the integrity of aquatic ecosystems*, N. Polunin (ed.). Cambridge University Press. (in press).

Brown, F. A. (1972). The "clocks" timing biological rhythms. *Amer. Scient.* 60:756–766.

Bruce, J. R. (1928). Physical factors on the sandy beach. Part I. Tidal, climatic and edaphic. *J. Mar. Biol. Ass. U.K.* 15:535–552.

Burger, J. (1991). Foraging behavior and effects of human disturbance on the piping plover (*Charadius melodus*). *J. Coast. Res.* 7:39–52.

Burger, J., and Gochfeld, M. (1991). Human activity influence and diurnal and nocturnal foraging of sanderlings (*Calidris alba*). *Condor* 93:259–265.

Burger, J., Gochfeld, M., and Niles, L. J. J. (1995). Ecotourism and birds in coastal New Jersey: Contrasting responses of birds, tourists and managers. *Environ. Conserv.* 22:56–65.

Bursey, C. R. (1978). Temperature and salinity tolerances of the mole crab, *Emerita talpoida* (Crustacea, Anomura). *Comp. Biochem. Physiol.* 61A:81–83.

Caddy, J. F., and Defeo, O. (2003). Enhancing or restoring the productivity of natural populations of shellfish and other marine invertebrate resources. *FAO Fish. Tech. Paper* 448.

Campbell, E. E. (1996). The global distribution of surf diatom accumulations. *Rev. Chilena Hist. Nat.* 69:495–501.

Campbell, E. E., and Bate, G. C. (1987). Factors influencing the magnitude of primary production of surf zone phytoplankton. *Estuar. Coast. Shelf Sci.* 24:741–750.

Cardoso, R., and Defeo, O. (2003). Geographical patterns in reproductive biology of the Pan-American sandy beach isopod *Excirolana braziliensis*. *Mar. Biol.* 143:573–581.

Cardoso, R., and Defeo, O. (2004). Biogeographic patterns in life history traits of the Pan-American sandy beach isopod *Excirolana braziliensis*. *Estuar. Coast. Shelf Sci.* 61:559–568.

Cardoso, R. S., and Veloso, V. G. (2003). Population dynamics and secondary production of the wedge clam *Donax hanleyanus* (Bivalvia: Donacidae) on a high-energy, subtropical beach of Brazil. *Mar. Biol.* 142:153–162.

Carlisle, J. G., Schott, J. W., and Abramson, N. J. (1960). The barred surf perch in southern California. *Calif. Dept. Fish Game Fish Bull.* 190.

Carter, R. W. G., Curtis, T. G. F., and Sheehy-Skeffington, M. J. (eds.) (1992). *Coastal dunes: Geomorphology, ecology and management for conservation*. Rotterdam: A. A. Balkema.

Claerboudt, M. (2004). Shore litter along sandy beaches of the Gulf of Oman. *Mar. Pollut. Bull.* 49:770–777.

Clark, B. M., Bennett, B. A., and Lamberth, S. J. (1996). Factors affecting spatial variability in seine net catches of fish in the surf zone of False Bay, South Africa. *Mar. Ecol. Prog. Ser.* 131:17–34.

Clark, J. R. (1966). *Coastal zone management handbook.* Florida: CRC Press.

Clutter, R. J. (1967). Zonation of nearshore mysids. *Ecology* 48:200–208.

Cockcroft, A. C., and McLachlan, A. (1986). Distribution of juvenile and adult penaeid prawns *Macropetasma africanus* (Balss) in Algoa Bay. *S. Afr. J. Mar. Sci.* 4:245–255.

Cockcroft, A. C., and McLachlan, A. (1993). Nitrogen budget for a high energy ecosystem. *Mar. Ecol. Prog. Ser.* 100:287–299.

Coe, W. R. (1955). Ecology of the bean clam *Donax gouldi* on the coast of southern California. *Ecology* 36:512–514.

Colclough, J. H., and Brown, A. C. (1984). Uptake of dissolved organic matter by a marine whelk. *Trans. Roy. Soc. S. Afr.* 45:169–176.

Colombini, I., and Chelazzi, L. (2003). Influence of allochthonous input on sandy beach communities. *Oceanogr. Mar. Biol. Ann. Rev.* 41:115–159.

Colombini, I., Aloia, A., Bouslama, M. F., El Gtari, M., Fallaci, M., Roncini, L., Scapini, F., and Chelazzi, L. (2002). Small-scale spatial and seasonal differences in the distribution of beach arthropods on the northern Tunisian coast: Are species evenly distributed along the shore? *Mar. Biol.* 140:1001–1012.

Connell, J. H. (1975). Some mechanisms producing structure in animal communities: A model and evidence from field experiments. In: *Ecology and evolution of communities*, M. C. Cody and J. M. Diamond (eds.), pp 460–491. Boston: Harvard University Press.

Contreras, H., Jaramillo, E., Duarte, C., and McLachlan, A. (2003). Population abundances, growth and natural mortality of the crustacean macroinfauna at two sandy beach morphodynamic types in southern Chile. *Rev. Chilena Hist. Nat.* 76:543–561.

Coull, B. C. (1988). Ecology of meiofauna. In: *Introduction to the study of meiofauna*, R. P. Higgins and H. Thiel (eds.), pp 18–38. Washington, DC: Smithsonian Institution Press.

Coull, B. C., and Palmer, M. A. (1984). Field experimentation in meiofaunal ecology. *Hydrobiologia* 118:1–19.

Crain, A. D., Bolten, A. B., and Bjorndal, K. A. (1995). Effects of beach nourishment on sea turtles: A review and research initiatives. *Restoration Ecol.* 3:95–104.

Crisp, J. J., and Williams, R. (1971). Direct measurement of pore-size distribution on artificial and natural deposits and prediction of pore space accessible to interstitial organisms. *Mar. Biol.* 10:214–226.

Croker, R. A. (1967). Niche diversity in five sympatric species of intertidal amphipods (Crustacea: Haustoriidae). *Ecol. Monogr.* 37:173–199.

Croker, S. A., and Hatfield, E. B. (1980). Space partitioning and interactions in an intertidal sand burrowing amphipod guild. *Mar. Biol.* 61:79–88.

Cubit, J. (1969). The behavior and physical factors causing migration and aggregation in the sand crab, *Emerita analoga* (Stimpson). *Ecology* 50:118–123.

Dahl, E. (1952). Some aspects of the ecology and zonation of the fauna of sandy beaches. *Oikos* 4:1–27.

Davies, J. L. (1977). *Geographical variation in coastal development.* New York: Longman.

De Alava, A. (1993). Interdependencias ecologicas entre dos bivalves simpatricos en una playa arenosa de la costa atlantica uruguaya. MSc thesis, CINVESTAV-IPN, Merida, Uruguay.

Defeo, O. (1993). The effect of spatial scales in population dynamics and modeling of sedentary fisheries: The yellow clam *Mesodesma mactroides* of a Uruguayan exposed sandy beach. Ph.D. thesis, CINVESTAV-IPN, Merida, Uruguay.

Defeo, O. (1996a). Recruitment variability in sandy beach macroinfauna: Much yet to learn. *Rev. Chilena Hist. Nat.* 69:615–630.

Defeo, O. (1996b). Experimental management of an exploited sandy beach bivalve population. *Rev. Chilena Hist. Nat.* 69:605–614.

Defeo, O. (1998). Testing hypotheses on recruitment, growth and mortality in exploited bivalves: An experimental perspective. *Can. J. Fish. Aquat. Sci.* 125S:257–264.

Defeo, O. (2003). Marine invertebrate fisheries in sandy beaches: An overview. *J. Coast. Res. S. I.* 35:56–65.

Defeo, O., and de Alava, A. (1995). Effects of human activities on long-term trends in sandy beach populations: The wedge clam *Donax hanleyanus* in Uruguay. *Mar. Ecol. Prog. Ser.* 123:73–82.

Defeo, O., and Cardoso, R. (2002). Macroecology of population dynamics and life history traits of the mole crab *Emerita brasiliensis* in Atlantic sandy beaches of South America. *Mar. Ecol. Prog. Ser.* 239:169–179.

Defeo, O., and Cardoso, R. (2004). Latitudinal patterns in abundance and life history traits of the mole crab *Emerita brasiliensis* on South American sandy beaches. *Divers. Distrib.* 10:89–98.

Defeo, O., and Gomez, J. (2005). Morphodynamics and habitat safety in sandy beaches: Life history adaptations in a supralittoral amphipod. *Mar. Ecol. Prog. Ser.* 293:143–153.

Defeo, O., and Lercari, D. (2004). Testing taxonomic resolution levels for ecological monitoring in sandy beach macrobenthic communities. *Aquatic Conserv: Mar. Freshw. Ecosyst.* 14:65–74.

Defeo, O., and McLachlan, A. (2005). Patterns, processes and regulatory mechanisms in sandy beach macrofauna: A multi-scale analysis. *Mar. Ecol. Prog. Ser.* 295:1–20.

Defeo, O., and Rueda, M. (2002). Spatial structure, sampling design and abundance estimates in sandy beach macrofauna: Some warnings and new perspectives. *Mar. Biol.* 140:1215–1225.

Defeo, O., Brazeiro, A., de Alava, A., and Riestra, G. (1997). Is sandy beach macrofauna only physically controlled? Role of substrate and competition in isopods. *Estuar. Coast. Shelf Sci.* 45:453–462.

Defeo, O., Gomez, J., and Lercai, D. (2001). Testing the swash exclusion hypothesis in sandy beach populations: The mole crab *Emerita brasiliensis* in Uruguay. *Mar. Ecol. Prog. Ser.* 212:159–170.

Defeo, O., Lercai, D., and Gomez, J. (2003). The role of morphodynamics in structuring sandy beach populations and communities: What should be expected? *J. Coast. Res.* 35S:352–362.

Degraer, S., Mouton, I., de Neve, L., and Vincx, M. (1999). Community structure and intertidal zonation of the macrobenthos on a macrotidal, ultra-dissipative sandy beach: Summer-winter comparison. *Estuaries* 22:742–752.

Degraer, S., Volckaert, A., and Vincx, M. (2003). Macrobenthic zonation patterns along a morphodynamic continuum of macrotidal, low bar/rip and ultradissipative sandy beaches. *Estuar. Coast. Shelf Sci.* 56:459–468.

Deidun, A., Azzopardi, M., Saliba, S., and Schembri, P. J. (2003). Low faunal diversity on Maltese sandy beaches: Fact or artefact? *Estuar. Coast. Shelf Sci.* 56:1–10.

De la Huz, R., Lastra, M., and Lopez, J. (2002). The influence of sediment grain size on burrowing, growth and metabolism of *Donax trunculus* L. (Bivalvia: Donacidae). *J. Sea Res.* 47:85–95.

De la Huz, R., Lastra, M., Junoy, J., Castellanos, C., and Vieitez, J. M. (2005). Biological impacts of oil pollution and cleaning in the intertidal zone of exposed sandy beaches: Preliminary study of the "Prestige" oil spill. *Estuar. Coast. Shelf Sci.* 65:19–29.

De Lancey, L. B. (1987). The summer zooplankton of the surf zone at Folly Beach, South Carolina. *J. Coast. Res.* 3:211–217.

Denny, M. W. (1988). *Biology and the mechanics of the wave swept environment*. New Jersey: Princeton University Press.

De Patra, K. D., and Levin, L. A. (1989). Evidence of the passive deposition of meiofauna into fiddler crab burrows. *J. Exp. Mar. Biol. Ecol.* 125:173–192.

De Ruyck, A. M. C., McLachlan, A., and Donn, T. E. (1991). The activity of three intertidal sand beach isopods (Flabellifera: Cirolanidae). *J. Exp. Mar. Biol. Ecol.* 146:163–180.

De Villiers, G. (1974). Growth, population dynamics, a mass mortality and arrangement of white sand mussels, *Donax serra* Roding, on beaches in the south-western Cape Province. *Investl. Rep. Div. Sea Fish. S. Afr.* 109:1–31.

Dexter, D. M. (1983). Community structure of intertidal sandy beaches in New South Wales, Australia. In: *Sandy beaches on ecosystems*, A. McLachlan and T. Erasmus (eds.), pp 461–472, The Hague: Dr W. Junk.

Dillery, D. C., and Knapp, L. V. (1969). Longshore movements of the sand crab *Emerita analoga* (Decapoda: Hippidae). *Crustaceana* 18:233–240.

Doing, H. (1985). Coastal foredune zonation and succession in various parts of the world. *Vegetatio* 61:65–75.

Donn, T. E. (1987). Longshore distribution of *Donax serra* in two log spiral bays in the Eastern Cape, South Africa. *Mar. Ecol. Prog. Ser.* 35:217–222.

Donn, T. E. (1990a). Zonation patterns of *Donax serra* Roding (Bivalvia: Donacidae) in southern Africa. *J. Coast. Res.* 6:903–911.

Donn, T. E. (1990b). Morphometics of *Donax serra* Roding (Bivalvia: Donacidae) populations with contrasting zonation. *J. Coast. Res.* 6:893–901.

Donn, T. E., and Els, S. F. (1990). Burrowing times of *Donax serra* from the south and west coasts of South Africa. *Veliger* 33:355–358.

Donn, T. E., Clark, D. J., McLachlan, A., and du Toit, P. (1986). Distribution and abundance of *Donax serra* Roding (Bivalvia: Donacidae) as related to beach morphology. I. Semilunar migrations. *J. Exp. Mar. Biol. Ecol.* 102:121–131.

Dugan, J. E., and Hubbard, D. M. (1996). Local variation in populations of the sand crab *Emerita analoga* on sandy beaches in southern California. *Rev. Chilena Hist. Nat.* 69:579–588.

Dugan, J. E., and McLachlan, A. (1999). An assessment of longshore movement in *Donax serra* Roding (Bivalvia: Donacidae) on an exposed sandy beach. *J. Exp. Mar. Biol. Ecol.* 234:111–124.

Dugan, J. E., and Hubbard, D. M. (2006). Ecological responses to coastal armoring on exposed sandy beaches. *Shore and Beach* (in press).

Dugan, J. E., Hubbard, D. M., and Wenner, A. M. (1994). Geographic variation in life history of the sand crab, *Emerita analoga* (Stimpson) on the California coast: Relationships to environmental variables. *J. Exp. Mar. Biol. Ecol.* 181:255–278.

Dugan, J. E., Hubbard, D. M., and Page, H. M. (1995). Scaling population density to body size: Tests in two soft-sediment intertidal communities. *J. Coast. Res.* 11:849–857.

Dugan, J. E., Hubbard, D. M., and Lastra, M. (2000). Burrowing abilities and swash behavior of three crabs, *Emerita analoga* Stimpson, *Blepharipoda occidentalis* Randall, and *Lepidopa californica* Efford (Anomura, Hippoidea), of exposed sandy beaches. *J. Exp. Mar. Biol. Ecol.* 255:229–245.

Dugan, J. E., Hubbard, D. M., McCrary, M. D., and Pierson, M. O. (2003). The response of macrofauna communities and shorebirds to macrophyte wrack subsidies on exposed sandy beaches of southern California. *Estuar. Coast. Shelf Sci.* 58S:25–40.

Dugan, J. E., Jaramillo, E., and Hubbard, D. M. (2004). Competitive interactions in macrofaunal animals of exposed sandy beaches. *Oecologia* 139:630–640.

Duncan, J. R. (1964). The effects of water table and tide cycle on swash-backwash sediment distribution and beach profile development. *Mar. Geol.* 2:186–197.

Dunnet, G. M. (1982). Oil pollution and seabird populations. *Phil. Trans. Roy. Soc. Lond.* 297B:413–427.

Du Preez, D., and Campbell, E. E. (1996a). Cell coatings of surf diatoms. *Rev. Chilena Hist. Nat.* 69:539–544.

Du Preez, D., and Campbell, E. E. (1996b). The photophysiology of surf diatoms: A review. *Rev. Chilena Hist. Nat.* 69:545–551.

Du Preez, H. H. (1984). Molluscan predation by *Ovalipes punctatus* (de Haan) (Crustacea: Brachyura: Portunidae). *J. Exp. Mar. Biol. Ecol.* 84:55–71.

Du Preez, H. H., McLachlan, A., Marais, J. F. K., and Cockcroft, A. C. (1990). Bioenergetics of fishes in a high-energy surf-zone. *Mar. Biol.* 106:1–12.

Dye, A. H. (1981). A study of benthic oxygen consumption on exposed sandy beaches. *Estuar. Coast. Shelf Sci.* 13:671–680.

Dyer, K. R. (1986). *Coastal and estuarine sediment dynamics.* New York: Wiley.

Ehlinger, G. S., and Tankersley, R. A. (2003). Larval hatching in the horseshoe crab, *Limulus polyphemus*: Facilitation by environmental cues. *J. Exp. Mar. Biol. Ecol.* 292:199–212.

Eleftheriou, A., and McIntyre, A. D. (1976). The intertidal fauna of sandy beaches: A survey of the Scottish coast. *Scottish Fish. Res. Rep.* 6:1–61.

Ellers, O. (1995a). Behavioral control of swash-riding in the clam *Donax variabilis*. *Biol. Bull.* 189:120–127.

Ellers, O. (1995b). Discrimination among wave-generated sounds by a swash-riding clam. *Biol. Bull.* 189:128–137.

Ellers, O. (1995c). Form and motion *of Donax variabilis* in flow. *Biol. Bull.* 189:138–147.

Emery, K. O., and Foster, J. F. (1948). Water tables in marine beaches. *J. Mar. Res.* 7:644–654.

Enright, J. T. (1972). A virtuoso isopod: Circalunar ryhthms and their tidal fine structure. *J. Comp. Physiol.* 77:41–162.

Ercolini, A., and Scapini, F. (1974). Sun compass and shore slope in the orientation of littoral amphipods (*Talitrus saltator* Montagu). *Monit. Zool. Ital. (NS)* 8:85–115.

Ercolini, A., Pardi, L., and Scapini, F. (1983). An optical directional factor in the sky might improve the direction finding of sandhoppers on the seashore. *Monit. Zool. Ital. (NS)* 17:313–317.

FAO (2002). *World agriculture: Towards 2015/2030, an FAO perspective.* UN Food Agriculture Organization, Rome.

Fairweather, P. G. (1990). Ecological changes due to our use of the coast: Research needs versus effort. *Proc. Ecol. Soc. Aust.* 16:71–77.

Fanini, L., Cantarino, C. L., and Scapini, F. (2005). Relationship between the dynamics of two *Talitrus saltator* populations and the impacts of activities linked to tourism. *Oceanologia* 47:93–112.

Fegley, S. R. (1988). A comparison of meiofaunal settlement onto the sediment surface and recolonisation of defaunated sandy sediment. *J. Exp. Mar. Biol. Ecol.* 123:97–113.

Fenchel, T. (1978). The ecology of micro- and meiobenthos. *Ann. Rev. Ecol. Syst.* 9:99–121.

Fenchel, J., and Staarup, B. J. (1971). Vertical distribution of photosynthetic pigments and the penetration of light in marine sediments. *Oikos* 22:172–182.

Fenchel, T. M., and Riedl, R. J. (1970). The sulfide system: A new biotic community underneath the oxidized layer of marine sand bottoms. *Mar. Biol.* 7:255–268.

Fincham, A. A. (1971). Ecology and population studies of some intertidal and sublittoral sand-dwelling amphipods. *J. Mar. Biol. Ass. U.K.* 51:471–488.

Fiori, S., Vidal-Martinez, V., Sima-Varez, R., Rodriguez-Canul, R., Aguirre-Macedo, M. L., and Defeo, O. (2004). Field and laboratory observations of the mass mortality of the yellow clam

Mesodesma mactroides in South America: The case of Isla del Jabali, Argentina. *J. Shellfish Res.* 23:451–455.

Fishelson, L. (1983). Population ecology and biology of *Dotilla sulcata* (Crustacea, Ocypodidae) on sandy beaches in the Red Sea. In: *Sandy beaches as ecosystems*, A. McLachlan and T. Erasmus (eds.), pp 643–654, The Hague: Dr W. Junk.

Folk, R. L. (1974). *Petrology of sedimentary rocks*. Texas: Hemphills.

Forward, R. B. (1986). Behavioural responses of a sand beach amphipod to light and pressure. *J. Exp. Mar. Biol. Ecol.* 98:65–113.

Freestone, A. L., and Nordstrom, K. F. (2001). Early development of vegetation in restored dune plant microhabitats on a nourished beach at Ocean City, New Jersey. *J. Coast. Conserv.* 7:105–116.

Frost, M. T., Attrill, M. J., Rowden, A. A., and Foggo, A. (2004). Abundance: occupancy relationships in macrofauna on exposed sandy beaches (patterns and mechanisms). *Ecography* 27:643–649.

Gauci, M. J., Deidun, A., and Schembri, P. J. (2005). Faunistic diversity of Maltese pocket sandy and shingle beaches: Are these of conservation value? *Oceanologia* 47:219–241.

Gamble, F. W., and Keeble, F. (1904). The bionomics of *Convoluta roscoffensis* with special reference to its green cells. *Quart. J. Microsc. Sci.* 47:44–57.

Gee, J. M., Warwick, R. M., Schaaning, M., Berge, J. A., and Ambrose, W. G. (1985). Effects of organic enrichment on meiofaunal abundance and community structure in sublittoral soft sediments. *J. Exp. Mar. Biol. Ecol.* 91:247–262.

Gerlach, S. A. (1954). Das Sublittoral der sandigen Meereskusten als Lebensraum einer Mikrofauna. *Kieler Meeresforsch.* 10:121–129.

Gheskiere, T., Hoste, E., Kotwicki, L., Degraer, S., Vanaverbeke, J., and Vincx, M. (2002). The sandy beach meiofauna and free-living nematodes from De Panne (Belgium). *Bull. Roy. Belgian Inst. Nat. Sci. Biol. Suppl.* 72:53–57.

Gheskiere, T., Hoste, E., Vanaverbeke, J., Vincx, M., and Degraer, S. (2004). Horizontal zonation patterns and feeding structure of marine nematode assemblages on a macrotidal, ultra-dissipative sandy beach (De Panne, Belgium). *J. Sea Res.* 52:211–226.

Gheskiere, T., Vincx, M., Urban-Malinga, B., Rossano, C., Scapini, F., and Degraer, S. (2005a). Nematodes from wave-dominated sandy beaches: Diversity, zonation patterns and testing of the isocommunities concept. *Estuar. Coast. Shelf Sci.* 62:365–375.

Gheskiere, T., Vincx, M., Weslawski, J. M., Scapini, F., and Degraer, S. (2005b). Meiofauna as descriptor of tourism induced changes at sandy beaches. *Mar. Environ. Res.* 60:245–265.

Gianuca, N. M. (1983). A preliminary account of the ecology of sandy beaches in Southern Brazil. In: *Sandy beaches as ecosytems*, A. McLachlan and T. Erasmus (eds.), pp 413–419, The Hague: Dr W. Junk.

Gibbs, R. J., Matthews, M. D., and Link, P. A. (1971). The relationship between sphere size and settling velocity. *J. Sed. Pet.* 41:7–18.

Gibson, D. J., Ely, J. S., and Looney, P. B. (1997). A Markovian approach to modeling succession on a coastal barrier island following beach nourishment. *J. Coast. Res.* 13:831–841.

Giere, O. (1993). *Meiobenthology: The microscopic fauna in aquatic sediments*. Berlin: Springer Verlag.

Gimenez, L., and Yannicelli, B. (2000). Longshore patterns of distribution of macroinfauna on a Uruguayan sandy beach: An analysis at different spatial scales and their potential causes. *Mar. Ecol. Prog. Ser.* 199:111–125.

Godfrey, R. J., and Godfrey, M. M. (1981). Ecological effects of off-road vehicles on Cape Cod. *Oceanus* 23:56–67.

Gomez, J., and Defeo, O. (1999). Life history of the sandhopper *Pseudorchestia brasiliensis* (Amphipoda) in sandy beaches with contrasting morphodynamics. *Mar. Ecol. Prog. Ser.* 182:209–220.

Gray, J. S. (1967). Substrate selection by the archiannelid *Protodrilus hypoleucus* Armenante. *J. Exp. Mar. Biol. Ecol.* 1:47–54.

Gray, J. S., and Pearson, T. H. (1982). Objective selection of sensitive species indicative of pollution-induced changes in benthic communities. I. Comparative morphology. *Mar. Ecol. Prog. Ser.* 9:111–119.

Greene, K. (2002). *Beach nourishment: A review of the biological and physical impacts*. Habitat Management Series No. 7, Atlantic States Marine Fisheries Commission, Washington DC.

Griffiths, C. L., Stenton-Dozey, J. M. E., and Koop, K. (1983). Kelp wrack and energy flow through a sandy beach. In: *Sandy beaches as ecosytems*, A. McLachlan and T. Erasmus (eds.), pp 547–556, The Hague: Dr W. Junk.

Gunter, G. (1979). Notes on sea beach ecology: Food sources on sandy beaches and localized diatom blooms bordering Gulf beaches. *Gulf Res. Rep.* 6:305–307.

Hacking, N. (1997). Sandy beach macrofauna of eastern Australia: A geographical comparison. Ph.D. thesis, University of New England, Australia.

Hager, R. P., and Croker, R. A. (1980). The sand burrowing amphipod *Amphiporeia virginiana* Schoemaker 1933 in the tidal plankton. *Can. J. Zool.* 58:860–864.

Hamm, L., Capobianco, M., Dette, H. H., Lechuga, A., Spanhoff, R., and Stive, M. J. F. (2002). A summary of European experience with shore nourishment. *Coast. Eng.* 47:237–264.

Hamner, W. M., Smyth, M., and Mulford, E. D. (1968). Orientation of the sand beach isopod *Tylos punctatus*. *Anim. Behav.* 16:405–409.

Haque, A. M., Szymelfenig, M., and Weslawski, J. M. (1996). Sandy littoral zoobenthos of the Polish Baltic coast. *Oceanologia* 38:361–378.

Harden-Jones, F. R. (1968). *Fish migrations*. London: Edward Arnold.

Harris, R. P. (1972). Horizontal and vertical distribution of the interstitial harpacticoid copepods of a sandy beach. *J. Mar. Biol. Ass. U.K.* 52:375–387.

Harris, S. A., da Silva, F. M., Bolton, J. J., and Brown, A. C. (1986). Algal gardens and herbivory in a scavenging sandy-beach nassariid whelk. *Malacologia* 27:299–305.

Hayes, W. B. (1974). Sand beach energetics: Importance of the isopod *Tylos punctatus*. *Ecology* 55:838–847.

Hayes, W. B. (1977). Factors affecting the distribution of *Tylos punctatus* (Isopoda, Oniscoidea) on beaches in southern California and northern Mexico. *Pacific Sci.* 31:165–187.

Heip, C., Vincx, M., and Vranken, G. (1985). The ecology of marine nematodes. *Oceanogr. Mar. Biol. Ann. Rev.* 23:399–489.

Herrnkind, W. F. (1972). Orientation in shore living arthropods, especially the sand fiddler crab. In: *The behaviour of marine animals*, H. E. Winn and B. L. Olla (eds.), New York: Plenum Press.

Hesp, P. (1991). Ecological processes and plant adaptations on coastal dunes. *J. Arid. Environ.* 21:165–191.

Heymanns, J. J., and McLachlan, A. (1996). Carbon budget and network analysis of a high-energy beach/surf-zone ecosystem. *Estuar. Coast. Shelf Sci.* 43:485–505.

Higgins, R. P., and Thiel, H. (eds.) (1988). *Introduction to the study of meiofauna*. Washington, DC: Smithsonian Institution Press.

Hockey, P. A. R., Siegfried, W. R., Crowe, A. A., and Cooper, J. (1983). Ecological structure and energy requirements of sandy beach avifauna in southern Africa. In: *Sandy beaches as ecosystems*, A. McLachlan and T. Erasmus (eds.), pp 507–521, The Hague: Dr W. Junk.

Hockin, D. C. (1982). The effects of sediment particle diameter upon the meiobenthic copepod community of an intertidal beach: A field and laboratory experiment. *J. Anim. Ecol.* 51:555–572.

Hodgson, A. N. (1982). Studies on wound healing and regeneration of the siphons of the bivalve *Donax serra* (Roding). *Trans. Roy. Soc. S. Afr.* 44:489–498.

Hodgson, A. N., and Brown, A. C. (1985). Contact chemoreception by the propodium of the sandy beach whelk *Bullia digitalis* (Gastropoda: Nassariidae). *Comp. Biochem. Physiol.* 82A:425–427.

Hodgson, A. N., and Trueman, E. R. (1985). The use of the foot in burrowing by the whelk *Bullia*. *J. Mollusc. Stud.* 51:101.

Hoeseler, V. (1988). Entstchung und Neutiger Zustand der jingen Duneninseln Memmert und Mellum sowie Forschungsprogram zur Besiedlung durch lInsekten und andere Glieder Flussen. *Drosera* 1/2:5–46.

Horrocks, J. A., and Scott, N. M. (1991). Nest site location and nest success in the hawksbill turtle *Eretmochelys imbricata* in Barbados, West Indies. *Mar. Ecol. Prog. Ser.* 69:1–8.

Hosier, P. E., Kochar, M., and Thayer, V. (1981). Off-road vehicle and pedestrian track effects on the sea-approach of hatchling loggerhead turtles. *Environ. Conserv.* 8:158–161.

Hubbard, D. M., and Dugan, J. E. (2003). Shorebird use of an exposed sandy beach in southern California. *Estuar. Coast. Shelf Sci.* 58S:41–54.

Hummel, H. (2003). Geographical patterns of dominant bivalves and a polychaete in Europe: No metapopulations in the marine coastal zone? *Helgoland. Mar. Res.* 56:247–251.

Huttel, M. (1986). Active aggregation and downshore migration in the trochid snail *Umbonium vestiarium* on a tropical sand flat. *Ophelia* 26:221–232.

Inoue, T., Suda, Y., and Sano, M. (2005). Food habits of fishes in the surf zone of a sandy beach at Sanrimatsubara, Fukuoka Prefecture, Japan. *Ichthyol. Res.* 52:9–14.

James, R. J. (1999). Cusps and pipis on a sandy ocean beach in New South Wales. *Aust. J. Ecol.* 24:587–592.

James, R. J., and Fairweather, P. G. (1996). Spatial variation of intertidal macrofauna on a sandy ocean beach in Australia. *Estuar. Coast. Shelf Sci.* 43:81–107.

Janssen, G., and Mulder, S. (2005). Zonation of macrofauna across sandy beaches and surf zones along the Dutch coast. *Oceanologia* 47:265–282.

Jaramillo, E., and McLachlan, A. (1993). Community and population responses of the macroinfauna to physical factors over a range of exposed sandy beaches in south-central Chile. *Estuar. Coast. Shelf Sci.* 37:615–624.

Jaramillo, E., McLachlan, A., and Coetzee, P. (1993). Intertidal zonation patterns of macroinfauna over a range of exposed sandy beaches in south-central Chile. *Mar. Ecol. Prog. Ser.* 101:105–118.

Jaramillo, E., McLachlan, A., and Dugan, J. E. (1995). Total sample area and estimates of species richness in exposed sandy beaches. *Mar. Ecol. Prog. Ser.* 119:311–314.

Jaramillo, E., Contreras, H., and Quinn, P. (1996). Macroinfauna and human disturbance in a sandy beach of south-central Chile. *Rev. Chilena Hist. Nat.* 69:655–663.

Jaramillo, E., Dugan, J., and Contreras, H. (2000). Abundance, population structure, tidal movement and burrowing rate of *Emerita analoga* (Stimpson 1857) (Anomura: Hippidae) at a dissipative and a reflective sandy beach in south central Chile. *PZNI Mar. Ecol.* 21:113–127.

Jaramillo, E., Avellanal, M. H., Gonzalez, M., and Kennedy, F. (2000). Locomotor activity of *Phalerisidia maculata* Kulzer (Coleoptera, Tenebrionidae) on Chilean sandy beaches. *Rev. Chilena Hist. Nat.* 73:67–77.

Jaramillo, E., Contreras, H., Duarte, C., and Avellanal, M. H. (2003). Locomotor activity and zonation of upper shore arthropods in a sandy beach of north central Chile. *Estuar. Coast. Shelf Sci.* 58S:177–197.

Jedrzejczak, M. F. (2002). Stranded *Zostera marina* L. vs wrack fauna community interactions on a Baltic sandy beach (Hel, Poland): A short-term pilot study. Part II. Driftline effects of succession changes and colonisation of beach fauna. *Oceanologia* 44:367–387.

Jedrzejczak, M. F. (2003). Predrying of stranded wrack material as an aspect of the litterbag techniques in the sandy beach studies. *Oceanolog. Hydrobiol. Stud.* 32:59–74.

Johannes, R. E. (1980). The ecological significance of the submarine discharge of groundwater. *Mar. Ecol. Prog. Ser.* 3:365–373.

Jones, D. A. (1972). Aspects of the ecology and behaviour of *Ocypode ceratophthalmus* (Pallas) and *O. kuhlii* de Haan (Crustacea: Ocypodidae). *J. Exp. Mar. Biol. Ecol.* 8:31–43.

Jones, D. A., and Hobbins, C. S. C. (1985). The role of biological rhythms in some sand beach cirolanid isopods. *J. Exp. Mar. Biol. Ecol.* 93:67–69.

Judd, F. W., Lonard, R. I., Everitt, J. H., and Villareal, R. (1989). Effects of vehicular traffic in the secondary dunes and vegetated flats of South Padre Island, Texas. In: *Coastal Zone 89*, 4634–4645. American Society of Civil Engineers, New York.

Karavas, N., Georghiou, K., Arianoutsou, M., and Dimopoulos, D. (2005). Vegetation and sand characteristics influencing nesting activity of *Caretta caretta* on Sekania beach. *Biol. Conserv.* 121:177–188.

Kennedy, F., Naylor, E., and Jaramillo, E. (2000). Ontogenetic differences in the circadian locomotor activity rhythm of the talitrid amphipod crustacean *Orchestoidea tuberculata. Mar. Biol.* 137:511–517.

Kensley, B. (1974). Aspects of the biology and ecology of the genus *Tylos* Latreille. *Ann. S. Afr. Mus.* 65:401–471.

Kikulawa, A., Kamezaki, N., and Hidetoshi, O. (1999). Factors affecting nesting beach selection by loggerhead turtles (*Caretta caretta*): A multiple regression approach. *J. Zool. Lond.* 249:447–454.

Klapow, L. A. (1972). Fortnightly molting and reproduction cycles in the sand beach isopod *Excirolana chiltoni. Biol. Bull.* 143:568–591.

Koch, R. (1983). Molecular connectivity index for assessing ecotoxicological behavior of organic compounds. *Toxicol. Env. Chem.* 6:87–96.

Komar, P. D. (1998). *Beach processes and sedimentation* (2d ed.). New Jersey: Prentice-Hall.

Kooijman, A. M., and de Haan, M. W. A. (1995). Grazing as a measure against grass encroachment in Dutch dry dune grassland: Effects on vegetation and soil. *J. Coast. Conserv.* 1:127–134.

Koop, K., and Griffiths, C. L. (1982). The relative significance of bacteria, meio- and macrofauna on an exposed sandy beach. *Mar. Biol.* 66:295–300.

Koop, K., and Lucas, M. I. (1983). Carbon flow and nutrient regeneration from the decomposition of macrophyte debris in a sandy beach microcosm. In: *Sandy beaches as ecosystems*, A. McLachlan and T. Erasmus (eds.), pp 249–262, The Hague: Dr W. Junk.

Koop, K., Newell, R. C., and Lucas, M. I. (1982). Biodegradation and carbon flow based on kelp debris (*Ecklonia maxima*) in a sandy beach microcosm. *Mar. Ecol. Prog. Ser.* 7:315–326.

Kotwicki, L., Szymelfenig, M., de Troch, M., Urban-Malinga, B., and Weslawski, J. M. (2005). Latitudinal biodiversity patterns of meiofauna from sandy littoral beaches. *Biodiv. Conserv.* 14:461–474.

Kyle, R., Robertson, W. D., and Binnie, S. L. (1997). Subsistence shellfish harvesting in Maputaland Reserve in northern Kwa-Zulu-Natal, South Africa: Sandy beach organisms. *Biol. Conserv.* 82:173–182.

Lasiak, T. A. (1981). Nursery grounds of juvenile teleosts: Evidence from the surf zone of Kings Beach, Port Elizabeth. *S. Afr. J. Zool.* 77:388–390.

Lasiak, T. A. (1983). Recruitment and growth patterns of juvenile marine teleosts caught at Kings Beach, Algoa Bay. *S. Afr. J. Zool.* 18:25–30.

Lasiak, T. A. (1984). Structural aspects of the surf zone fish assemblages at Kings Beach, Algoa Bay, South Africa: Short-term fluctuations. *Estuar. Coast. Shelf Sci.* 18:347–360.

Lasiak, T. A., and McLachlan, A. (1987). Opportunistc utilization of mysid shoals by surf-zone teleosts. *Mar. Ecol. Prog. Ser.* 37:1–7.

Lastra, M., and McLachlan, A. (1996). Spatial and temporal variations in recruitment of *Donax serra* Roding (Bivalvia: Donacidae) on an exposed sandy beach of South Africa. *Rev. Chilena Hist. Nat.* 69:631–639.

Lastra, M., Dugan, J. E., and Hubbard, D. M. (2002). Burrowing and swash behavior of the Pacific mole crab *Hippa pacifica* (Anomura, Hippidae) in tropical sandy beaches. *J. Crustac. Biol.* 22:53–58.

Laudien J., Flint, N. S., van der Bank, F. H., and Brey, T. (2003). Genetic and morphological variation in populations of the surf clam *Donax serra* (Roding) from southern African sandy beaches. *Biochem. Syst. Ecol.* 31:751–772.

Layman, C. A. (2000). Fish assemblage structure of the shallow ocean surf-zone on the eastern shore of Virginia barrier islands. *Estuar. Coast. Shelf Sci.* 51:201–213.

Leber, K. M. (1982). Seasonality of macroinvertebrates on a temperate, high wave energy sandy beach. *Bull. Mar. Sci.* 32:86–98.

Lenanton, R. C. J., Robertson, A. I., and Hansen, J. A. (1982). Nearshore accumulations of detached macrophytes as nursery areas for fish. *Mar. Ecol. Prog. Ser.* 9:51–57.

Lercari, D., and Defeo, O. (1999). Effects of freshwater discharge in sandy beach populations: The mole crab *Emerita brasiliensis* in Uruguay. *Estuar. Coast. Shelf Sci.* 49:457–468.

Lercari, D., and Defeo, O. (2003). Variation of a sandy beach macrobenthic community along a human-induced environmental gradient. *Estuar. Coast. Shelf Sci.* 58S:17–24.

Lewin, J., and Schaeffer, L. T. (1983). The role of phytoplankton in surf ecosystems. In: *Sandy beaches as ecosystems*, A. McLachlan and T. Erasmus (eds.), pp 381–389, The Hague: Dr W. Junk.

Malan, D. E., and McLachlan, A. (1991). *In situ* benthic oxygen fluxes in a nearshore coastal marine system: A new approach to quantify the effect of wave action. *Mar. Ecol. Prog. Ser.* 73:69–81.

Manning, M. L., and Lindquist, N. (2003). Helpful habitant or pernicious passenger: Interactions between an infaunal bivalve, an epifaunal hydroid and three potential predators. *Oecologia* 134:415–422.

Maros, A., Louveaux, A., Godfrey, M. H., and Girondt, M. (2003). *Scapteriscus didactylus* (Orthoptera, Gryllotalpidae), predator of leatherback turtle eggs in French Guiana. *Mar. Ecol. Prog. Ser.* 249:289–296.

Marques, J. C., Goncalves, S. C., Pardal, M-A., Chelazzi, L., Fallaci, M., Bouslama, M. F., El Gtari, M., Charfi-Cheikhrouba, F., and Scapini, F. (2003). Biology, population dynamics and secondary production of the sandhopper *Talitrus saltator* (Montagu) (Amphipoda, Talitridae) at Lavos (Western part of Portugal), Collelungo (western coast of Italy) and Zouaraa (north-west coast of Tunisia): A comparative study of Atlantic and Mediterranean populations. *Estuar. Coast. Shelf Sci.* 58S:127–148.

Marsden, I. D. (1991). Kelp-sandhopper interactions on a sandy beach in New Zealand. I. Drift composition and distribution. *J. Exp. Mar. Biol. Ecol.* 152:61–74.

Marsh, B. A., and Branch, G. M. (1979). Circadian and circatidal rhythms of oxygen consumption in the sandy beach isopod *Tylos granulatus* (Krauss). *J. Exp. Mar. Biol. Ecol.* 37:77–89.

Martinez, M. L., and Psuty, N. P. (2004). *Coastal dunes: Ecology and conservation*. New York: Springer.

McArdle, S., and McLachlan, A. (1991). Dynamics of the swash zone and effluent line on sandy beachs. *Mar. Ecol. Prog. Ser.* 76:91–99.

McArdle, S., and McLachlan, A. (1992). Sand beach ecology: Swash features relevant to the macrofauna. *J. Coast. Res.* 8:398–407.

McDermott, J. J. (1983). Food web in the surf zone of an exposed sandy beach along the mid Atlantic coast of the United States. In: *Sandy beaches as ecosystems*, A. McLachlan and T. Erasmus (eds.), pp 529–538, The Hague: Dr W. Junk.

McFarland, W. N. (1963). Seasonal changes in the numbers and biomass of fishes from the surf at Mustang Island, Texas. *Publ. Inst. Mar. Sci. Texas* 9:91–105.

McGwynne, L. E. (1991). The microbial loop: Its role in a diatom-enriched surf zone. Ph.D. thesis, University of Port Elizabeth, South Africa.

McGwyne, L., and McLachlan, A. (1992). *Ecology and management of sandy coasts*. University of Port Elizabeth, South Africa, Institute for Coastal Research, Report No. 30.

McLachlan, A. (1978). A quantitative analysis of the meiofauna and the chemistry of the redox potential discontinuity zone in a sheltered sandy beach. *Estuar. Coast. Mar. Sci.* 7:275–290.

McLachlan, A. (1979). Volumes of seawater filtered by Eastern Cape sandy beaches. *S. Afr. J. Sci.* 75:75–79.

McLachlan, A. (1980a). Intertidal zonation of macrofauna and stratification of meiofauna on high energy sandy beaches in the Eastern Cape, South Africa. *Trans. Roy. Soc. S. Afr.* 44:213–222.

McLachlan, A. (1980b). The definition of sandy beaches in relation to exposure: A simple rating system. *S. Afr. J. Sci.* 76:137–138.

McLachlan, A. (1982). A model for the estimation of water filtration and nutrient regeneration by exposed sandy beaches. *Mar. Environ. Res.* 6:37–47.

McLachlan, A. (1983). Sandy beach ecology: A review. In: *Sandy beaches as ecosystems*, A. McLachlan and T. Erasmus (eds.), pp 321–380, The Hague: Dr W. Junk.

McLachlan, A. (1989). Water filtration by dissipative beaches. *Limnol. Oceanogr.* 34:774–780.

Mclachlan, A. (1991). Ecology of coastal dune fauna. *J. Arid Environ.* 21:229–243.

McLachlan, A. (1985). The biomass of macro- and interstitial fauna on clean and wrack-covered beaches in Western Australia. *Estuar. Coast. Shelf Sci.* 21:587–599.

McLachlan, A. (1996). Physical factors in benthic ecology: Effects of changing sand particle size on beach fauna. *Mar. Ecol. Prog. Ser.* 131:205–217.

McLachlan, A. (1998). Interactions between two species of *Donax* on a high energy beach: An experimental approach. *J. Mollusc. Stud.* 64:492–495.

McLachlan, A. (2001). Coastal beach ecosystems. In: *Encyclopedia of biodiversity*, S. Levin (ed.), pp 741–751, New York: Academic Press.

McLachlan, A., and Bate, G. C. (1984). Carbon budget for a high energy surf zone. *Vie Milieu* 34:67–77.

McLachlan, A., and Dorvlo, A. (2005). Global patterns in sandy beach macrofauna communities. *J. Coast. Res.* 21:674–687.

McLachlan, A., and Dorvlo, A. (2006). Global patterns in sandy beach macrobenthic communities: Biological factors. *J. Coast. Res.* (in press).

McLachlan, A., and Erasmus, T. (eds.) (1983). *Sandy beaches as ecosystems*. The Hague: Dr W. Junk.

McLachlan, A., and McGwynne, L. E. (1986). Do sandy beaches accumulate nitrogen? *Mar. Ecol. Prog. Ser.* 34:191–195.

McLachlan, A., and Harty, B. (1981). Effects of oil on water filtration by exposed sandy beaches. *Mar. Pollut. Bull.* 12:374–378.

McLachlan, A., and Hesp, A. (1984). Faunal response to the morphology and water circulation of a sandy beach with cusps. *Mar. Ecol. Prog. Ser.* 19:133–144.

McLachlan, A., and Jaramillo, E. (1995). Zonation on sandy beaches. *Oceanogr. Mar. Biol. Ann. Rev.* 33:305–335.

McLachlan, A., and Romer, G. (1990). Trophic relations in a high energy beach and surf zone ecosystem. In: *Trophic relations in the marine environment*, M. Barnes and R. Gibson (eds.), pp 356–371, Aberdeen: University Press.

McLachlan, A., and Turner, J. (1994). The interstitial environment of sandy beaches. *PZNI Mar. Ecol.* 15:177–211.

McLachlan, A., and Young, N. (1982). Effects of low temperatures on the burrowing rates of four sandy beach molluscs. *J. Exp. Mar. Biol. Ecol.* 65:275–284.

McLachlan, A., Erasmus, T., and Furstenberg, J. P. (1977). Migrations of sandy beach meiofauna. *Zool. Afr.* 12:257–277.

McLachlan, A., Wooldridge, T., and van der Horst, G. (1979a). Tidal movements of the macrofauna on an exposed sandy beach in South Africa. *J. Zool. Lond.* 188:433–442.

McLachlan, A., Dye, A. H., and van der Ryst, P. (1979b). Vertical gradients in the fauna and oxidation of two exposed sandy beaches. *S. Afr. J. Zool.* 14:43–47.

McLachlan, A., Wooldridge, T., and Dye, A. H. (1981a). The ecology of sandy beaches in southern Africa. *S. Afr. J. Zool.* 16:219–231.

McLachlan, A., Erasmus, T., Dye, A. H., Wooldridge, T., van der Horst, G., Rossouw, G., Lasiak, T. A., and McGwynne, L. E. (1981b). Sand beach energetics: An ecosystem approach towards a high energy interface. *Estuar. Coast. Shelf Sci.* 13:11–25.

McLachlan, A., Dye, A. H., and Newton, B. (1981c). Simulation of the interstitial system of exposed sandy beaches. *Estuar. Coast. Shelf Sci.* 12:267–278.

McLachlan, A., Cockroft, A. C., and Malan, D. E. (1984). Benthic faunal response to a high energy gradient. *Mar. Ecol. Prog. Ser.* 16:51–53.

McLachlan, A., Eliot, I. G., and Clark, D. J. (1985). Water filtration through reflective microtidal beaches and shallow sublittoral sands and its implications for an inshore ecosystem in Western Australia. *Estuar. Coast. Shelf Sci.* 21:91–104.

McLachlan, A., de Ruyck, A., du Toit, P., and Cockcroft, A. (1992). Groundwater ecology at the dune-beach interface. In: *Proceedings of the first international conference on groundwater ecology*, J. A. Stanford and J. J. Simons (eds.), pp 209–216, American Water Resources Assn. Technical Publication Series TPS-92.2.

McLachlan, A., Jaramillo, E., Donn, T. E., and Wessels, F. (1993). Sandy beach macrofauna communities and their control by the physical environment: A geographical comparison. *J. Coast. Res.* 15S:27–38.

McLachlan, A., Jaramillo, E., Defeo, O., Dugan, J., de Ruyck, A., and Coetzee, P. (1995). Adaptations of bivalves to different beach types. *J. Exp. Mar. Biol. Ecol.* 187:147–160.

McLachlan, A., Dugan, J. E., Defeo, O., Ansell, A. D., Hubbard, D. M., Jaramillo, E., and Penchaszadeh, P. (1996a). Beach clam fisheries. *Oceanogr. Mar. Biol. Ann. Rev.* 34:163–232.

Mclachlan, A., Kerley, G., and Rickard, C. (1996b). Ecology and energetics of slacks in the Alexandria coastal dunefield. *Landscape Urban Plan.* 34:267–276.

McLusky, D. S., and Stirling, A. (1975). The oxygen consumption of *Donax incarnatus* and *D. spiculum* from tropical beaches. *Comp. Biochem. Physiol.* 51A:942–947.

Mendelssohn, I. A., Hester, M. W., Marteferrante, F. J., and Talbot, F. (1991). Experimental dune building and vegetative stabilization in a sand-deficient barrier island setting on the Louisiana coast. *J. Coast. Res.* 7:137–149.

Middaugh, D. P., Kohl, H. W., and Burnett, L. E. (1983). Concurrent measurements of intertidal environmental variables and embryo survival for the California grunion *Leuresthes tenuis*, and Atlantic silverside, *Menidia menidia* (Pisces: Atherinidae). *Calif. Fish. Game.* 69:89–96.

Modde, T., and Ross, S. T. (1981). Seasonality of fish occupying a surf zone habitat in the Gulf of Mexico. *Fish. Bull.* 78:911–922.

Moffet, M. D., McLachlan, A., Winter, P. E. D., and de Ruyck, A. M. C. (1996). Impact of trampling on beach fauna. *J. Coast. Conserv.* 4:87–90.

Moore, C. G. (1979). The zonation of psammolittoral harpacticoid copepods around the Isle of Man. *J. Mar. Biol. Ass. U.K.* 59:711–724.

Moore, S. L., Gregorio, D., and Carreon, M. (2001). Composition and distribution of beach debris in Orange County, California. *Mar. Pollut. Bull.* 42:241–245.

Morioka, S. A., Ohno, A., Kohno, H., and Taki, Y. (1993). Recruitment and larval survival of milkfish *Chanos chanos* larvae in the surf zone. *Japan. J. Ichthyol.* 40:247–260.

Mortimer, J. A. (1995). Factors influencing beach selection by nesting sea turtles. In: *Biology and conservation of sea turtles*, K. A. Bjorndal (ed.), pp 45–51, Washington, DC: Smithsonian Institution Press.

Munro, A. L. S., Wells, J. B. J., and McIntyre, A. D. (1978). Energy flow in the flora and meiofauna of sandy beaches. *Proc. Roy. Soc. Edinb.* 76B:297–315.

Myers, J. P., Ruiz, G. R., Walters, J. R., and Pitelka, F. A. (1982). Do shorebirds depress their prey? *Wader Study Group Bull.* 35:31.

Nardi, M., Morgan, E., and Scapini, F. (2003). Seasonal variation in the free-running period in two *Talitrus saltator* populations from Italian beaches differing in morphodynamics and human uses. *Estuar. Coast. Shelf Sci.* 58S:199–206.

National Research Council (1995). *Beach nourishment and protection.* Washington DC: National Academy Press.

Naylor, E. (1988). Clock-controlled behaviour in intertidal animals. In: *Behavioural adaptations to intertidal life.* G. Chelazzi and M. Vannini (eds.), pp 1–14, New York: Plenum Press.

Naylor, E., and Kennedy, F. (2003). Ontogeny of behavioural adaptations in beach crustaceans: Some temporal considerations for integrated coastal zone management and conservation. *Estuar. Coast. Shelf Sci.* 58S:169–175.

Naylor, E., and Rejeki, S. (1996). Tidal migrations and rhythmic behaviour of sand beach Crustacea. *Rev. Chilena Hist. Nat.* 69:475–484.

Neff, M. N., and Anderson, J. W. (1981). *Response of marine animals to petroleum and specific petroleum hydrocarbons.* Basking, UK: Applied Science Publishers.

Nel, P. (1995). The effect of sand particle size on sandy beach macrofauna. MSc thesis, University of Port Elizabeth, South Africa.

Nel, P. (2001). Physical and biological factors structuring sandy beach macrofauna communities. Ph.D. thesis, University of Cape Town, South Africa.

Nel, P., McLachlan, A., and Winter, P. E. D. (2001). The effect of grain size on the burrowing of two *Donax* species. *J. Exp. Mar. Biol. Ecol.* 265:219–238.

Newell, R. C. (1979). *Biology of intertidal animals* (3rd ed.). Kent: Marine Ecological Surveys.

Nicholas, W. L., and Trueman, J. W. H. (2005). Biodiversity of marine nematodes in Australian sandy beaches from tropical and temperate regions. *Biodiv. Conserv.* 14:823–839.

Nicholls, R. J., and Leatherman, S. P. (1996). Adapting to sea level rise: Relative sea-level trends to 2100 for the United States. *Coast. Mgmnt.* 24:301–324.

Nordstrom, K. F. (1994). Developed coasts. In: *Coastal evolution*, R. W. G. Carter and C. Woodroffe (eds.), pp 477–509, New York Cambridge University Press.

Nordstrom, K. F. (2000). *Beaches and dunes on developed coasts.* UK Cambridge: Cambridge University Press.

Nordstrom, K. F., and Mauriello, M. N. (2001). Restoring and maintaining naturally-functioning landforms and biota on intensively developed barrier islands under a no-retreat scenario. *Shore and Beach* 69:19–28.

Nordstrom, K., Psuty, N., and Carter, B. (1990). The study of coastal dunes. In: *Coastal dunes: Form and process*. K. Nordstrom, N. Psuty, and B. Carter (eds.), pp 1–14, New York: John Wiley & Sons.

Nordstrom, K., Jackson, N. L., Bruno, M. S., and de Batts, H. A. (2002). Municipal initiative for managing dunes in coastal residential areas: A case study of Avalon, New Jersey, USA. *Geomorphology* 47:137–152.

Noy-Meir, I. (1979). Structure and function of desert ecosystems. *Israel J. Bot.* 28:1–19.

Odendaal, F. J., Turchin, P., Hoy, G., Wickens, P., Wells, J., and Schroeder, G. (1992). *Bullia digitalis* (Gastropoda) actively pursues moving prey while swash riding. *J. Zool., Lond.* 228:103–113.

Pardi, L. (1954). Über die Orientierung von *Tylos latreilli* (Aud. & sav.) (Isopoda terrestria). *Z. Tier-psychol.* 11:175–181.

Pardi, L. (1955). L'orientamento diurno di *Tylos latreilli* (Aud. & Sav.). *Boll. Inst. Mus. Univ. Torino* 4:167–196.

Pardi, L., and Papi, F. (1952). Die sonne als Kompass bei *Talitrus saltator* (Montagu) (Amphipoda, Talitridae). *Naturniss.* 39:262–263.

Pardi, L., and Scapini, F. (1983). Inheritance of solar direction finding in sandhoppers; mass-crossing experiments. *J. Comp. Physiol.* 151:435–440.

Patin, S. A. (1982). *Pollution and the biological resources of the oceans*. London: Butterworths.

Pauly, D. (1997). Small scale fisheries in the tropics: Marginality, marginalization, and some implications for fisheries management. In: *Global trends in fisheries management*, E. K. Pitch, P. D. Huppert, and M. D. Sisseuweiwe (eds.), pp 40–49, American Fisheries Society Symposium 20, Bethesda, Maryland.

Pearse, A. S., Humm, H. J., and Wharton, G. W. (1942). Ecology of sandy beaches at Beaufort, North Carolina. *Ecol. Monogr.* 12:135–190.

Pearson, T. H., and Rosenberg, R. (1978). Macrobenthic succession in relation to organic enrichment and the marine environment. *Oceanogr. Mar. Biol. Ann. Rev.* 16:229–311.

Pessanha, A. L. M., and Araujo, F. G. (2003). Spatial, temporal and diel variations of fish assemblages at two sandy beaches in Sepetiba Bay, Rio de Janeiro, Brazil. *Estuar. Coast. Shelf Sci.* 57:817–828.

Peterson, C. H. (1991). Intertidal zonation of marine invertebrates in sand and mud. *Am. Sci.* 79:236–249.

Peterson, C. H., and Bishop, M. J. (2005). Assessing the environmental impacts of beach nourishment. *BioScience* 55:887–895.

Peterson, C. H., and Lipcius, R. N. (2003). Conceptual progress towards predicting quantitative ecosystem benefits of ecological restorations. *Mar. Ecol. Prog. Ser.* 264:297–307.

Peterson, C. H., Hickerson, D. H. M., and Johnson, G. G. (2000). Short-term consequences of nourishment and bulldozing on the dominant large invertebrates of a sandy beach. *J. Coast. Res.* 16:368–378.

Phillips, R. C., and Menez, E. G. (1988). Seagrasses. *Smithsonian Contributions to Marine Science* 34, Washington, DC.

Pigram, J. (1983). *Outdoor recreation and resource management*. New York: St Martin's Press.

Pollock, L. W., and Hummon, W. D. (1971). Cyclic changes in interstitial water content, atmospheric exposure and temperature in a marine beach. *Limnol. Oceanogr.* 16:522–535.

Poxton, M. G., and Nasir, N. A. (1985). The distribution and population dynamics of 0-group plaice (*Pleuronectes platessa* L.) on nursery grounds in the Firth of Forth. *Estuar. Coast. Shelf Sci.* 21:845–857.

Rakocinski, C. F., Heard, R. W., and Le Croy, S. E. (1996). Responses by macrobenthic assemblages to extensive beach restoration at Perdido Key, Florida, USA. *J. Coast. Res.* 12:326–353.

Rafaelli, D., and Mason, C. F. (1981). Pollution monitoring with meiofauna using the ratio of nematodes to copepods. *Mar. Pollut. Bull.* 12:158–160.

Ranwell, D. S. (1972). *Ecology of salt marshes and sand dunes.* London: Chapman & Hall.

Reise, K. (1985). *Tidal flat ecology: An experimental approach to species interactions.* Berlin: Springer-Verlag.

Remane, A. (1933). Verteilung und Organisation der benthonischen Mikrofauna der Kieler Bucht. *Wiss. Meeresunters., Abt. Kiel.* NF21:161–221.

Ricciardi, A., and Bourget, E. (1999). Global patterns of macroinvertebrate biomass in marine intertidal communities. *Mar. Ecol. Prog. Ser.* 185:21–35.

Rickard, C. A., McLachlan, A., and Kerley, G. I. H. (1994). The effects of vehicular and pedestrian traffic on dune vegetation in South Africa. *Ocean Coast. Mgmnt.* 23:225–247.

Riedl, R. J. (1971). How much seawater passes through sandy beaches? *Int. Rev. Ges. Hydrobiol.* 56:923–946.

Riedl, R. J., and Machan, R. (1972). Hydrodynamic patterns in lotic intertidal sands and their bioclimatological implications. *Mar. Biol.* 13:179–209.

Riedl, R. J., Huang, N., and Machan, R. (1972). The subtidal pump: A mechanism of water exchange by wave action. *Mar. Biol.* 13:210–221.

Rijusdorp, A. D., van Straten, M., and van der Veer, H. W. (1985). Selective tidal transport of North Sea plaice larvae *Pleuronectes platessa* in coastal nursery areas. *Trans. Am. Fish. Soc.* 114:461–430.

Roberts, J. D. (1984). Terrestrial egg deposition and direct development in *Arenophryne rotunda* Tyler, a myobatrachnid frog from coastal sand dunes at Shark Bay, W. A. *Austr. Wildlife Res.* 11:191–200.

Roberts, D., Rittschof, D., Gerhart, D. J., Schmidt, A. R., and Hill, L. G. (1989). Vertical migration of the clam *Mercenaria mercenaria* (L.) (Mollusca: Bivalvia): Environmental correlates and ecological significance. *J. Exp. Mar. Biol. Ecol.* 126:271–280.

Robertson, A. I., and Hansen, J. A. (1981). Decomposing seaweed: A nuisance or a vital link in coastal food chains? CSIRO Marine Lab. Res. Rep., Perth, Australia, 75–83.

Robertson, A. I., and Lucas, J. S. (1983). Food choice, feeding rates, and the turnover of macrophyte biomass by a surf-zone inhabiting amphipod. *J. Exp. Mar. Biol. Ecol.* 72:99–124.

Robertson, A. I., and Lenanton, R. C. J. (1984). Fish community structure and food chain dynamics in the surf-zone of sandy beaches: The role of detached macrophyte detritus. *J. Exp. Mar. Biol. Ecol.* 84:265–283.

Robertson, D. A., and Pfeiffer, W. (1982). Deposit feeding by the ghost crab *Ocypode quadrata. J. Exp. Mar. Biol. Ecol.* 56:165–177.

Rodil, I. F., and Lastra, M. (2003). Environmental factors affecting benthic macrofauna along a gradient of intermediate sandy beaches in northern Spain. *Estuar. Coast. Shelf Sci.* 61:37–44.

Rodriguez, J. G., Lastra, M., and Lopez, J. (2003). Meiofauna distribution along a gradient of sandy beaches in northern Spain. *Estuar. Coast. Shelf Sci.* 58S:63–69.

Romer, G. (1986). Faunal assemblages and food chains associated with surf-zone phytoplankton blooms. MSc thesis, University of Port Elizabeth, South Africa.

Romer, G., and McLachlan, A. (1986). Mullet grazing on surf diatom accumulations. *J. Fish Biol.* 28:93–104.

Ross, S. W., and Lancaster, J. E. (2002). Movements and site fidelity of two juvenile fish species using surf zone nursery habitats along the southeastern North Carolina coast. *Environ. Biol. Fish.* 63:161–172.

Ross, S. T., McMichael, R. H., and Ruple, D. L. (1987). Seasonal and diel variation in the standing crop of fishes and macroinvertebrates from a Gulf of Mexico surf zone. *Estuar. Coast. Shelf Sci.* 25:391–412.

Rumbold, D. G., Davis, P. W., and Perretta, C. (2001). Estimating the effect of beach nourishment on *Caretta caretta* (loggerhead sea turtle) nesting. *Restor. Ecol.* 9:304–310.

Ruple, D. L. (1984). Occurrence of larval fishes in the surf zone of a northern Gulf of Mexico barrier island. *Estuar. Coast. Shelf Sci.* 18:191–208.

Sakamato, I. (1990). Use of respiration in the sandy beach or on the tidal flat: 1. Permeable sandy beach. *Mar. Pollut. Bull.* 23:123–130.

Salvat, B. (1964). Les conditions hydrodynamiques interstitielles des sediments meubles inter-tidaux et la repartition verticale de la fauna endogee. *C. R. Acad. Sci. Paris* 259:1576–1579.

Sanders, H. L., Grassle, J. F., Hampson, G. R., Morse, L. S., Garner-Price, S., and Jones, C. C. (1980). Anatomy of an oil spill: Long-term effects from the grounding of the barge Florida off west Falmouth, Massachusetts. *J. Mar. Res.* 38:265–280.

Sanduli, R., and De Nicola, M. (1991). Responses of meiobenthic communities along a gradient of sewage pollution. *Mar. Pollut. Bull.* 22:463–467.

Sastre, M. P. (1985). Aggregated patterns of dispersion in *Donax denticulatus. Bull. Mar. Sci.* 36:220–224.

Savidge, W. B., and Taghon, G. L. (1988). Passive and active components of colonization following two types of disturbance in an intertidal sandflat. *J. Exp. Mar. Biol. Ecol.* 115:137–155.

Scapini, F. (1988). Heredity and learning in animal orientation. *Monit. Zool. Ital. (NS)* 22:203–234.

Scapini, F. (1995). Heredity, individual experience, canalization: Sandhoppers as a case study. *Polsk. Arch. Hydrobiol.* 42:559–568.

Scapini, F., and Fasinella, D. (1990). Genetic determination and plasticity in the sun orientation of natural populations of *Talitrus saltator. Mar. Biol.* 107:141–145.

Scapini, F., Ugolini, A., and Pardi, L. (1985). Inheritance in solar direction finding in sandhoppers. II. Differences in arcuated coastlines. *J. Comp. Physiol.* 56A:729–735.

Scapini, F., Buiatti, M., and Ottaviano, O. (1988). Phenotypic plasticity in sun orientation of sand-hoppers. *J. Comp. Physiol.* 163A:739–747.

Scapini, F., Lagar, M. C., and Mexxetti, M. C. (1993). The use of slope and visual information in sandhoppers: Innateness and plasticity. *Mar. Biol.* 115:545–553.

Scapini, F., Buiatti, M., De Matthaeis, E., and Mattoccia, M. (1995). Orientation behaviour and heterozygosity of sand hopper populations in relation to stability of beach environments. *J. Evol. Biol.* 8:43–52.

Scapini, F., Fallaci, M., and Mezzetti, M. C. (1996). Orientation and migration in sandhoppers. *Rev. Chilena Hist. Nat.* 69:553–563.

Scapini, F., Aloia, A., Bouslama, M. F., Chelazzi, L., Colombini, I., El Gtari, M., Fallaci, M., and Marchetti, G. M. (2002). Multiple regression analysis of the sources of variation in orientation of two sympatric sandhoppers, *Talitrus saltator* and *Talorchestia brito*, from an exposed Mediterranean beach. *Behav. Ecol. Sociobiol.* 51:403–414.

Schratzberger, M., Gee, J. M., Rees, H. L., Boyd, S. E., and Wall, C. M. (2000). The structure and taxonomic composition of sublittoral meiofauna assemblages as an indicator of the status of marine environments. *J. Mar. Biol. Ass. U.K.* 80:969–980.

Schneider, D. C. (1985). Migratory shorebirds: Resources depletion in the tropics? *Ornith. Monogr.* 30:546–558.

Schoeman, D. S., and Richardson, A. J. (2002). Investigating biotic and abiotic factors affecting the recruitment of an intertidal clam on an exposed sandy beach using a generalized additive model. *J. Exp. Mar. Biol. Ecol.* 276:67–81.

Schoeman, D. S., McLachlan, A., and Dugan, J. E. (2000). Lessons from a disturbance experiment in the intertidal zone of an exposed sandy beach. *Estuar. Coast. Shelf Sci.* 50:869–884.

Schoeman, D. S., Wheeler, M., and Wait, M. (2003). The relative accuracy of standard estimations for macrofaunal abundance and species richness derived from selected intertidal transect designs used to sample exposed sandy beaches. *Estuar. Coast. Shelf Sci.* 58S:5–16.

Schwinghamer, P. (1981). Characteristic size distributions of integral benthic communities. *Can. J. Fish. Aquatic Sci.* 38:1255–1263.

Senta, T., and Hirai, A. (1981). Seasonal occurrence of milkfish fry at Tanegashima and Yakushima in southern Japan. *Jap. J. Ichth.* 28:45–51.

Senta, T., and Kinoshita, I. (1985). Larval and juvenile fishes occurring in surf zones of western Japan. *Trans. Am. Fish. Soc.* 114:609–618.

Seymour, M. K. (1971). Burrowing behaviour in the European lugworm *Arenicola marina* (Polychaeta; Arenicolidae). *J. Zool., Lond.* 164:93–113.

Shepherd, R. A., Knott, B., and Eliot, I. G. (1988). The relationship of juvenile southern mole crabs *Hippa australis* Hale (Crustacea: Anonura: Hippidae) to surficial swash water circulation over several diurnal spring-tide cycles during winter condition on a microtidal sandy beach. *J. Exp. Mar. Biol. Ecol.* 121:209–226.

Shesma, D. J., Barron, K. M., and Ellis, J. T. (2002). Retention of beach sands by dams and debris basins in Southern California. *J. Coast. Res.* SI 36:662–674.

Shiells, G. M., and Anderson, K. J. (1985). Pollution monitoring using the nematode/copepod ratio: A practical application. *Mar. Pollut. Bull.* 16:62–68.

Short, A. D. (1996). The role of wave height, period, slope, tide range and embaymentization in beach classifications: A review. *Rev. Chilena Hist. Nat.* 69:589–604.

Short, A. D. (ed.) (1999). *Handbook of beach and shoreface morphodynamics.* London: John Wiley.

Short, A. D., and Wright, L. D. (1983). Physical variability of sandy beaches. In: *Sandy beaches as ecosystem*, A. McLachlan and T. Erasmus (eds.), pp 133–144, The Hague: Dr W. Junk.

Silva, M. D., Araujo, F. G., de Azevedo, M. C. C., and Santos, J. N. D. (2003). The nursery function of sandy beaches in a Brazilian tropical bay for O-group anchovies (Teleostei: Engraulidae): Diel, seasonal and spatial patterns. *J. Mar. Biol. Ass. U.K.* 84:1229–1232.

Soares, A. (2003). Sandy beach morphodynamics and macrobenthic communities in temperate, subtropical and tropical regions: A macroecological approach. Ph.D. thesis, University of Port Elizabeth, South Africa.

Soares, A. G., McLachlan, A., and Schlacher, T. (1996). Disturbance effects of stranded kelp on populations of the sandy beach bivalve *Donax serra* (Roding). *J. Exp. Mar. Biol. Ecol.* 205:165–186.

Soares, A. G., Callahan, R. K., and De Ruyck, A. M. C. (1998). Microevolution and phenotypic plasticity in *Donax serra* Roding (Bivalvia: Donacidae) on high energy sandy beaches. *J. Mollusc. Stud.* 64:407–421.

Soares, A. G., Scapini, F., Brown, A. C., and McLachlan, A. (1999). Phenotypic plasticity genetic similarity and evolutionary inertia in changing environments. *J. Mollusc. Stud.* 65:136–139.

Southward, A. J. (1982). An ecologist's view of the implications of the observed physiological and biochemical effects of petroleum compounds on marine organisms and ecosystems. *Phil. Trans. Roy. Soc. Lond.* 297B:241–255.

Starfish (2005). www.starfish.ch/reef/reptiles.html

Steele, J. H. (1976). Comparative studies of beaches. *Phil. Trans. Roy. Soc. Edinb.* 274B:401–415.

Steele, J. H., McIntyre, A. D., Edwards, R. R. C., and Trevallion, A. (1970). Interrelationships of a young plaice population with its invertebrate food supply. In: *Animal populations in relation to their food resources*, A. Watson (ed.), *British Ecol. Soc. Symp.* 10:375–388, Oxford: Blackwell.

Stanley, S. M. (1970). *Relation of shell form to life habits of the Bivalvia (Mollusca).* Geological Society of America Mem 125, Baltimore, Maryland.

Stenton-Dozey, J. M. E. (1989). Physiology and energetics of the sandy-beach bivalve *Donax serra* Roding, with special reference to temperature and chlorine tolerance. Ph.D. thesis, University of Cape Town South Africa.

Stephen, A. C. (1931). Notes on the biology of certain lamellibranchs on the Scottish coast. *J. Mar. Biol. Ass. U.K.* 17:277–300.

Strydom, N. A. (2003). Occurrence of larval and early juvenile fishes in the surf zone adjacent to two intermittently open estuaries, South Africa. *Environ. Biol. Fish.* 66:349–359.

Strydom, N. A., and d'Hotman, B. D. (2005). Estuary-dependence of larval fishes in a non-estuary associated South African surf zone: Evidence for continuity of surf assemblages. *Estuar. Coast. Shelf Sci.* 63:101–108.

Suda, Y., Inone, T., and Uchida, H. (2002). Fish communities in the surf zone of a protected sandy beach at Doigahama, Yamaguchi Prefecture, Japan. *Estuar. Coast. Shelf Sci.* 55:81–96.

Suda, Y., Shiino, S., Nagata, R., Fuzawa, T., Hiwatari, T., Kohata, K., Hamaoka, S., and Watanabe, M. (2005). Revision of the ichthyofauna of reflective sandy beach on the Okhotsk coast of northern Hokkaido, Japan, with notes on the food habits of some fish. *Proc. 20th Intnl. Symp. Okhotsk Sea.* 23–28.

Sumich, J. L. (1999). An introduction to the biology of marine life (2nd ed.). Boston: McGraw-Hill.

Swart, D. H. (1983). Physical aspects of sandy beaches: A review. In: *Sandy beaches as ecosystems*, A. McLachlan and T. Erasmus (eds.), pp 5–44, The Hague: Dr W. Junk.

Swart, D. H., and Reyneke, P. G. (1988). The role of driftsands at Waenhuiskrans, South Africa. *J. Coast. Res. S. I.* 3:97–102.

Swedmark, B. (1964). The interstitial fauna of marine sand. *Biol. Rev.* 39:1–42.

Takahashi, K., and Kawaguchi, K. (1998). Diet and feeding rhythm of the sand-burrowing mysids *Archeomysis kakubai* and *A. japonica* in Otsuchi Bay, northeastern Japan. *Mar. Ecol. Prog. Ser.* 162:191–199.

Takahashi, K., Hirose, T., and Kawaguchi, K. (1999). The importance of intertidal sand-burrowing peracarid crustaceans as prey for fish in the surf-zone of a sandy beach in Otsuchi Bay, northeastern Japan. *Fish Sci.* 65:856–864.

Talbot, M. M. B. (1986). The distribution of surf diatoms *Anaulus birostatus* in relation to the nearshore circulation in an exposed beach/surf ecosystem. Ph.D. thesis, University of Port Elizabeth, South Africa.

Talbot, M. M. B., Bate, G. C., and Campbell, E. E. (1990). A review of the ecology of surf zone diatoms, with special reference to *Anaulus australis*. *Oceanogr. Mar. Biol. Ann. Rev.* 28: 155–175.

Tamaki, A. (1987). Comparison of resistivity to transport by wave action in several polychaete species on an intertidal sand flat. *Mar. Ecol. Prog. Ser.* 37:181–189.

Tarazona, J., and Parendes, C. (1992). Impacto de los eventos El Niño sobre las commundades bentonicas de playa arenosa durante 1976–1986. Paleo ENSO Records International Symposium, 299–303, Lima, Peru.

Tinley, K. L. (1985). *Coastal dunes of South Africa*. South African CSIR, National Scientific Programmes Report No. 109.

Trueman, E. R., and Ansell, A. D. (1969). The mechanism of burrowing into soft substrates by marine animals. *Oceanogr. Mar. Biol. Ann. Rev.* 7:315–366.

Trueman, E. R., and Brown, A. C. (1976). Locomotion, pedal retraction and extension, and the hydraulic system of *Bullia* (Gastropoda: Nassariidae). *J. Zool., Lond.* 178:365–384.

Trueman, E. R., and Brown, A. C. (1985). Dynamics of burrowing and pedal extension in *Donax serra* (Mollusca: Bivalvia). *J. Zool., Lond.* 207:345–355.

Trueman, E. R., and Brown, A. C. (1987). Locomotory function of the pedal musculature of the nassariid whelk, *Bullia. J. Mollusc. Stud.* 53:287–288.

Trueman, E. R., and Brown, A. C. (1989). The effect of shell shape on the burrowing performance of species of *Bullia* (Nasariidae: Gastropoda). *J. Mollusc. Stud.* 55:129–131.

Turner, I. L. (1993). The total water content of sandy beaches. *J. Coast. Res.* S. I. 15:11–26.

Turner, R. K., Lorenzoni, I., Beaumart, N., Bateman, I. J., Langford, I. H., and McDonald, A. L. (1998). Coastal management for sustainable development: Analyzing environmental and socio-economic changes on the U.K. coast. *Geogr. J.* 164:269–281.

Ugolini, A., Scapini, F., and Pardi, L. (1986). Interaction between solar orientation and landscape visibility in *Talitrus saltator* (Crustacea: Amphipoda). *Mar. Biol.* 90:449–460.

Ullberg, J., and Olafsson, E. (2003). Free-living marine nematodes actively choose habitat when descending from the water column. *Mar. Ecol. Prog. Ser.* 260:141–149.

United Nations (1998). *World population prospects: The 1998 revision.* New York, United Nations Secretariat, Department of Economic and Social Affairs, Population Division.

Vanaverbeke, J., Steyaert, M., Vanreusel, A., and Vincx, M. (2003). Nematode biomass spectra as descriptors of functional changes due to human and natural impact. *Mar. Ecol. Prog. Ser.* 249:157–170.

Vandermeulen, J. H. I. (1982). Some conclusions regarding long-term biological effects of some major oils spills. *Phil. Trans. Roy. Soc. Lond.* 297B:335–351.

Van der Merwe, D., and Van der Merwe, D. (1991). Effects of off-road vehicles on the macrofauna of a sandy beach. *S. Afr. J. Sci.* 87:210–213.

Van der Merwe, D., and McLachlan, A. (1987). Significance of free-floating macrophytes in the ecology of a sandy beach surf zone. *Mar. Ecol. Prog. Ser.* 38:53–63.

Van der Merwe, D., and McLachlan, A. (1991). The meiofauna of a coastal dune slack. *J. Arid Environ.* 21:213–228.

Van Heerdt, P. F., and Morzer Bruyns, M. F. (1960). A biocoenological investigation of the yellow dune region of Terschelling. *Tijdschr. Entomol.* 103:225–275.

Vanini, M. (1980). Notes on the behaviour of *Ocypode ryderi* (Crustacea, Brachyura). *Mar. Behav. Physiol.* 7:171–184.

Vassallo, P., and Fabiano, M. (2005). Trophodynamic variations on microtidal north Mediterranean sandy beaches. *Oceanologia* 47:351–364.

Veloso, V. G., and Cardoso, R. S. (1999). Population biology of the mole crab *Emerita brasiliensis* (Decapoda: Hippidae) at Urca Beach, Brazil. *J. Crustac. Biol.* 19:147–153.

Veloso, V. G., Silva, E. S., Cactano, C. II. S., and Cardoso, R. S. (2005). Comparison between the macroinfauna of urbanized and protected beaches in Rio de Janeiro State, Brazil. *Biol. Conserv.* 127:44, 510–515.

Vincx, M. (1996). Meiofauna in marine and freshwater sediments. In: *Methods for the examination of organismal diversity in soils and sediments,* G. S. Hall (ed.), pp 187–195, Wallingfort, UK: CAB International.

Von Blaricom, G. R. (1982). Experimental analyses of structural regulation in a marine sand community exposed to oceanic swell. *Ecol. Monogr.* 52:283–305.

Wade, B. A. (1967). Studies on the biology of the West Indian beach clam *Donax denticulatus* Linne. 1. Ecology. *Bull. Mar. Sci.* 17:149–174.

Warwick, R. W. (1984). Species size distribution in marine benthic communities. *Oceologia* 61:32–41.

Watt-Pringle, P., and Strydom, N. A. (2003). Habitat use by larval fishes in a temperate South African surf zone. *Estuar. Coast. Shelf Sci.* 58:765–774.

Webb, J. E. (1991). Hydrodynamics, organisms and pollution of coastal sands. *Ocean Shore Mgmnt.* 16:23–51.

Webb, P., and Wooldridge, T. (1990). Diel horizontal migration of *Mesopodopsis slabberi* (Crustacea: Mysidacea) in Algoa Bay, southern Africa. *Mar. Ecol. Prog. Ser.* 62:73–77.

Webb, P., Perissinotto, R., and Wooldridge, T. H. (1987). Feeding of *Mesopodopsis slabberi* (Crustacea, Mysidacea) on naturally occurring phytoplankton. *Mar. Ecol. Prog. Ser.* 38:115–123.

Wenner, A. M. (1988). Crustaceans and other invertebrates as indicators of beach pollution. In: *Marine organisms as indicators*, D. F. Sale and G. S. Keppel (eds.), pp 199–229, New York: Springer-Verlag.

Weslawski, J. M., Urban-Malinga, B., Kotwicki, L., Opalinski, K. W., Szymmelfenig, M., and Dutkowski, M. (2000). Sandy coastlines: Are there conflicts between recreation and natural values? *Oceanol. Stud.* 29:5–18.

Whittle, K. J., Hardy, R., Mackie, P. R., and McGill, A. S. (1982). A qualitative assessment of the sources and fate of petroleum compounds in the marine environment. *Phil. Trans. Roy. Soc. Lond.* 297B:193–218.

Wieser, W., and Schiemer, F. (1977). The ecophysiology of some marine nematodes of Bermuda: Seasonal aspects. *J. Exp. Mar. Biol. Ecol.* 26:97–106.

Wilce, R. T., and Quinlan, A. V. (1983). Fouling of the sandy beaches of Nahant Bay (Massachusetts) by an abnormal free-living form of the macro-alga *Pilagella littoralis* (Phaeophyta). II. Population characteristics. In: *Sandy beaches as ecosystems*, A. McLachlan and T. Erasmus (eds.), pp 285–296, The Hague: Dr W. Junk.

Williams, J. A. (1978). The annual pattern of reproduction of *Talitrus saltator* (Crustacea: Amphipoda). *Mar. Biol.* 184:231–244.

Wilson, D. P. (1955). The role of micro-organisms in the settlement of *Ophelia bicornis* Savigny. *J. Mar. Biol. Ass. U.K.* 34:513–543.

Wilson, W. H. (1981). Sediment-mediated interactions in a densely populated infaunal assemblage: The effects of the polychaete *Abarenicola pacifica*. *J. Mar. Res.* 39:735–748.

Wolcott, T. G., and Wolcott, D. L. (1984). Impact of off-road vehicles on macronivertebrates on a mid-Atlantic beach. *Biol. Conserv.* 29:217–240.

Woodin, S. A. (1977). Algal gardening behavior by neriid polychaetes: Effects on soft-bottom community structure. *Mar. Biol.* 44:39–42.

Woodin, S. A. (1981). Disturbances and community structure in a shallow-water sand flat. *Ecology* 62:1052–1066.

Wooldridge, T. (1983). Ecology of beach and surf zone mysid shrimps in the Eastern Cape, South Africa. In: *Sandy beaches as ecosystems*, A. McLachlan and T. Erasmus (eds.), pp 449–460, The Hague: Dr W. Junk.

Wooldridge, T. (1989). The spatial and temporal distribution of mysid shrimps and phytoplankton accumulations in a high energy surfzone. *Vie Milieu* 39:127–133.

World Resources Institute (1998). *World Resources 1998–1999*. Oxford University Press, New York, USA.

Wynberg, R. P., and Branch, G. M. (1997). Trampling associated with bait-collection for sand prawns *Callianassa kraussi* Stebbing: Effects on the biota of an intertidal sandflat. *Environ. Conserv.* 24:139–148.

Wynberg, R. P., and Brown, A. C. (1986). Oxygen consumption of the whelk *Bullia digitalis* (Dillwyn) at reduced oxygen tensions. *Comp. Biochem. Physiol.* 85A:45–47.

Yamamoto, M., Makino, H., Kagawa, T., and Tominaga, O. (2004). Occurrence and distribution of larval and juvenile Japanese flounder *Paralichthys olivaceus* at sandy beaches in eastern Hiuchi-Nada, central Seto Inland Sea, Japan. *Fish. Sci.* 70:1089–1097.

Yannicelli, B., Palacios, R., and Gimenez, L. (2002). Swimming ability and burrowing time of two cirolanid isopods from different levels of exposed sandy beaches. *J. Exp. Mar. Biol. Ecol.* 273:73–88.

Yu, O. H., Suh, H. L., and Shirayama, Y. (2003). Feeding ecology of three amphipod species *Synchelidium leuorostralum*, *S. trioostegitum* and *Gigantopsis japonica* in the surf zone of a sandy shore. *Mar. Ecol. Prog. Ser.* 258:189–199.

Appendix A: Measures of Beach Type

A

In addition to the three important indices covered in Chapter 2, DFV (dimensionless fall velocity, or Dean's Parameter), RTR (relative tide range), and BI (beach index), a number of other indices have been developed and used by ecologists to describe or categorize beaches. These are as follows.

Exposure Rating (McLachlan 1980)

Beaches are given a score on a 20-point scale of exposure based on wave action, surf-zone width, percent very fine sand, median particle diameter, depth of reduced layers, and presence of fauna with stable burrows. To a large extent this scale has been superseded by the beach morphodynamic models.

Beach State Index (McLachlan *et al.* 1993)

$$BSI = \log(DFV.tide/0.8 + 1)$$

Here, *tide* is maximum spring tide range in meters. This index ranges beaches on a scale of 0 to 4 from microtidal reflective to macrotidal dissipative and can be used to compare beaches from regions subject to different tide ranges. It is not as effective as BI but has been widely used in the past and differs from BI only in using breaker height rather than slope and sand fall velocity rather than particle size.

Beach Deposit Index (Soares 2003)

$$BDI = (1/\tan B)(a/Mz)$$

Here, *tan B* is the beach slope, *a* is 1.03 millimeters, and *Mz* is the mean sand particle size in millimeters. This index has only been used by Soares, who found it useful for microtidal beaches. It focuses on sedimentary factors and does not take tide range into account and thus should not be used for comparisons between regions.

Area (McLachlan and Dorvlo 2005)

$$Area = \log(tide/slope)$$

Here, *tide* is maximum spring tide range in meters, and *slope* is beach face slope. This is in effect a measure of beach width or intertidal area. Where the actual beach width has been measured there is no need for this index.

It is not recommended that any of these indices should be used in future work, save to compare with previous work based on them. Rather, the indices described in Chapter 2 and the features summarized in the conclusion thereof should be adequate for defining any beach.

Appendix B: The Chemical Environment of Sediments

This section is modified from an appendix to the Japanese edition, and was originally compiled and translated by Dr. Yasuhiro Hayakawa. The chemical environment in sediments generally depends on the oxidizing/reducing conditions and the supply of organic materials. A chemical formula of $(CH_2O)_{106}(NH_3)_{16}H_3PO_4$ is representative of marine particulate organic matter (POM) based on marine phytoplankton and detritus with an elemental ratio carbon : nitrogen : phosphorus = 106 : 16 : 1, which is known as the Redfield ratio. Under oxidizing conditions, where dissolved oxygen as molecular O_2 is abundant — as in beaches exposed to waves — organic matter is decomposed in the course of *oxidative decomposition* by aerobic bacteria (as shown in the following) to carbon dioxide, nitrate, phosphate, and water.

$$(CH_2O)_{106}(NH_3)_{16}H_3PO_4 + 138O_2 \rightarrow 106CO_2 + 16HNO_3 + H_3PO_4 + 122H_2O$$
POM + oxygen $\rightarrow$ carbon dioxide + nitrate + phosphate + water

These materials, produced by decomposition, are dissolved in seawater as ions. If ammonia is produced during decomposition, it is changed under oxidizing conditions into nitrate by nitrifying bacteria in the *nitrification* process.

$$2NH_3 + 3O_2 \rightarrow 2NO_2^- + 2H^+ + 2H_2O$$
ammonia + oxygen $\rightarrow$ nitrite + hydrogen ions + water

and

$$2NO_2^- + O_2 \rightarrow 2NO_3^-$$
nitrite + oxygen $\rightarrow$ nitrate

If hydrogen sulfide is transported from the deeper substratum to the surface, where oxygen is sufficient, it is converted as follows into sulfur or sulfate by sulfur bacteria.

$$2H_2S + O_2 \rightarrow 2S + 2H_2O$$
sulfide + oxygen $\rightarrow$ sulfur + water

and

$$2S + 2H_2O + 3O_2 \rightarrow 4H^+ + 2SO_4^{2-}$$

sulfur + water + oxygen → hydrogen ions + sulfate

Those elements of carbon, nitrogen, and phosphorus derived from oxidative decomposition of organic matter and nitrification under oxidizing conditions are dissolved in seawater as bicarbonate, nitrate, and phosphate, respectively. They are recycled by organisms. These oxidizing reactions consume oxygen, which may bring about reducing conditions with a rapid decrease in the redox potential from positive to negative. The sediment layer where the redox potential changes rapidly is called the redox potential discontinuity (RPD) zone. The oxidizing zone above the RPD can extend to a depth of meters from the surface of exposed sands, whereas the RPD may occur just millimeters under the surface in fine sheltered sediments. The reducing zone below the RPD is toxic to most living organisms.

Decomposition of organic matter by anaerobic sulfate-reducing bacteria under reducing conditions with little dissolved oxygen uses sulfate ions in seawater instead of oxygen molecules to produce carbon dioxide, ammonia, phosphate, sulfide, and water.

$$(CH_2O)_{106}(NH_3)_{16}H_3PO_4 + 53SO_4^{2-} \rightarrow 106CO_2 + 16NH_3 + H_3PO_4 + 53S^{2-} + 106H_2O$$

POM + sulfate → carbon dioxide + ammonia + phosphate + sulfide + water

Most ammonia produced is dissociated to ammonium ions (NH_4^+) in seawater, with relatively low toxicity to living organisms. However, the portion of nondissociated ammonia (NH_3) that is toxic could increase when water temperature and pH increase. Sulfide ions produced are also toxic, but can combine with metals to make less toxic forms. For instance, ferrous sulfide (FeS) is produced by combination of sulfide with iron in the interstitial water, turning the bottom black due to the color of this compound. High production rates of sulfides can lead to increasing free sulfide ions that do not combine with metals and then make the sediment smell of hydrogen sulfide.

In the reducing zone, *denitrification* by denitrifying bacteria can convert nitrogen compounds into gaseous nitrogen to be removed to the atmosphere. Hydrogen donors in these reactions are from organic matter. The resultant hydroxide ions produced in denitrification tend to make the water alkaline. Thus, the reducing zone in sediments plays an important role in the recycling of nitrogen.

$$2NO_2^- + 3H_2 \rightarrow N_2 + 2OH^- + 2H_2O$$

nitrite + hydrogen → gaseous nitrogen + hydroxide ion + water

and

$$2NO_3^- + 5H_2 \rightarrow N_2 + 2OH^- + 4H_2O$$

nitrate + hydrogen → gaseous nitrogen + hydroxide ion + water

The chemical environment in sediments is characterized by active processes driven by both benthos and bacteria. Organic matter supplied there is decomposed by biochemical processes into various inorganic compounds. The chemical environment in sediments thus has a significant influence on the global cycles of these materials.

Index